Advances in Surface Engineering
Volume III: Engineering Applications

Advances in Surface Engineering

Volume III: Engineering Applications

Edited by

P.K. Datta
University of Northumbria at Newcastle, UK

J.S. Burnell-Gray
University of Northumbria at Newcastle, UK

THE ROYAL
SOCIETY OF
CHEMISTRY
Information
Services

The Proceedings of the Fourth International Conference on Advances in Surface Engineering held at The University of Northumbria at Newcastle on 14–17 May 1996.

The front cover illustration is taken from the contribution by S.J. Bull and A.R. McCabe, p.209.

Special Publication No. 208

ISBN 0-85404-757-3

A catalogue record for this book is available from the British Library

Published by The Royal Society of Chemistry,
Thomas Graham House, Science Park, Milton Road,
Cambridge CB4 4WF, UK

Printed and bound by
Bookcraft (Bath) Ltd

Preface

Advances in Surface Engineering is based on the Proceedings of the *4th International Conference on Advances in Surface Engineering* which was hosted by the University of Northumbria's *Surface Engineering Research Group* between 14–17th May 1996.

Acknowledgements

The editors wish to thank Professor Gilbert Smith, the Vice Chancellor of the University of Northumbria for opening the conference. The editors are grateful to Professor Van de Voorde for giving the opening technical keynote talk.

The editors wish to express their gratitude for the support extended by:

The Department of Trade and Industry, The Institute of Materials, The Institute of Corrosion, The Royal Society of Chemistry, Multi-Arc (UK) Ltd, Buehler Ltd, Gearing/Micromaterials, Mats (UK), Tech Vac and Woodhead Publishing.

Special thanks are due to Professor Cryan head of the *School of Engineering* at the *University of Northumbria at Newcastle*.

The support and encouragement of many colleagues at the *University of Northumbria at Newcastle* and friends in other universities, is gratefully acknowledged.

The human commitment to any conference or book is substantial and often not fully acknowledged. In this regard the work of Kath Hynes, the secretaries and the technicians from the *School of Engineering* and the members of the *Surface Engineering Research Group* (*SERG*) should be fully recognized.

Finally special commendation is reserved for Dan Smith who administered the *4th International Conference on Advances in Surface Engineering* (*4ICASE*) and David Griffin of *SERG* who desk-top published the conference proceedings.

P. K. Datta
J. S. Burnell-Gray

Surface Engineering Research Group
School of Engineering
University of Northumbria at Newcastle

Contents
Volume III Engineering Applications

Introduction 1

Section 3.1 Biomedical

3.1.1 Surfaces in Implantable Ceramics Technology
A. Ravagliok, A. Krajewski and P. Vincenzini 19

3.1.2 In-Vitro Evaluation of Corrosion Resistance of Nitrogen Ion Implanted
Titanium Simulated Body Fluid
M. Subbaiyan, T. Sundararajian, S. Rajeswari, U. K. Mudali,
K. G. M. Nair and N. S. Thampi 26

3.1.3 Pitting Corrosion Studies on Nitrogen Implanted 316L SS for
Biomedical Applications
M. Subbaiyan, K. M. Veerabadran, N. S. Thampi, K. Krishnan,
U. K. Mudali and R. K. Dayal 38

Section 3.2 Aerospace

3.2.1 Characterization of High Temperature Metal Matrix
Composite Coatings
A. R. Marder 51

3.2.2 Simulation of Failure Mechanisms in Al_2O_3/Al Metal Matrix
Composites and the Effect of Fibre/Matrix Interface Bond Strength
Using Finite Element Modelling Techniques
D. Griffin, A. Daadbin, J. S. Burnell-Gray and P. K. Datta 70

3.2.3 High Temperature Air Oxidation Performance of Modified Aluminide
Coatings on a Nickel-Based Superalloy
W. Y. Chan, J. S. Burnell-Gray and P. K. Datta 78

3.2.4 Taguchi Optimization of an LPPS MCrAlY Coating for Use as a
Bond Coat for Thermal Barrier Coatings
R. E. Jones, J. Cawley, D. S. Rickerby and J. Green 99

3.2.5 Joining of Ferritic ODS Alloys for High Temperature Application
M. W. Carroll and N. J. Wood 115

Section 3.3 Automotive

3.3.1   Volume Production of Thermal Sprayed Coatings in the Automotive
         Industry
         K. Harrison and A. R. Nicoll 131

3.3.2   Development of Coatings for Lubricant-free Ball Bearings and of an
         Adapted Test Bench for the Evaluation of the Coated Ball Bearings
         O. Knotek, E. Lugscheider, F. Löffler and M. Möller 147

Section 3.4 Cutting Tools and Manufacture

3.4.1   Successful Products through Surface Engineering
         M. Sarwar 159

3.4.2   Indirect Monitoring of Carbide Tool Wear in the Facing of Low
         Carbon Steels Through Measurements of the Roughness of
         Machined Surfaces
         K. N. Strafford, J. Audy and L. Blunt 167

3.4.3   Potential of Physical Vapour Deposited Coatings of a Cermet
         for Interrupted Cutting
         G. E. D'Errico and E. Guglielmi 179

3.4.4   Sliding Wear Behaviour of TiN-Based PVD Coatings on
         Tool Steels
         C. Martini, E. Lanzoni and G. Palombarini 187

3.4.5   Performance of Physical Vapour Deposited Coatings on a
         Cermet Insert in Turning Operations
         G. E. D'Errico, R. Calzavarini and B. Vicenzi 197

3.4.6   Coating of Silicon Carbide Matrix Composites for Industrial
         Use in Corrosive Environments
         S. J. Bull and A. R. McCabe 209

3.4.7   Implementation of TiAlN and CrN Coatings and Ion Implantation
         in the Modern Plastics Moulding Industry
         E. J. Bienk and N. J. Mikkelsen 218

3.4.8   Thin Film Characterization Methods
         F. Löffler 224

Section 3.5 Power Generation

3.5.1   Studies of Surface Films on High Ni-Cr-Mo Alloys in Simulated SO_2
         Scrubber Environment
         N. Rajendran, G. Latha and S. Rajeswari 247

3.5.2   Surface Engineering of Composite and Graded Coatings for
        Resistance to Solid Particle Erosion at Elevated Temperatures
        M. M. Stack and D. Pena 260

3.5.3   A Study of the Corrosion Resistance and Microstructure of a
        Laser Treated Layer on Duplex and Austenitic Stainless Steels
        A. Neville and T. Hodgkiess 277

3.5.4   The Use of Thermal Spray Coatings in the South African Power
        Industry
        R. G. Wellman 294

Section 3.6 Marine

3.6.1   Properties of the Passive Films on SuperAustenitic Stainless
        Steels in Sea Water
        G. Latha, N. Rajendran and S. Rajeswari 303

3.6.2   Scheme of Coating and Surface Treatment for Improving Corrosion
        Resistance with Application to Marine Structures
        M. M. El-Gammal 315

3.6.3   Enhancement of Tooling Life Using Surface Engineering
        Technologies
        K. N. Strafford, T. P. Wilks, C. Subramanian and W. McMillan 325

3.6.4   Progress in Coated Cutting Tool Science and Technology
        K. N. Strafford 364

        Contributor Index 399

        Subject index 403

Contents
Volume I Fundamentals of Coatings

Introduction

Section 1.1 High Temperature Corrosion

1.1.1   The Oxidation Behaviour of Lanthanum Implanted Stainless Steels
*F. J. Ager, M. A. Respaldiza, A. Paúl, J. A. Odriozola, C. Luna,
J. Botella, J. C. Soares and M. F. da Silva*

1.1.2   The Oxidation and Hot Corrosion Properties of Rare-Earth Modified Cr/Al
Coatings on Fe- and Ni-Base Alloys
J. Kipkemoi and D. Tsipas

1.1.3   The Effect of H_2O, O_2 in Atmosphere and Cr, Si in Steel on the
Oxidation Behavior of Ferritic Stainless Steels
H. Fujikawa and Y. Shida

1.1.4   The Influence of Processing Route on the Oxidation/Sulphidation
of a Ti-48Al-2Nb-2Mn Intermetallic Alloy at 750 and 900°C
*H. L. Du, P. K Datta, J. Leggett, J. R. Nicholls, J. C. Bryar
and M. H. Jacobs*

1.1.5   Sulphidation Properties of Ti-Al Intermetallic Compounds at High
Temperatures in H_2S-H_2 Atmospheres
T. Yoshioka and T. Narita

1.1.6   High Temperature Corrosion of Electroless Nickel Film Formed
on Iron in Flowing Chlorine
T. Jiangping, L. Jian and L. Maoshen

1.1.7   The Effect of Coating Thickness on the Thermal Conductivity of
CVD and PVD Coatings
K. J. Lawson, J. R. Nicholls and D. S. Rickerby

Section 1.2 Aqueous Corrosion

1.2.1   Electrochemical Studies and Characterisation of Electrodeposited
Zinc-Chromium Alloy Coatings
M. R. El-Sharif, C. U. Chisholm, Y. J. Sut and L. Feng

1.2.2 Comparative Study of Corrosion Properties of Zinc Composites
in Acidic and Alkaline Environments
D. Sneddon, M. R. El-Sharif, C. U. Chisholm and A. M. DeSilva

1.2.3 Aluminium-Magnesium Corrosion Resistant Coatings
K. R. Baldwin, R. I. Bates, R. D. Arnell and C. J. E. Smith

1.2.4 Effect of Plasma and Conventional Gas Nitriding on Anodic Behaviour
of Iron and Low-Alloy Steel
J. Mankowski and J. Flis

Section 1.3 Wear

1.3.1 Modelling the Beneficial Effects of Oxidation on the Sliding
Wear of Metals and Alloys
F. H. Stott, J. Jiang and M. M. Stack

1.3.2 Oxidation and Erosion Resistance of Amorphous Bright Chromium
Electrodeposited Coatings
R. P. Baron and A. R. Marder

1.3.3 The Effect of Chromium Carbide Coatings on the Abrasive Wear of
AISI 44OC Martensitic Stainless Steel
A. B. Smith, A. Kempster, J. Smith and R. Benham

1.3.4 Abrasion Wear Behaviour of Squeeze Cast SiC Particulates
Reinforced Al-2024 Composites
S. M. Skolianos, G. Kiourtsidis and L. Escourru

1.3.5 Influence of Thickness of Electroless Nickel Interlayer on the
Tribological Behaviour of Mild Steel Coated With Titanium Nitride
C. Subramanian and K. N. Strafford

1.3.6 Tribological-Structure Relationships of Alumina-Based, Plasma
Sprayed Coatings
L. C. Erickson, T. Troczynski and H. M. Hawthorne

1.3.7 The Tribological Characteristics of a Detonation Gun Coating of
Tungsten Carbide Under High Stress Abrasion
M. L. Binfield and T. S. Eyre

1.3.8 The Influence of the Substrate Bias Voltage on the Adhesion of
Sputtered Nb Coatings and Correlation with the Hardness Data
L. P Ward and P. K. Datta

Section 1.4 Fatigue and Other Failures

1.4.1   Rolling Contact Fatigue Performance of HVOF Coated Elements
         R. Ahmed and M. Hadfield

1.4.2   Fatigue Failure and Crack Growth Response of an Electroless Ni-B
         Coated Low Carbon Steel
         P. B. Bedingfield, A. L. Dowson and P. K. Datta

1.4.3   Observations and Analysis of a 200nm Aluminium Film on Silicon
         Using Ultra Micro-Indentation
         A. J. Bushby and M. V. Swain

1.4.4   Depth-Profile Elemental Concentration, Microhardness and Corrosion
         Resistance of Nitrided Chromium Steels
         J. Flis, J. Mankowski, T. Bell and T. Zakroczymski

1.4.5   Failure Modes in the Scratch Adhesion Testing of Thin Coatings
         S. J. Bull

1.4.6   Strain Rate Sensitivity of Steel 03X26H6T
         M. Fouad, T. M. Tabaghia and S. Khalil

1.4.7   Elasticity vs Capillarity, or How Better to Control Wetting
         on Soft Materials
         M. E. R. Shanahan and A. Carré

1.4.8   The Influence of the Electrodeposition Parameters on the Gas
         Sensitivity and Morphology of Thin Films of Tetrabutylammonium
         BIS(1,3-Dithiol-2-Thione-4,5-Dithiolate) Nickelate
         J. R. Bates, P. Kathirgamanathan and R. W. Miles

         Contributor Index

         Subject index

Contents
Volume II Process Technology

Introduction

Section 2.1 PVD and CVD

2.1.1   Recent Developments in Magnetron Sputtering Systems
         R. D. Arnell and P. J. Kelly

2.1.2   Investigation of Ti$_2$N Films Deposited Using an Unbalanced
         Magnetron Sputtering Coating System
         *S. Yang, M. Ives, D. B. Lewis, J. Cawley, J. S. Brooks and
         W-D. Münz*

2.1.3   Evaluation of Defects in TiN Thin Films by Various
         Electrochemical Techniques
         S. Nitta and Y. Kimura

2.1.4   Annealing of Defects in PVD TiN: a Positron Annihilation
         Spectroscopy Study
         *S. J. Bull, A. M. Jones, A. R. McCabe, A. Saleh and
         P. Rice-Evans*

Section 2.2 Thermal, Plasma, Weld and Detonation

2.2.1   Spouted Bed Reactor for Coating Processes: a Mathematical Model
         G. Mazza, I. Sanchez, G. Flamant and D. Gauthier

2.2.2   Surface Engineering by Boron and Boron-Containing Compounds
         *G. Kariofillis, C. Salpistis, H. Stapountzis, J. Flitris
         and D. Tsipas*

2.2.3   Study on Carburizing/Boriding Compound Chemical Heat Treatment and its
         Application
         P-Z. Wang, L-Y. Shan, Z-Y. Ni and P. Ye

2.2.4   An Integrated Study of the Structure and Durability of Vacuum
         Furnace Fused Coatings
         T. Hodgkiess and A. Neville

2.2.5   Effects of Rare Earth Elements on the Plasma Nitriding of
         38CrMoAl Steel
         J. P. Shandong, H. Dong, T. Bell, F. Chen, Z. Mo,
         C. Wang and Q. Pen

2.2.6   APNEP – A New Form Of Non-Equilibrium Plasma, Operating at Atmospheric
         Pressure
         N. P. Wright

2.2.7   Scanning Reference Electrode Study of the Corrosion of Plasma
         Sprayed Stainless Steel Coatings
         D. T. Gawne, Z. Dou and Y. Bao

2.2.8   Thin Chromium Coatings Processed by HVOF-Spraying
         E. Lugscheider, P. Remer and H. Reymann

2.2.9   Dilution Control in Weld Surfacing Applications
         J. N. DuPont and A. R. Marder

2.2.10   Wire Explosion Coating
         M. G. Hocking

Section 2.3 Laser and EB

2.3.1   (Nickel Base Alloy + Cast Tungsten Carbide Particles) Composite
         Coating by Laser Cladding and the Wear Behaviour
         Wang Peng-zhu, Ma Xiong-dong, Ding Gang, Qu Jing-xin
         Shao He-sheng and Yang Yuan-sheng

2.3.2   Electron Beam Surface Hardening of 42CrMo4 Steel
         T. Jokinen and I. Meuronen

Section 2.4 Peening, Solar and Other

2.4.1   Controlled Shot Peening. Scope of Application and
         Assessment of Benefits
         G. J. Hammersley

2.4.2   Applications of High Solar Energy Density Beams in
         Surface Engineering
         A. J. Vázquez, G. P. Rodríguez and J. J. de Damborenea

2.4.3   Processing and Properties of Thermoplastic Coatings on Metals
         D. T. Gawne and Y. Bao

2.4.4   Ultrasonic Flow Polishing
         A. R. Jones and J. B. Hull

Section 2.5 Electrochemical and Electroless

2.5.1 The Influence of Formic Acid and Methanol on the Deposition of
Hard Chromium from Trivalent Chromium Electrolytes
S. K. Ibrahim, A. Watson and D. T. Gawne

2.5.2 Electrodeposition of Multilayer Zn-Ni Coatings from a
Single Electrolyte
M. R. Kalantary, G. D. Wilcox and D. R. Gabe

2.5.3 High Nickel Content Zinc-Nickel Electrodeposits from
Sulphate Baths
C. E. Lehmberg, D. B. Lewis and G. W. Marshall

2.5.4 The Electrodeposition of Ni-P-SiC Deposits
D. B. Lewis, B. E. Dodds and G. W. Marshall

2.5.5 Electroplated Iron Layers Containing Nitrogen
V. A. Stoyanov, J. D. Di-agieva and A. I. Pavlikianova

2.5.6 The Effects of Bath Ageing on the Internal Stress within Electroless
Nickel Deposits and Other Factors Influenced by the
Ageing Process
C. Kerr, D. Barker and F. Walsh

Contributor Index

Subject index

Introduction

J. S. Burnell-Gray and P. K. Datta

SURFACE ENGINEERING RESEARCH GROUP, UNIVERSITY OF NORTHUMBRIA AT NEWCASTLE, UK

1 SCOPE OF *ADVANCES IN SURFACE ENGINEERING*

Advances in Surface Engineering is aimed at reviewing and documenting the recent advances in research and application of this relatively newly emerging technology, the problems that remain to be solved and the directions of future research and development in surface engineering.

The three volumes incorporate both science and technical research papers. They demonstrate how SE technologies are continuously increasing the level of performance of components, devices and structures through the creation of high performance surfaces.

What distinguishes *Advances in Surface Engineering* is an attempt, though limited, to characterise the coatings/engineered surfaces in terms of their fundamental structural entities and to understand their behaviour and properties using the principles of material science and physics. This knowledge of coating structures together with an understanding of the mechanisms of degradation processes that operate on surfaces[1-3], has allowed the development of precisely designed surfaces/coatings with enhanced degrees of corrosion resistance, wear resistance and biocompatibility.

More broadly, *Advances in Surface Engineering* provides a lens for viewing fundamental changes in the SE and corrosion and wear management professions. In this era of advanced manufacturing technologies and virtual networks, most factors of production are available globally. Capital essentially flows freely; machines can be bought or their capacity rented; technology and technological mastery are readily transferable. What increasingly sets research, and its application apart are knowledge and expertise – *intellectual*, as opposed to physical, assets.

New and more powerful ways to think about problems and actions are the prized output from the *4th International Conference on Advances in Surface Engineering*. New theories enable engineers to conceptualize their activities in novel ways and initiate more effective programmes of action. New theoretical inspiration – ranging across a broad spectrum – can also help academic researchers redefine their efforts. Particularly prized is a new methodology or way of thinking that totally transforms the shape of a field and the leveraged efforts of hundreds of researchers.

Advances in Surface Engineering is structured in 3 volumes. *Volume 1* concerns fundamental aspects of corrosion and wear. *Volume 2* gives an appreciation of SE technologies. *Volume 3* deals with applications of surface engineering to selected industrial sectors – areas central to surface engineering and holding particular promise for improvements in existing and emerging surface engineering techniques.

The *Introduction* to *Advances in Surface Engineering* aims to provoke new debate and comprehension to devise a coherent and integrated framework for tackling the enduring engineering problem of understanding and controlling corrosion and wear exploiting SE as a sustained business asset. Embedded in the *Introduction* are attempts to identify:

1. What were the important issues in SE in 1992 (the time of the *3rd International Conference on Advances in Surface Engineering*)?
2. What has happened in SE since 1992?
3. What are the important SE issues in 1997?
4. What are likely to be the important issues – particularly relating to management – in SE in 2010?

Finally, we take the opportunity to present a synopsis of the activities of the University of Northumbria's (UNN) *Surface Engineering Research Group (SERG)*.

2 SURFACE ENGINEERING

2.1 Background

The growing importance of surface engineering is due to the realization that modified, treated and coated surfaces can prevent degradation processes more effectively, particularly those which originate at surfaces[4]. This applies to a wide range of engineering applications, as exemplified by the wear coatings listed in Table 1.

It is now widely recognized – see for instance the applications cited in Table 1 – that the successful exploitation of these processes and coatings may enable the use of simpler, cheaper and more easily available substrate materials, with substantial reduction in costs, minimization of demands for strategic materials and improvement in fabricability and performance. In demanding situations where the technology becomes constrained by surface-related requirements, the use of specially developed coating systems may represent the only real possibility for exploitation[4,6–9].

Table 1 *Industries and components using thermally applied wear coatings[5]*

Aero gas turbines	Land-based turbines	Others
* Turbine and compressor blades, vanes	• Turbine and compressor buckets, vanes, nozzles	♦ Feed rolls
* Gas path seals	• Piston rings (IC engine)	♦ Pump sleeves
* Mid-span stiffeners	• Hydroelectric valves	♦ Shaft sleeves
* Z-notch tip shroud	• Boiler tubes	♦ Gate valves, seats
* Combustor and nozzle assemblies	• Wear rings	♦ Rolling element bearings
* Blade dovetails	• Gas path seals	♦ Dies and moulds
* Flap and slat tracks	• Impeller shafts	♦ Diesel engine cylinder
* Compressor stators	• Impeller pump housings	♦ Hip joint prostheses
		♦ Hydraulic press sleeves
		♦ Grinding hammers
		♦ Agricultural knives

SCIENCE
ENGINEERING
thermal efficiency
fuel(s)
contact
aggressive environment
?
temperature, differential temperature
time
wear
corrosion
stress
corrosion mechanisms
kinetics/thermodynamics
wear mechanisms
environmental barrier
thermal barrier
diffusion barrier
wear resistance
thickness, weight
process & reproducibility
quality
treatment & process(es)
ageing
repair, tolerances
joining
cost, other & ?
structure modelling
diffusion modelling
defects
fracture mechanisms
deformation mechanisms
graded interfaces
coating/treatment layers
substrate
interfacial science
interfacial stresses
adhesion
epitaxy
substrate prep & finish
adhesion, bonding
machining damage
shape & complexity
microstructure
crystal structure
phases
heterogeneity
thermal properties
mechanical properties
other & ?
coating/treatment, effects
cost & availability
specific mechanical properties
specifications
heat treatment
thermal capabilities
machinability
other & ?

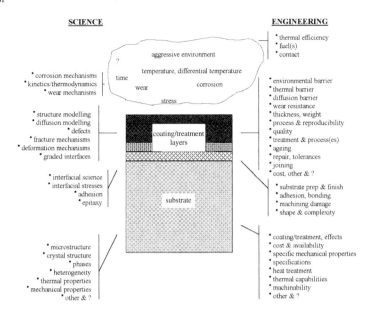

Figure 1 *Aspects of surface engineering research*[1]

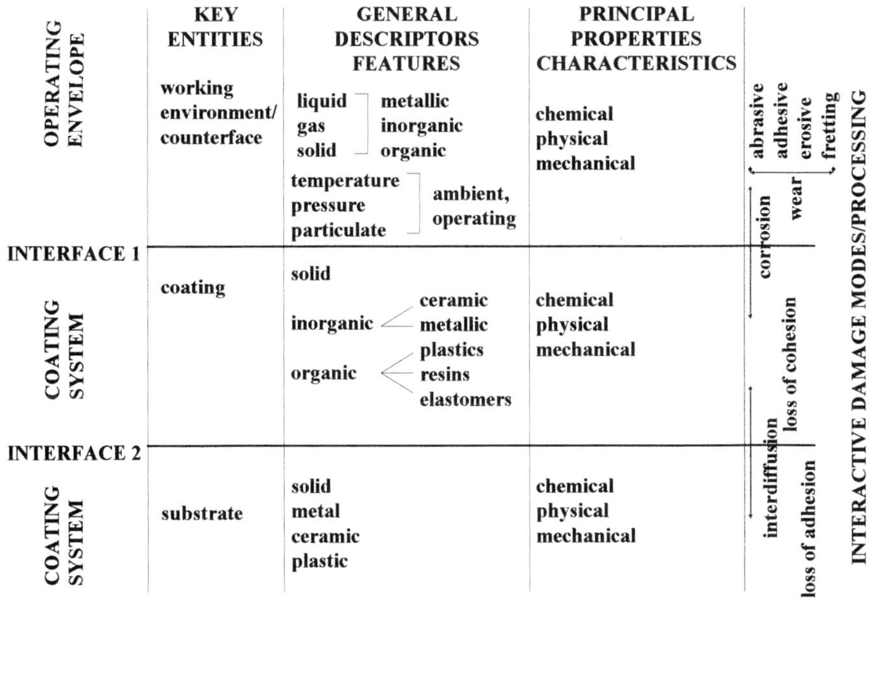

OPERATING ENVELOPE
KEY ENTITIES
GENERAL DESCRIPTORS FEATURES
PRINCIPAL PROPERTIES CHARACTERISTICS
working environment/ counterface
liquid
gas
solid
metallic
inorganic
organic
temperature
pressure
particulate
ambient, operating
chemical
physical
mechanical
abrasive
adhesive
erosive
fretting
wear
corrosion
INTERFACE 1
COATING SYSTEM
coating
solid
inorganic
ceramic
metallic
organic
plastics
resins
elastomers
chemical
physical
mechanical
loss of cohesion
INTERFACE 2
COATING SYSTEM
substrate
solid
metal
ceramic
plastic
chemical
physical
mechanical
interdiffusion
loss of adhesion
INTERACTIVE DAMAGE MODES/PROCESSING

Figure 2 *Generalised features of a working coating system*[6]

Surface engineering produces surfaces with a unique combination of bulk and surface properties resulting in the creation of a high performance composite material. However, the biggest benefit that flows from the use of surface engineering lies in the ability to create new surfaces with highly non-equilibrium structures.

2.2 Corrosion- and Wear-Related Failures

The basic features of a simple wear-or corrosion-resistant coating system are shown in Figures 1 and 2. Refering to Figure 2, selective interaction is required at *Interface 1* to provide a wear- or corrosion-resistant surface. Selective interaction of a different kind is required at *Interface 2* to obtain adequate adhesion. Such interaction(s) must not lead to the removal of coating constituents and/or their dilution by interdiffusion across *Interface 2*. The requirement for prolonged sustainability of the corrosion- and wear-resistance of the coatings imposes additional constraints on the design of the coating system. The coatings must contain a reservoir of elements to sustain the required selective interactions at the surface (*Interface 1*). Other constraints on the coating design may flow from the necessity of a surface to resist a number of degradation processes occurring at the same time. To satisfy these requirements at both surfaces or to prevent conjoint actions of different modes of degradation, multicomponent/ multiphase and multilayered coatings are required. Even so, adequate coating systems can now be designed and produced using intelligent combinations of various processes[10,11].

2.3 Surface Engineering Technologies

Surface engineering techniques generally consist of surface treatments where the compositions/structures or the mechanical properties of the existing surface are modified, or a different material is deposited to create a new surface. Surface engineering is essential in the application and exploitation of high performance engineering components. This is especially true in relation to both the rising costs of advanced performance structural materials and the increasingly high life-cycle costs associated with high performance systems. Table 2 illustrates the market share for various types of surface finish[12].

Deposition procedures, include traditional electrodeposition and chemical conversion coating, together with thermal spraying – where a plasma or electric arc melts a powder or wire source, and droplets of molten material are sprayed on to the surface to produce a coating; PVD, in which a vapour flux is generated by evaporation, sputtering or laser ablation; and CVD, where reaction of the vapour phase species with the substrate surface produces a coating.

Surface treatments include:
* mechanical processes that work-harden the surface – e.g. shot-peening;
* thermal treatments which harden the surface by quenching constituents in solid solution – e.g. laser or electron beam heating;
* diffusion treatments which modify the surface composition – e.g. carburizing and nitriding;
* chemical treatments that remove material or change the composition by chemical reactions – e.g. etching and oxidation; and
* ion implantation – see Table 3 for applications – where the surface composition is modified by accelerating ions to high energies and implanting them in the near-surface[4].

Table 4 lists characteristics of a coating which are important in relation to quality assurance.

Table 2 *Surface finishing – value by industry sectors*[12]

Coating	Size (£M)	Share (%)
Organic	1,450	43
Plating	705	21
Galvanizing	355	10
Surface heat treatment	325	10
Hard facing	100	3
Anodizing	65	2
Tin plating	40	1
Vitreous enamelling	40	1
PVD & CVD	25	1
Others	250	8
Total	3,355	100

Comparisons of certain of the above techniques are contained within Table 5 and examples of deposition and treatment technologies used in the aerospace industry are given in Table 6.

2.4 State of the Art and Future Developments

Since the early 1980s there has been a continuing and rapid development of advanced surface engineering practices for the optimization of corrosion and wear resistance. It is now possible to produce coatings of novel composition and microstructure in multilayer/multicomponent format as appropriate to the design specification, by a variety of sophisticated physical and chemical processes, including hybrid technologies. The paradox concerning compatibility between the environment, coating and substrate is no longer a problem. At the level of research scientists and engineers, efforts must be made to enhance and systematize understanding of the various process technologies. For instance in PVD, one such issue – which offers a distinct competitive advantage – concerns plasma densities and their importance

Table 3 *Industrial exploitation of ion implantation*[13]

Material	Application (specific examples)	Typical results
Cemented WC	Drilling (printed circuit board, dental burrs etc)	Four times normal life, less frequent breakage and better end product
Ti–6Al–4V	Orthopaedic implants (artificial hip and knee joints)	Significant (400 times) lifetime increase in laboratory tests
M50, 52100 steel	Bearings (precision bearings for aircraft)	Improved protection against corrosion, sliding wear and rolling contact fatigue
Various alloys	Extrusion (spinnerets, nozzles and dies)	Four to six times normal performance
D2 steel	Punching and stamping (pellet punches for nuclear fuel, scoring dies for cans)	Three to five times normal life

Table 4 *Characteristic properties of a coating*

Structural	Mechanical	Physical	Chemical
Composition	Adhesion	Specific heat	Chemical stability
Density	Cohesion	Thermal expansion	Environmental compatibility
Porosity	Hardness		Corrosion resistance
Phase contents	Modulus		Biocompatibility
Crystallinity	Ductility		
Grain size	Strength		
Amorphosity	Fracture toughness		
Defect structures	Internal stress		
Dislocations	Wear resistance		
Vacancies	Friction coefficient		
	Deformation mode		

in facilitating the repeatable manufacture of advanced surface engineered artefacts with outstanding properties and performances[10,11]. However, a further critical problem with regard to quality assessment is the precise significance of measured properties and characteristics – e.g. hardness – in relation to actual coating performance. Here insight is needed into the consequence of particular hardness levels in relation to wear performance, so that a coating engineered to a given hardness could be expected to offer a predetermined design wear life.

At the fundamental level there is a need to understand structure/property relationships in, for example coatings, so that surface engineered systems can be designed from conception to develop desired properties. This also requires a better understanding of the degradation processes which need to be prevented/minimized by the designed surface. In this regard there is considerable scope for the creation of tailored coatings of chosen composition, structure and properties – including multilayer/multicomponent format – by highly adaptable PVD and CVD technologies (Table 7).

Table 5 *Comparison between five surface engineering processes[14]*

Process	Resistance to wear	Risk of distortion	Resistance to impact	Convenience	Range of materials
Plasma spraying – atmospheric	High	Low	Low	Very good, gun is offered to the work	Extensive
Plating	High	Low	Medium	Low, work is processed in a bath	Low
Welding	Medium	High	Good	Good	Medium
CVD/PVD ion deposition	High	Low	Good	Low vacuum chamber required	Good
Cladding	Low	Low	Good	Good	Low

Table 6 *Surface engineering technologies used in the aerospace industry[15]*

Technique	Material	Requirement	Application
Mechanical treatments, eg peening	Steels, titanium-based and nickel-based alloys	Improved mechanical and wear properties	Compressor blade roots
Paints	Phenolic and epoxy polyurethanes	Cosmetic, corrosion and wear, earthing, emissivity and infrared	Shafts, discs, blading
Polishing	Steels, titanium-based and nickel-based alloys	Cosmetic, salvage and repair efficiency	Aerofoil surfaces on vanes and blades
Electrochemical	Tribomet, chromium	Corrosion and wear, salvage and repair	Bearing chambers, stator vanes
Thermal spraying (D-gun, flame spraying, plasma spraying)	Al/Si polyester, WC/Co, CuNiIn	Corrosion and wear, salvage and repair, seals, net-shapes	Snubbers, gas-path seals, combustor cans
Thermochemical	Nitrogen and carbon into steels	Improved mechanical properties	Shafts and gears
Pack aluminizing	Nickel-based alloys	Corrosion/oxidation	Aerofoil surfaces on vanes and blades

2.5 Quality Assurance

Quality assurance of surface engineered coatings and surfaces is a major issue particularly for the coating users and producers[10].

In the absence of a definitive knowledge of the structure/property relationships in coatings deposited on a surface, only an empirical approach can be adopted[10]. A coating can be described in terms of its characteristic properties (Table 4).

One approach which is being increasingly adopted is to define the functionality of the coating for a particular application in terms of a sub-set of the properties listed above. For example, to achieve quality assurance of a load-bearing prosthesis, consideration can be given, in the first instance, to the sub-set of properties consisting of adhesion, strength, wear resistance, friction and biocompatibility. Similarly corrosion resistant coatings can be quality assured by addressing parameters such as adhesion, residual stress, ductility, K_{Ic} , fatigue/crack growth resistance and chemical stability.

Quality assurance of deposition technology[17], as well as manufactured and surface engineered artefacts to reliably impart specified measurable performances, is central to surface engineering. Effort needs to be made to develop expert systems to design and select coating/treatment systems and hence define the appropriate process technology.

These skills and competencies must be applied within the construct of business realities – achieving sustained world-class competitive advantage.

Table 7 *Future surface engineering activities*[4,16]

1. Surface engineering of non-ferrous metals
2. Surface engineering of polymers and composites
3. Surface engineering of ceramics
4. Mathematical modelling of surface engineered components
5. Surface engineering in material manufacture
6. Statistical process control in surface engineering
7. Non-destructive evaluation of surface engineered components
8. Duplex or hybrid surface engineering technologies and design, eg:
* laser treatment of thermal and plasma spray coatings
* ion beam mixing and ion-assisted coatings
* hot isostatic pressing of overlay coatings
* thermochemical treatment of pre-carburized steels
* thermochemical treatment of pre-laser hardened steels
* CVD treatment of pre-carburized steels
* PVD treatment of pre-nitrided steels
* ion implantation of pre-nitrided steels

3 MANAGEMENT ISSUES

Hard and anecdotal findings from research into the strategic management of technology and the exploitation of technological innovation suggest that, for leading firms in a wide variety of industries, developing advanced technologies *per se* is rarely the constraining challenge in technological innovation. Rather, the challenge is innovation in the market. In this regard quality management, management of change, strategic management, innovation and knowledge management are as important as technological issues.

3.1 Quality

Bench-marking is a means of measuring and comparing performance and may be defined as, "the continuous process of measuring products, services and practices against the toughest competitors or those companies recognized as industry leaders". The emphasis should be on understanding how surface technologists carry out their activities; learning how other groups excel in surface engineering; and then adapting and reinterpreting what has been learnt in a way that makes for competitive advantage. Instead of aiming to improve only against previous performance, technologists should use bench-marking to inject an element of imagination into the quest for progress, while simultaneously and objectively scrutinizing established processes. Inevitably, the perspective is international[18,19].

As part of the normal research work an external audit could be carried out, this may comprise a survey of the controls surrounding researchers' daily records – before and after implementation of quality improvements. Assessments of the following areas could be among the audit's key functions:

* assurance of daily task completion;
* periodic review by management;
* verifiability of results claimed against physical evidence;
* evidence of consistency/quality of results claimed; and
* indication of appropriate coordination with clients.

Items for further investigation might also include:
- How is control exercised by management – formally or informally?
- Is control proportional to risk, exposure and objectives?
- Are guidelines for the delegation of authority suitable for the organization?

A standard part of the audit would be an analysis of the effectiveness of researchers' time management. The *day in the life of* technique consists of unobtrusively observing the activities of selected researchers over a period of several days. The allotment of their – *value added* or *non-value added* – time should be reviewed with both the researcher and manager. For *non-value added* activities the root causes of the operation would be investigated and minimized[20].

3.2 Change

Management of strategic and organizational change must address three key questions:
1. How is change initiated and implemented in relatively successful organizations?
2. What is the rôle of the management at the service provider and end-user companies in initiating change?
3. What is the contribution of management development in the implementation of organizational and strategic change?

Change management must at the same time stimulate innovation and provide mechanisms for dealing with uncertainty in knowledge-intensive SE business environments. This might be accomplished by systematically analyzing the organization's values, needs, interests and relationships, and productively applying the insights gained. An important aim of this process is to determine – using multiple levels of comparison – specific factors that might lead to competitive advantage.

Individual organizations and consortia need to assess complementary plant networks – here the objective should be to provide qualitative insight and analytical tools to facilitate the development and adaptation of plant networks in response to diverse and changing markets, manufacturing costs and technological capabilities across countries[21].

3.3 Strategy

During times of tumult strategic alliances become increasingly attractive. Research has indicated that to make strategic alliances work, success does not necessarily come from the structural or systems aspects of the alliance, but rather from the quality of relationships, the degree of trust, mutual commitment and the flexibility of attitude brought to the relationship. General assessments of the management challenges associated with creating and managing strategic alliances must consider the dynamics of cross-functional partnerships and a specific evaluation of the rôle of international alliances in high-tech SE industries. Alliances need to be analyzed from three perspectives:
a. direct economic costs and benefits;
b. historical evolution; and
c. external networks.

Growing reliance on strategic alliances has prompted investigations into how alliance partners utilize complementary technological capabilities to create products that neither could develop individually and what drives companies to choose strategic alliances and joint ventures over internal R&D and licensing as approaches to augmenting technological capabilities. Also being explored are how technology-based companies develop and exploit strategically valuable knowledge assets and the managerial and technical processes by which novel technologies are applied to the development of complex new products. Other studies are suggesting ways in which senior executives can nurture superior product development performance and advancing the notion that technological innovation is often less a constraint than the need to innovate in the marketplace. Likewise, research suggests the principal challenge in entrepreneurship is building skills such as selling ideas and products and applying theories of competitive advantage, not communicating new knowledge or developing new theories that better explain entrepreneurship. Of particular interest is how technological knowledge is transferred across company boundaries and the rôle collaborative development might play in countering a company's core rigidities, i.e. deeply in-grained, but out-dated, technological capabilities.

Strategic decision-making may be considered as essentially a technique for making judgements when the outcome depends in part on the actions of others, it involves systematic analysis consistent with the commonplace expedient of putting oneself in another's shoes, and interactive decision and value analysis. The process may involve clinical, statistical and theoretical research focused on major commitments, notably investments and disinvestments.

The impact of technology on industry structure – such as investments in SE technologies – need to be evaluated as part of coherent business strategies and be viewed as strategic necessities rather than attempts to gain sustainable advantage. In considering technological change and competitive strategy it is necessary to explore the dynamic links between and strategic consequences of technological change and shifts in organizational structure and competitive advantage. Research attempts to answer questions such as:

* How does a firm identify opportunities to create value from technological change and substitution?
* How can firms create sustainable profits in the face of SE technological change and resulting adjustments in competitive dynamics?
* What methods are available to managers for evaluating highly uncertain projects and the value of developing specific competencies and capabilities?
* How might competitors' capabilities be assessed?
* How does technological change affect competitive dynamics and redefine industry structures?
* Under what conditions should a firm invest in a new technology?
* How do competitive dynamics affect the evolution of technology and attendant quality standards?

The merging of computers, telecommunications and SE – and the blurring of functional and technological boundaries within the SE industry – forces managers to pursue a wide range of topics, including the competitive dynamics within and between quality standards, the rôle of alliances in SE technology, and the relationship between technology choices, scope of a firm (i.e. degree of horizontal and vertical integration) and financial decisions.

Research might profitably be conducted into the determinants of superior SE process development performance and process development strategies within, for instance, the aerospace and automotive industries. Detailed qualitative and quantitative data could be collected on the histories, strategies and performance of surface engineered artefacts. Statistical analysis could then be used to identify how such factors as organizational structure, project strategy, organizational capabilities and technological environment influence process development lead times, productivity and costs. Such a study would be expected to shed light on the special challenges that attend the management of R&D projects and building of development capabilities in SE-based industries; in addition this would yield significant insights into the potential strategic rôle of SE technologies in these important industries over the coming decade[21].

3.4 Innovation

Innovation – the successful exploitation of new ideas – is the key to sustained competitiveness. British university research groups have responded to the Government's "Science, Engineering and Technology" and "Competitiveness" White Papers, and especially that aspect which is concerned with the application of new techniques and ways of working that improve the effectiveness of individuals and organizations. The scope of several research programmes covers the rôle of innovative management in the achievement of sustained improvement in the bottom-line performance of commercial and industrial businesses. In particular emphasis is placed on the human and organizational processes and conditions that contribute to this. Such "research on business in business" is designed to increase knowledge and understanding of these crucial elements and to encourage their exploitation by industry[22].

Organizations frequently manage innovation and development activities not only as single projects but also as a cohesive set of related projects. Research is currently underway in a number of organizations to compare the different methods used for managing project sets. The aim is to identify "best practice" methods and to examine the link between the strategic aims of the organization and the formation of the goals and objectives of the projects[23].

3.5 Knowledge

In new lean, business processes where non-value added tasks have been eliminated, IT (information technology) can facilitate manufacturing philosophies, e.g. JIT (just in time). However, complex problem areas still exist in re-engineering companies. These problems will occur with large and small companies, and basic IT cannot address them. They include the handling of incomplete, conflicting and vague data, the discovery of knowledge in massive data sets, the interpretation of legislation and inter-organizational contracts, the management of change, and the re-application of an expert's accrued experience and expertise. *Knowledge-based systems* (KBSs) provide a series of techniques that can help to assess, manage and ameliorate these problems. There have been a number of reported successes where KBSs have added value to business processes, and in fact made business re-engineering possible[24].

A *knowledge-based system* refers to any assemblage which incorporates a level of expertise and experience, which can be used to address new situations in an intelligent way. The knowledge may derive from human *experts*, research papers and reports, or computer systems. KBS technology – e.g. expert systems, neural networks, data mining and artificial intelligence – applied to SE can offer industry significant benefits:
- intelligent decision-making support, making safer, faster and more effective decisions;
- consistency of approach and assurance of quality standards;
- better service delivery, increased productivity and improved cost control;
- dissemination of scarce expertise across the organization;
- a valuable training tool for new engineers and managers; and
- developing a way of making complex situations more transparent for the decision-maker and linking shop-floor activity to commercial transactions.

The technical focus might be concentrated in the fields of knowledge-based SE and corrosion and wear management, and it is in these areas that developed solutions to complex industrial problems should be sought[25].

4 THE *SURFACE ENGINEERING RESEARCH GROUP*

4.1 Background

The *Surface Engineering Research Group* (*SERG*) comprises a Director (Prof. P. K. Datta), an Assistant Director (Dr J. S. Burnell-Gray), 5 academic consultants, 2 Research Fellows, 2 Research Associates and 4 Research Students. It has five core functions: *Research, Education, Technology Transfer, Consultancy* and *Training*.

SERG firmly believes that surface engineering is one of the several keys to UK industry gaining a world-class competitive advantage. The *Group's* objectives are to identify outstanding pivotal research issues, promote the take-up of SE research results in industry and influence UK and European Union research policy to help create the next generation of manufacturing systems.

Our technical focus is concentrated in the fields of surface engineering, corrosion and wear, and it is in these areas that we seek to develop commercial applications as solutions to complex industrial problems. The research focus of the laboratory is addressed along with partners in for instance Rolls-Royce, Multi-Arc (UK), Chromalloy UK and Johnson Matthey. *SERG's* research portfolio also supports cross-fertilization of the work of other groups.

The development of successful applications in corrosion and wear management is not a straightforward task as it requires the synergistic combination of expertise on coatings deposition, surface engineering, corrosion and wear engineering, as well as corrosion/wear theory. However, the rewards of SE are across-the-board improvements in value, quality, customer support and productivity.

4.2 Research Portfolio

It is within this demanding and continually changing framework – also see Figure 3 – that the *Group's* research activities are based. Added-value to the component in relation to enhanced mechanical, thermal, chemical, electrical or optoelectrical attributes, as well as fitness-for-purpose and the minimization of life cycle costs, are the driving forces for the justification and adoption of surface engineering practices. In this regard the *Group's* research has not only contributed to the characterization and understanding of the functional behaviour of coatings, but also at a more fundamental level has involved the systematic application and testing of scientific principles with a view to creating entirely new types of surfaces with novel properties. More pragmatically this research effort has in addition contributed to the achievement of more reliable coatings with reproducible properties. This reproducibility aids prediction of the performance, e.g. mechanisms and time-dependent behaviour, of surface-modified artefacts using surface analytical techniques, system modelling, interfacial simulation and NDE – all of which are essential if engineers are to fully exploit the potentials of the discipline.

The scope of research programmes – from basic, through strategic/pre-competitive to applied – is outlined in Table 8. Other examples of applied surface engineering research lie not only in the field of aero engine turbine blades, but also in natural gas production and combustion, and orthopaedics.

The *Surface Engineering Research Group* advocates the principles of surface engineering and provides local and national industry with a world-class facility for the study of corrosion and wear control using surface engineering technologies. Since the early 1980s *SERG* has gained an international reputation in the area of surface engineering and also acts as a regional teaching facility in corrosion and wear prevention using the latest surface engineering technologies involving surface analysis, surface modification and coatings deposition. *SERG* offers a well-founded mechanical engineering workshop and corrosion laboratory comprising computer-aided machining, non-traditional machining, and high temperature gaseous and molten salt, as well as aqueous corrosion facilities.

Table 8 *SERG's research portfolio*

Current and Recent Programmes
- *Design and Optimization of High Temperature (HT) Protective Coatings* concerns the design and development of HT degradation resistant MCrAlYX-type coating systems capable of withstanding corrosion in coal gasifier atmospheres typically containing significant Cl_2 or S_2 potentials. The project formed part of the EPSRC Rolling Programme in Surface Engineering jointly pursued by ourselves, Hull University and Sheffield Hallam University.
- *Interfacial Modification of MMCs* aiming to improve HT mechanical properties and corrosion resistance of Ti-Ti alloy matrix/SiC fibres. Interfacial modelling is used to quantitatively and qualitatively describe diffusion and corrosion mechanisms. HT studies are performed in atmospheres designed to simulate aero-engine compressors. Part of the EPSRC Rolling Programme with UNN support.
- *Optimization of Electroless Deposition Processes for Protective Coatings* studying and exploiting selected parameters (bath chemistry, and coating composition and morphology) of electroless Ni-B and Ni-P coatings with varying concentrations of B and P, as a function of coating condition and determining the resultant effect on corrosion, wear and fatigue behaviour. Part of the EPSRC Rolling Programme.
- *Design and Development of Pt-Aluminide Coatings for Improved Corrosion Resistance* determining a reference database and enhancing Pt-modified aluminide coatings, deposited with or without Ta, on superalloy substrates designed for use as gas turbine blades. Hot corrosion is monitored in simulated gas turbine environments. Sponsored by EPSRC, Rolls-Royce plc, C-UK Ltd. and Johnson Matthey.
- *Improved High Temperature Resistant Silicon Nitride-Silicon Carbide Composites* concerning the characterization and optimization of Si_3N_4-SiC composites during exposure to replicated combustion environments. The EU provided funds for this research programme, jointly pursued by ourselves, Limerick University, T&N Technology and British Gas.
- *High Temperature Corrosion of Car Engine Valve and Valve Seat Materials* involving the study and selection of appropriate Cr_2O_3- and Al_2O_3-forming alloys, oxide dispersion-containing and other mechanically alloyed materials, to optimize durability in simulated car engine atmospheres. Funding by EPSRC and British Gas.
- *Surface Engineered (Diamond-Like Carbon) Prostheses* embraces a wide range of activities, viz: depositing and characterizing DLC coatings, biocompatibility tests, and fatigue and wear studies of coated prostheses. Funding is from DTI Link. Partners include 3M, Teer Coatings and the Royal Victoria Infirmary (Newcastle).
- *Studies of Electroless Coatings* concerning the deposition and characterization of electroless coatings on a number of substrates. The aim was to produce coatings with a set of specific properties such as high corrosion resistance, texture, appearance and good adhesion. The project was sponsored by NPL.

Completed Programmes
* *Coatings Technology*
  - Development of Ta and Nb ion-plated coatings as novel load-bearing (human) implant materials

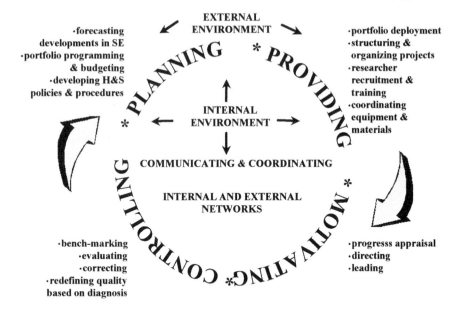

Figure 3 *SERG's research management functions*

- Electrolytic deposition of Co-Sn alloys for the electrical industry
* *Aqueous Corrosion*
  - Corrosion of implant systems in bodily fluids
  - Corrosion of electroless coating systems
  - Passivation in electrolytically deposited alloys
* *Environmental Cracking*
  - Stress corrosion cracking (SCC) in titanium alloys used in submarine hulls
  - Development of software and hardware to study and analyse crack propagation
  - SCC of nodular cast irons in heavy goods vehicle (HGV) suspensions
* *High Temperature Degradation*
  - Chloridation of binary alloys, MCrAlX-type and MCrAlY-type coatings alloys
  - Sulphidation/oxidation of advanced engineering ceramics, Si_3N_4/SiC composites, CoCrAlYX-type coatings alloys, refractory metals, HfN and Nb_2N
  - Oxidation of titanium alloys and MMCs for aerospace applications

---

4.3 Future Challenges, Development and Plans

Plans to consolidate and expand on the success enjoyed by *SERG* centre on:

Research
- ◆ Expanding into interfacial modelling and interfacial engineering;
- ◆ moving into the area of pack cementation;
- ◆ diversifying into deposition (PVD) control and optimization;
- ◆ assessing the viability of PVD catalyse;

- moving into new pre-treatments, e.g. pre-nitridation and pre-carburization;
- extending electroless deposition into cobalt alloys;
- moving away from sulphidation towards chloridation and erosion/corrosion;
- introducing the concept of knowledge-based systems;
- contingency funding to refurbish/maintain existing equipment;
- buying a mass spectrometer;
- constructing a burner rig; and
- further exploiting IPR.

Education
- Offering the facility of split PhDs, interchange of research fellows and travelling professors;
- providing a surface engineering module to existing postgraduate and EPSRC IGDS courses; and
- developing an IGDS in Research Management.

Technology Transfer
- Presenting the 5th International Conference on Advances in Surface Engineering in 2000;
- hosting the 1998 Institute of Corrosion's Corrosion Science Symposium;
- hosting the 2nd European Workshop in Surface Engineering Technologies for SMEs in 1998;
- promoting in-house courses in corrosion, wear and surface engineering based on EPSRC's Rolling Programme in Surface Engineering commencing 1998; and
- further enhancing consultancy and contractual services to industry.

References

1. J. S. Burnell-Gray and P. K. Datta (editors), 'Surface Engineering Casebook' Woodhead Publishing, November 1996.
2. K. N. Strafford, P. K. Datta and J. S. Gray (editors), 'Surface Engineering Practice: Processes, Fundamentals and Applications in Corrosion and Wear', Published by Ellis Horwood, 1990.
3. P. K. Datta and J. S. Gray (editors), Conf Proc 3rd Int Conf Advances in Surface Engineering, Newcastle upon Tyne, May 1992. Published by The Royal Society of Chemistry, 1993, 'Surface Engineering Vol I: Fundamentals of Coatings', 'Surface Engineering Vol II: Engineering Applications', 'Surface Engineering Vol III: Process Technology and Surface Analysis'.
4. V. Sankaran, 'Surface Engineering – A Consultancy Report', Advances in Materials Technology: Monitor, Issue 24/25, February 1992.
5. P. Sahoo, *Powder Metallurgy International*, 1993, **25**, 73.
6. K. N. Strafford and S. Subramanian, *J. Materials Processing Technology*, 1995, **53**, 393.
7. Reference 2, Keynote paper, Chapter 3.1.1, p. 397.
8. J. S. Burnell-Gray and P. K. Datta (editors), 'Quality Issues in Surface Engineering', To be published by Woodhead Publishing 1997.
9. J. S. Burnell-Gray, Internal reports, MBA course, University of Sunderland , 1995-1997.

10. Correspondence with K. N. Strafford, University of South Australia.
11. Correspondence with K. N. Strafford, University of South Australia.
12. D. Hemsley, *Engineering*, 1994, **235**, 25.
13. P. Sioshansi, *Thin Solid Films*, 1984, **118**, 61.
14. I. H. Hoff, *Welding and Metal Fabrication*, 1995, **63**, 266.
15. D. S. Rickerby and A. Matthews, 'Advanced Surface Coatings: A Handbook of Surface Engineering', Published by Blackie, 1991.
16. T. Bell, *J. Phys D: Appl Phys*, 1992, **25**, A297.
17. C. Subramanian, K. N. Strafford, T. P. Wilks, L. P. Ward and W. McMillan, *Surface and Coatings Technology*, 1993, **62**, 529.
18. A. van de Vliet, *Management Today*, January 1996, 56.
19. F. C. Allan, *Special Libraries*, 1993, **84**, 123.
20. S. J. Burns, *Internal Auditor*, 1991, **48**, 56.
21. Anon, Harvard Business School, Internet home pages: www.hbs.edu/research/summaries/lec.html, accessed 21st October 1996.
22. Anon, Economic & Social Research Council, Internet home pages: www.bus.ed.ac.uk:8080/ESRC-Innovation.html, accessed 23rd October 1996.
23. Anon, Cranfield University, School of Management, Internet home pages: www.cranfield.ac.uk/som/res/default.html, accessed 18th October 1996.
24. Anon, Ulster University, Northern Ireland Knowledge Engineering Laboratory, Internet home pages: www.nikel.infj.ulst.ac.uk/nintro.htm, accessed 23rd October 1996.
25. Anon, London University, London Business School, Internet home pages: http://www.lbs.lon.ac.uk/om/research.html#tech: accessed 18th October 1996.

Section 3.1 Biomedical

3.1.1
Surfaces in Implantable Ceramics Technology

A. Ravagliok A. Krajewski and P. Vincenzini

IRTEC-CNR. FAENZA, ITALY

1 INTRODUCTION

The very favourable biochemical, biological and biophysical responses of ceramic surfaces in connection with living tissues that justify their use as coatings are outlined. Bioceramic coatings are indeed among the most suitable options to make bioactive and bioconductive the surface of a prosthetic device for the tissue to be regenerated, by assuring bone growth and the joining to the device. The paper will especially focus on:

1. available materials,
2. physico-chemical aspects of coating/substrate interface,
3. coating/tissue interactions,
4. applicational trends and case studies.

2 COATING, WHY?

An ongoing approach to overcome the mechanical limitations of ceramics and bioactive glasses for load-bearing applications in surgical implants is to apply the material as a coating on a mechanically tough substrate. Specific metal alloys and medical grade Alumina and Zirconia are the commonly used substrates.

When applied on metals, the main goal of the coating is to form a protective barrier against the surrounding physiological media to avoid their chemical interaction. Indeed, the occurrence of ion release from metal alloys due to redox reactions with physiological media is well known. This affects the physical integrity of the alloy and, at the same time, a detrimental action may possibly follow from the release of toxic ions.

When applied on medical grade high strength ceramics, the main goal of the coating is to confer suitable surface properties in order that a good adhesion be established between the implant and the surrounding tissues. At this purpose, it is common to apply substances capable of (i) stimulating bone growth (bioactive substances), (ii) favouring osteointegration by the direct linkage of growing bone onto the implant surface, and (iii) promoting bone growth all around the implant surfaces (osteoconduction).

3 MATERIALS

The correct choice of the coating material is dependent on the specific function of the substrate to be coated and on the specific type of interaction we expect for the coating surface towards host tissues. Table I collects the most prevailing solutions adopted today to coat different implantable substrates.

Various techniques are available to apply inert or active coatings, the appropriate selection of a particular technique depending on the nature of both the substrate and the coating material and on the expected function. The most common coating method for the application of thick biomedical coatings is plasma spraying; practically all alumina and hydroxyapatite coatings, and some bioglass coatings, are applied by this method. Use of isostatic hot pressing also has been reported. When thinner coatings are required, most suitable techniques appear to be sol-gel deposition, electrophoretic coating, ion beam sputter deposition, reactive chemical vapour deposition and radio frequency magnetron sputtering[1]. Bioactive glasses are mostly deposited by methods such as: enamelling (glazing), flame spraying, rapid immersion and sol-gel deposition. Ion implantation also is being increasingly used to modify surfaces of surgical implants. For example a thin antifriction film on titanium substrate is obtained by N^{5+} implantation, whereas a thin TiC film is obtained by C^{4+} implantation.

4 THE COATING-SUBSTRATE INTERFACE

Physical and chemical interactions are established at the coating/substrate interface. When applying bioglass coatings, a permeation of the glass along the grain boundaries of the metal or ceramic substrate generally occurs. Any open porosity in the substrate enhances this effect. A further, perhaps more critical aspect is the partial dissolution of the substrate in the biological glass, thus altering its composition and properties. Particularly dangerous has been found to be the dissolution of alumina in the bioglass coating, which affects negatively the bioactivity of the coating. Other changes in the substrate may be related to possible effects of dissolution/reprecipitation at the coating/interface. For example, when zirconia substrates are used, a certain amount of monoclinic phase may be formed at the coating/substrate interface because of reprecipitation of dissolved zirconia during cooling whose stabilizer (e.g. Y_2O_3) remained dissolved into the glass. Plasma spraying hydroxyapatite involves structural transformations

Table 1 *Presently used coating systems according to substrate material*

Substrate material	Bioactive coating	Bioinert coating
316l stainless steel	Hydroxyapatite Biological glass	Alumina
Co-Cr-Mo alloy	Hydroxyapatite Biological glass	Alumina
Titanium alloys	Biological glass Hydroxyapatite Tricalciumphosphate	Alumina TiN TiC
Alumina	Biological glass	-
Zirconia	Biological glass	-

of the starting hydroxyapatite phase, in particular if nonstoichiometric powders are used, with the possible crystallization of high temperature phases such as tetracalciumphosphate (TeCP), whereas α- and β-tricalciumphosphate (TCP) and/or CaO may form as secondary phases from non-stoichiometric hydroxyapatite. Because of its solubility in physiological fluids, CaO may be particularly dangerous after implantation as it causes microcavities in the material which may be detrimental to the mechanical integrity of the implant. On the other hand, TCP is resorbable; thus an excess amount of this may also be deleterious to the mechanical integrity of the coating. If a substantial amount of α-TCP is formed, microcracks may be generated during cooling from the deposition temperature as a consequence of the very large mismatch of the thermal expansion coefficients between α-TCP and hydroxyapatite (the thermal expansion coefficient for α-TCP is about five times than for hydroxyapatite). By heating the coated implant up to a maximum temperature of 900°C for typically 2 hours (not always this procedure may be applied to metallic substrates) a recovery of the starting crystalline form may be obtained involving a drastic decrease, up to the complete disappearance, of α-TCP. Operating in wet (by steam) atmosphere also a noticeable decrease of β-TCP occurs accompanied by the reconstitution of some hydroxyapatite, but with the formation of undesirable amounts of free CaO.

When substrates based on titanium alloys are used, titanium ion interdiffusion in the hydroxyapatite crystalline lattice may occur, β-TCP and CaO may possibly form and $CaTiO_3$ may possibly crystallize by further reaction of CaO with the TiO_2 resulting from metallic titanium oxidation with the H_2O formed during the transformation from hydroxyapatite to β-TCP[2]. Even traces of Ti_3P were detected at the interface. With alumina coatings, allotropic phase transformations may occur that weaken the coating microstructure. Doping with small amounts of MgO is a suitable measure to reduce the effect. Indeed MgO is known to react with the most unstable alumina phases to give rise to $MgAlO_4$ spinel which segregates at the grain boundaries, thus strengthening the texture.

Zirconia coatings also show problems. At the high deposition temperature of the plasma some evolution of the stabilizer occurs with formation of different zirconia phases and stabilizer segregation at grain boundaries.

5 THE BONE BONDING CAPABILITY

Several requirements should be fulfilled which affect the bonding capability of the coated surface with the surrounding tissue. Among these:

1. the flux of ions from the coating to the tissue should be intense enough to allow a progressive tissue reacting reconstruction, whereas a too high ion release may be inappropriate,
2. suitable electrical surface charge and possibly a uniform distribution of the dipoles present over all the surface has to be realized,
3. complete absence of ions which are not recognized by the body as metabolytes, particularly trivalent cations such as B^{3+}, Al^{3+}, etc. is required.

A common characteristic of bioactive materials is a time-dependent modification of the surface which is in contact with the surrounding tissue after implantation. Kinetic processes (biokinetics) that give rise to the modification obviously differ for different materials and generally occur by more than one stage. A list of the main sequence in the initial stage of the

Table 2 *Physiological responses to a foreign body in the initial stages*

Action	Mediator	Activity	Reaction time
Concentration gradient	Ions	Exchange	Immediate
Surface activity	Protiens	reactive coating	Some hours
Integration	Cells	Cyte/blast cells	Some days

physiological response of a tissue to a foreign body is reported in the Table 2.

Biological glasses undergo several intermediate restructuring stages of their surface involving the production of a surface silica gel onto which hydroxyapatite is deposited. During this transformation process, that corresponds to the first action of Table 2, a substantial ionic flux is established which inhibits any further interaction with the surrounding tissue. Only when a sufficiently thick film of hydroxyapatite deposited onto the surface is completed, a further step may start involving bonding of the coating surface to the bone. A list of the complete sequence of events on a biological glass is reported in the Table 3.

For all bioactive materials the mechanism of bone bonding occurs through various steps common to all of them. The first one is protein adhesion.

Different surfaces have different protein adsorption properties. Indeed any material is characterized by its own specific distribution of surface charges. Therefore microstructure, nanoporosity and chemistry play an important role on the adsorption properties of material surface.

The adsorption of a complex molecular ensemble such as a protein, resides on the reciprocal correspondence of charge distribution on the material surface and on the distribution of the functional chemical groups of opposite charge in the protein molecule. In such a way a bidimensional matching of the type "key-look" is established at the surface.

It generally has been observed that the steps involved in protein deposition occur very

Table 3 *Listing of the sequence of interfacial reaction involved in forming a bond between tissue and biological glass (rearrangement from L. L. Hench listing 1995)*

ln(time) (hours)	Reaction stages	Description
0	1	Fast exchange of cations for H+ and formation of Si–OH bonds
1	2	Release of more complex ions
	3	Polycondensation $Si\text{-}OH + Si\text{-}OH = Si\text{-}O\text{-}Si + H_2O$
	4	Adsorption in an amorphous way of Ca^{2+}, PO_4^{3-}, CO_3^{2-} ions
2	5	Crystallization of hydroxyl carbonate apatite (HCA)
	6	Adsorption of biological moieties in HCA layer (proteins, etc.)
10	7	Action of macrophages
20	8	Attachment of stem cells
100	9	Differentiation of stem cells
	10	Generation of matrix
	11	Crystallisation of matrix

quickly on negatively charged surfaces. Negatively charges surface can also promote the activation of the Hageman's factor by rendering the adsorbed proteins more susceptible to limited proteolysis[3]. The proteins involved in this adhesion process are albumin, fibronectin, fibrinogen and collagen. Usually, cells adhere to the material surface through the interaction of fibronectin. It is also possible for the cells to adhere directly through their own cellular membrane (i.e. without the interposition of fibronectin) if a suitable charge distribution is established at the surface of the material. The transformation of *Zymogens* into active enzymes follows from the adsorption process of proteins[4]. Associated to the proteins adsorbed on the material surface there are also immuno-gamma-globulins (IgG), hormones such as bone growth factors, and activated enzymes which influence cell differentiation and proliferation.

The interaction between osteoblasts receptors and the corresponding protein ligands on the surface contribute to the cellular adhesion. Anyway, the selection of the cellular species that will preferentially adapt on a surface is strictly related to the specific type of proteins preferentially adsorbed by each material.

Bioinert materials are such that the step sequence of the reconstructive intervention leads to the formation of fibrous tissue. The occurrence of ions subjected to metabolism such as Mg^{2+}, Ca^{2+}, K^+, Na^+ in the physiological liquid near the surface of the material involves higher or lower cell activity according to their relative amounts. This comes from the possible conditioning of the cell membrane potential according to the Ca^{2+}/Mg^{2+} and Na^+/K^+ cellular pumps.

It has been also demonstrated[5] that the occurrence of silicate ions may favour a more rapid bone growth, thus suggesting their possible role in tissue repair and ostheogenesis. The higher the crystallinity of the implanted bioactive substance, the higher is the stability of the constituting chemical species and, consequently, the lower is their release in the biological environment. Occurrence of an amorphous phase involves higher solubility, hence an higher rate of ion release from the substance.

When implanted in the bone, hydroxyapatite is in its thermodynamical condition of highest chemical stability and therefore it behaves as a bioactive compound with poor solubility and therefore scarce osteoinduction. Any other bioactive compound (i.e. TCP) will undergo modification, at least within the near-interface region. This chemical rearrangement involves a motion of small quantities of ions to be exchanged with the environmental physiological liquids, consequently a high osteoinduction takes place with no good connection between neoforming bone and material surface, until a more stable formation of a hydroxyapatite layer has occurred. TCP implanted in the bone tends to transform into hydroxyapatite. The only noticeable difference between the ceramic hydroxyapatite and the hydroxyapatite phase that forms by reaction at the surface of a biological glass resides in the morphology of the microstructure that does not correspond to the one developing by nature as mineralogical phase into the bone.

6 THE BONE/CERAMIC COATING INTERFACE

Apart from TiC and TiN coatings used for joints, for which do not exist specific problems concerning interaction between the ceramic layer and the biological environment substantially made of sinovial liquid, major drawbacks are encountered when dealing with ceramic coatings on prostheses intended for use in contact with bone. One of the major problems arises from the high flux of mechanical load, also transient, the cross section of the coating has to withstand.

The organism response to this mechanical load exercise is to strengthen the bone tissue in order to enable it to withstand the applied stress. The problem arises when the threshold of the maximum tolerable load (stress) for the cells is exceeded, this leading to their progressive death. If the applied stress remains tolerable, coatings are subjected to a first phagocytic step that may result in a thinning of the coating. For bioinert ceramic coatings, an excessive thinning is a remote possibility, whereas bioactive ceramics, and particularly hydroxyapatite coatings, may undergo a very substantial reduction in their thickness depending especially on the occurrence of substantial amounts of amorphous phases or other biodegradable components.

Porosity plays an important role in emphasizing the solubilization process. Its role also is important in the subsequent step of bone reconstruction in the environment of the prosthesis. Indeed, if microcracks occur at the coating/substrate interface the physiological liquid may permeate into the open porosity, cross the whole coating and interact with the substrate. In addition, osteoblasts may enter the pores. This, on one side introduces a beneficial vitalization effect on the material of the coating, but, on the other side, may cause, at the same time, remodelling of the texture of the ceramic material and local weakening.

When the plasma spray technique is applied to deposit hydroxyapatite coatings, extreme care has to be put to avoid an excess of amorphous phases and of other phosphatic components coming from the chemical disproportion of the powders caused by excessive impact and thermal energies and related microstructural damages. A substantial amount of drawbacks rises from non-stoichiometrically well balanced powders, by unsuitable grain size and shape, and from a bad use of the coating technique.

7 SOME CONSIDERATIONS ABOUT ONGOING APPLICATIONS

Nowadays, increasing interest is placed on bioactive coatings such as hydroxyapatite and biological glasses for applications involving surfaces capable to anchor to the bone tissue, possibly avoiding use of surgical cements. Indeed, previous experience with hip joint stems plasma sprayed with alumina coatings resulted in somewhat dubious post-operative results in several cases.

Actual findings indicate that use of coatings should be restricted to prostheses that have to withstand mechanical loads that increase-decrease with time in relatively long terms: i.e. mechanical shocks or transients should be minimal. Further on, no substantial shear stresses should affect the cross section of the coating.

On the other hand, the use of hydroxyapatite coatings on hip joint stems resulted in substantially positive results as demonstrated by several casistics of surgical interventions. Nevertheless not all is clear and the need exists for further improvement or re-thinking of the present approach to biomedical coatings.

Coatings, either bioinert or bioactive, are also applied on prostheses to be infibulated in hollow bone, either short or long bone. An example of advantageous use of hydroxyapatite coatings is in odontology. Here the currently used titanium alloy implants show a satisfactory bonding to bone in deep bone regions, whereas chronic phlogosis can be favoured in correspondence of the collar region. This derives from the poor adhesion of the metal to bone and a gap forms between the gingival crown and the root of the tooth. Bacteria colonies thus are easily developed owing to the presence of saliva and organic residues in the gap. A solution for the need of more suitable materials resides in bioactive ceramic coatings for their ability to induce the reconstruction of the gingival tissue all around the prosthesis.

One recent application consists in making weakly bioactive the rim plate of auricular prostheses made with bioinert ceramic (alumina or zirconia) for total ossicular replacement. This has been realized by the use of a biological glass coating[6] whose role is to induce the timpanic tissue to bond to the prosthesis surface thus realizing its fixation.

References.

1. J. G. C. Wolke, K van Dick, H.G. Schaeken, K. DeGroot and A. Jansen, *J. Biom Mat. Res.*, 1994, **28**, 1477.
2. C. Chai and B. Ben-Nissan, *J. Austral. Ceram. Soc.*, 1993, **29**, 81.
3. L. Vroman, *Bull. N. Y. Acad. Med.*, 1988, **64**, 352.
4. S. Sepatnekar et al, J. *Biom. Mat. Res.*, 1995, **29**, 247.

3.1.2

In-Vitro Evaluation of Corrosion Resistance of Nitrogen Ion Implanted Titanium Simulated Body Fluid

M. Subbaiyan[1], T. Sundararajian[1], S. Rajeswari[1], U. Kamachi Mudali[2], K. G. M. Nair[2] and N. S. Thampi[2]

[1]DEPARTMENT OF ANALYTICAL CHEMISTRY, UNIVERSITY OF MADRAS, GUINDY CAMPUS, MADRAS – 600 025,INDIA

[2]METALLURGY AND MATERIALS GROUP, INDIRA GANDHI CENTRE FOR ATOMIC RESEARCH, KALPAKKAM – 603 102, INDIA

1 INTRODUCTION

The favourable local tissue response to titanium and its alloy Ti6Al4V, has promoted its widespread use in dentistry, implantology and orthopaedic surgery. Owing to higher corrosion resistance as well as enhanced fatigue strength, this alloy(Ti 6Al4V) has been considered as an interesting alternative to previously used stainless steels and cobalt-chromium alloys. However, in the long term use the presence of vanadium in this alloy may cause problems, although vanadium in Ti 6Al4V alloy is very stable and the amount of ion release found in the normal situations is always lower than the toxic level. Toxicity of vanadium is well known, and it can be aggravated when an implant is fractured and the material undergoes fretting[1]. This phenomenon may cause local irritation of the tissues surrounding the implant or in long term implantations, the patient may be affected by the metallic ions released by the implants[2].

In order to avoid this potential risk several solutions have been proposed. One among them is titanium (commercial purity, CP) which can be used as counterpart for this alloy. However, it is not inert when incorporated into the aggressive medium of the human body. Titanium releases the corrosion products into the surrounding tissues and fluids[3-9]. Moreover, in articular prosthesis like hip and knee joints, fretting and wear may take place, causing an increase in wear of titanium and its alloys, and the degradation of ultra high molecular weight polyethylene attached to them. This phenomenon is more prone to occur with the titanium and its alloy than with the other currently used materials such as stainless steels and cobalt-chromium alloys[10-12].

Considerable attention has been paid in the recent past on the surface treatment and finishing of these materials to reduce these adverse phenomena. Particularly, nitriding by physical vapour nitridation and ion implantation have been used to improve the tribological behaviour of the materials[13-16]. However, less attention was paid towards the evaluation of rate of ion release and the corrosion resistance of the modified material. This is partially due to the difficulty in evaluating corrosion resistance of Ti and its alloys.

Yu et al[17] found that nitrogen ion implanted Ti6Al4V at the dose rate of 3×10^{17} Ions/ cm^2 reduces the passive current of the specimen in the bio brine solution. In addition to improvement in the mechanical properties (surface hardness, wear and fretting resistance), considerable changes in the electrochemical behaviour were also observed. Nitrogen ion implantation on theTi6Al4V alloy at different doses influencing the corrosion resistance has also been reported[18-19]. At higher doses large numbers of TiN and carbonitride formation were noticed which increased the wear resistance[20]. However, the role of these precipitates in influencing the corrosion resistance of the materials is not yet established.

The aim of the present study is to optimize the conditions of nitrogen ion implantation on commercially pure titanium, and to correlate the implantation parameters to the corrosion resistance in comparison with the unimplanted specimen in a Ringer's solution by electrochemical techniques namely, OCP-time measurements, cyclic polarization and potentiotransient studies. X-ray photoelectron spectroscopic studies were carried out to analyze the surface concentration and chemical state of the elements in the as-implanted and passive film of implanted and unimplanted specimens.

2 EXPERIMENTAL

Titanium (CP, supplied by Midhani, INDIA) samples of 8.5 mm diameter were ground upto 1000 grit SiC paper, and the final polish was done subsequently by 5 and 1 micron diamond paste. The polished samples were ultrasonically cleaned with acetone and rinsed with deionized water. Nitrogen ion was implanted on the finished specimens in a TAM SAMES 150 kV accelerator at the energy of 70 keV, and with the doses of 5×10^{15}, 1×10^{16}, 4×10^{16}, 7×10^{16}, and 1×10^{17} ions/cm^2. During implantation the pressure at the target chamber was maintained at 1×10^{-6} torr.

The implanted samples were mounted on to a sample holder to expose only the implanted portion of the specimen to the test solution during electrochemical experiments. The design of the sample mount is illustrated in Figure 1. The other side of the specimen was attached to a brass rod by applying silver paste. The contact region was covered by Teflon tape and over that water insoluble nonconducting varnish was uniformly coated and cured for a few hours.

Ringer's solution (9 g/l NaCl, 0.43 g/l KCl, 0.24 g/l CaCl$_2$ and 0.2 g/l NaHCO$_3$) was chosen to simulate the body fluid condition. The pH of the solution was adjusted to 7.4, and the temperature was maintained at $37.4 \pm 1°C$. The solution was continuously deaerated using purified nitrogen gas throughout the experiment. Open circuit potential-time measurements

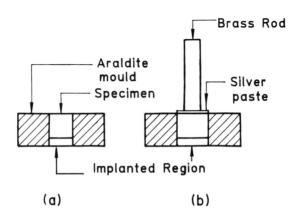

Figure 1 *Specimen preparation for (a) implantation and (b) electrochemical study*

and cyclic polarization technique was used to study the effect of nitrogen ion implantation and variation in the doses.

2.1 OCP-Time Measurements

The specimens were immersed in the Ringer's solution and the potential was monitored as a function of time with respect to saturated calomel electrode (SCE) as the reference, until the potential of the specimen reached a stable value.

2.2 Cyclic Polarization Measurements

Cyclic polarization studies were carried out using Wenking ST72 potentiostat. Platinum foil was used as the counter electrode. The working electrode (specimen) was allowed to reach the steady state OCP, before imposing the dynamic potential. A scan rate of 1mV/s was then applied from 250 mV below the OCP to 2000 mV, and subsequently reversed to obtain the cyclic polarization curve.

2.3 PotentioTransient Studies

Potentiotransient studies were carried out by applying the potential of 1.5 V from the OCP of the working electrode (specimen) immersed in the test solution, and the specimen was kept at the same potential for 3 hrs. The current density of the specimen was noted as a function of time at this potential until the potential was released to OCP condition. Again the OCP of the specimen was noted with respect to time to understand the stability of a stable passive film.

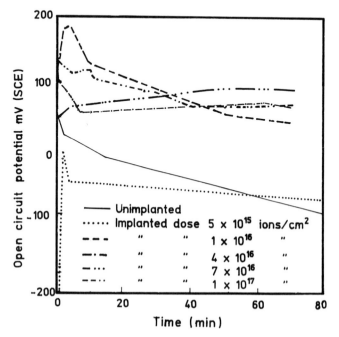

Figure 2 *Variation of OCP with respect to doses as a function of time*

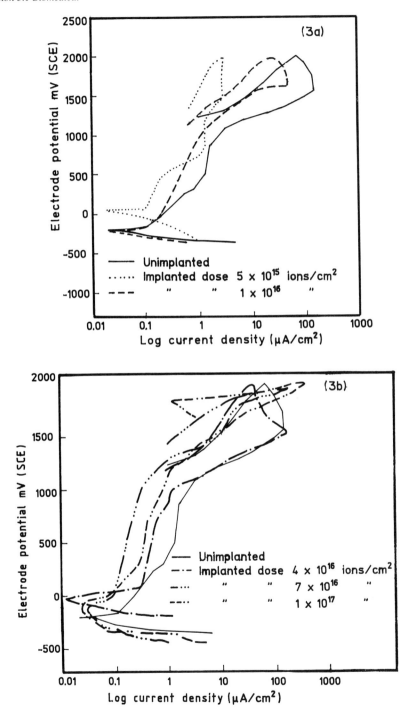

Figure 3 *(a) and (b) Cyclic polarization curves for unimplanted and implanted specimens*

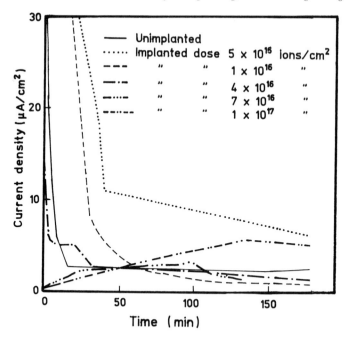

Figure 4 *Variation in the current density as a function of time to the specimens at a constant impressed potential of 1.5 V*

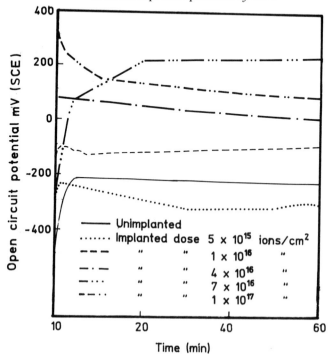

Figure 5 *Variation of OCP as a function of time after the impressed potential (1.5 V) studies*

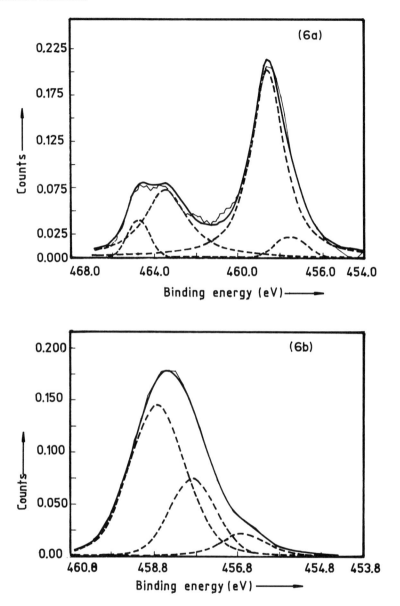

Figure 6 *(a) and (b) XPS spectra of Ti 2p electrons of implanted-passivated and as-implanted specimens*

2.4 XPS Measurements

XPS study was performed to identify the variations in the concentration of the elements in the modified surface, and their respective chemical state in the passive film. For this purpose, the implanted (optimum dose) and unimplanted specimens were passivated in the Ringer's solution for 1 hour at the impressed potential of 1000 mV (SCE). After passivation the specimens were rinsed with deionized water. Along with these specimens, as implanted

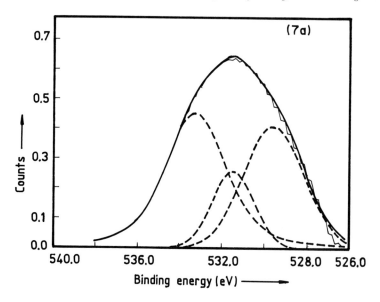

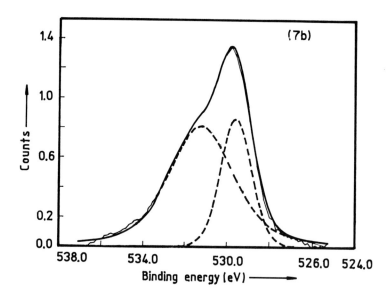

Figure 7 *(a) and (b) XPS spectra of O 1s electrons of implanted-passivated and unimplanted-passivated specimens*

specimen (optimum dose) was also analyzed for comparison purposes.

XPS analysis was carried out with a VG SCIENTIFIC ESCA LAB Mark 11 using Al Kα source (energy 1486.6 eV). The binding energies given in this paper are relative to fermi level and all values were corrected with respect to C 1 s spectrum. The resolution of the instrument for Au $4f_{7/2}$ electrons with Al Kα was 0.8 eV under the experimental condition and the base pressure was 7.4×10^{-9} mbar during spectral acquisition.

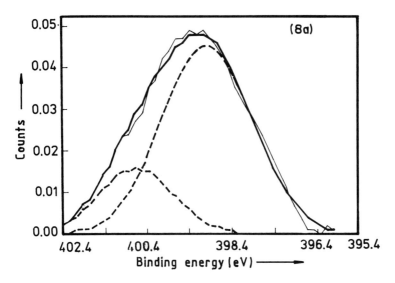

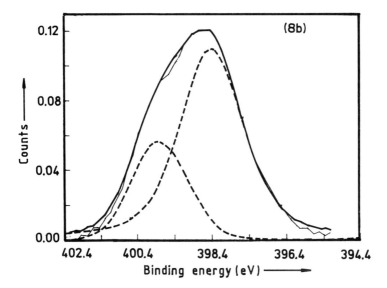

Figure 8 *(a) and (b) XPS spectra of N 1s electrons of implanted-passivated and as-implanted specimens*

3 RESULTS AND DISCUSSION

3.1 Open Circuit Potential-Time Measurements

The OCP of the specimens in the Ringer's solution measured as a function of time for implanted and unimplanted specimens are given in Figure 2. The OCP was found to shift towards the noble direction with increase in the dose rate upto 7×10^{16} Ions/cm^2 and it was found to decrease at 1×10^{17} ions/cm^2. The OCP of unimplanted specimen with time shifted towards the active direction indicating the dissolution of the passive film. A similar trend was observed with specimens implanted at lower doses (5×10^{15} and 1×10^{16} ions/cm^2). However, the attainment of the stable passive film was achieved in a shorter duration compared to the unimplanted specimen. For the specimen implanted at 7×10^{16} ions/cm^2, the OCP was not only shifted towards noble direction but also there was rapid attainment of the stable potential indicating the formation of the stable passive film compared to the other specimens. The shift of OCP towards the noble direction indicates a decrease in the number of the anodic sites. At doses higher than 7×10^{16} ions/cm^2 the OCP shifted towards active direction indicating an increase in the number of anodic sites.

3.2 Cyclic Polarization Studies

Figure 3a and 3b are cyclic polarisation curves showing the current density of the passive film which varied with the variation of the doses. Implantation at lower doses (5×10^{15} and 1×10^{16} ions/cm^2) did not improve the corrosion resistance while the doses 7×10^{16} ions/cm^2 showed a reduction in the passive current density. Also, the dose of 1×10^{17} ions/cm^2 was found to be detrimental, as evident from an increase in the current density implying an increase in the rate of dissolution. From these results it is inferred that at a dose of 7×10^{16} ions/cm^2 a stable passive film arising from sufficient titanium nitride was formed on the implanted region. The decrease in the corrosion resistance at 1×10^{17} Ions/cm^2 is attributed to the fomation of carbonitrides which was reported by Hohmuth and Rauschenbach[21]. The rapid increase in the current density at 1.2V is due to oxygen evolution at the anode (specimen) and thickening of the passive film. In implanted specimens this phenomenon also shifted to noble potential[22,23].

3.3 Potentiotransient Studies

Figure 4 shows the current density versus time curves at the appled potential of 1.5V for the specimens implanted at different doses and for the unimplanted reference specimen. The unimplanted specimen showed a higher current density at initial stages, which was directly proportional to the rate of ion release from the specimens, the sudden decrease to the constant value reveals that the ion release was controlled. The specimens implanted at low doses (5×10^{15}, 1×10^{16} ions/cm^2) showed much higher current density compared to the unimplanted specimen. Though the specimen which was implanted at dose 1×10^{16} ions/cm^2 showed lower current density at the end of the experiment, the time taken to reach the steady state was comparatively larger than the unimplanted specimens. However, the lower current density and lesser time taken to attain the steady state for the specimens implanted at higher doses (4×10^{16}, 7×10^{16}, 1×10^{17} ions/cm^2) indicated an excellent corrosion resistance. These observations reveal that the passive film formed on the specimens at doses from 4×10^{16} ions/cm^2 and above are highly stable, particularly the specimen implanted at the dose 7×10^{16} ions/cm^2 showed a maximum corrosion resistance.

Figure 5 shows the open circuit potential versus time measured after the potentiotransient studies of the specimen. These measurements were done to understand the kinetics of the passive film formation. The results show a similar trend as that the of fresh specimens for low doses but the potential is shifted to still nobler direction with higher doses of implantation. Among these the OCP of the specimen implanted at the dose 7×10^{16} ions/cm^2 showed the shift towards the more noble direction than any other specimen. It may be due to the sufficient

TiN formation on the modified layer and formation of a tenacious passive film on the specimen which prohibited further dissolution.

3.4 XPS Measurements

3.4.1 Ti 2p Spectra. The high resolution spectra for Ti 2p (Figures 6a and 6b) correspond to implanted-passivated and as-implanted specimens respectively. In implanted-passivated specimen the peak can be deconvoluted into four peaks. The main peaks at 458.6 and 464.7 eV correspond to the Ti^{4+} ($2p_{3/2}$) and Ti^{4+} ($2p_{1/2}$) respectively[24]. These results indicated that the passive film mainly consisted of TiO_2 oxides. The difference between these two peaks was reported to be 5.8 eV for Ti^{4+} compounds[25]. Since separation between the two peaks was 6.0 eV, the presence of some other compound along with TiO_2 is expected. A peak at 463.5 eV was observed in implanted-passivated specimens and no reports are available to account for this peak, and it needs further investigations.

Apart from this, the peak observed at 457.7 eV corresponds to the presence of oxynitrides[26]. The spectra of as implanted specimen showed asymmetric broadening to the lower binding energy side which can be separated as three peaks. The peak at, 456.9 eV corresponds to the TiN^{1+x} indicating the presence of non-stoichiometric titanium nitride in the implanted region. Along with that, the peaks found at 457.9 and 458.8 eV correspond to Ti^{3+} and Ti^{4+} ($2p_{3/2}$) respectively. Also the titanium enrichment was observed in implanted and passivated specimen than the as-implanted specimen.

3.4.2 O 1s Spectra. Figures 7a and 7b show the O1s spectra of implanted-passivated and unimplanted-passivated specimens respectively. The spectra of implanted-passivated specimen can be separated into three peaks. The peaks at 529.5, 531.4 and 533.3 eV correspond to O^{2-} (O-Ti), physisorbed OH and surface contamination (O-C) respectively[28]. However the unimplanted-passivated specimens showed only two peaks 529.5 and 531.2 eV, which are corresponding to O-Ti bond and physisorbed oxygen respectively. More over, intensity of oxygen on the surface is higher in unimplanted-passivated specimen than the implanted-passivated specimen.

3.4.3 N 1s Spectra. Figures 8a and 8b show the N 1s spectra of implanted-passivated and as-implanted specimens. The implanted-passivated specimen showed two peaks at 398.3 and 399.8 eV and the as-implanted specimen showed two peaks at the binding energy of 398.8 and 400.6 eV. According to the earlier report, the peak for the stoichiometric TiN is at 396.6 eV and the N-C bond will vary from 400.6 to 401.8 eV[29]. In this case, the peak at 399.8 eV in the implanted-passivated specimen and the peak at 400.6 eV in the as-implanted specimen are due to the N-C bond and probably due to the appearence of carbonitrides. The peak at 398.3 eV for implanted-passivated and 398.8 eV for as-implanted specimens are probably due to non-stoichiometric titanium nitride formation.

3.5 General Discussion About the Corrosion Resistance of Nitrogen Ion Implanted Titanium

The nitrogen ion implanted titanium showed marked changes in the corrosion resistance with variation in doses. Among the various doses, the doses between 4×10^{16} to 7×10^{16} ions/cm^2 were found to be the optimum and more than that would be detrimental. This can be confirmed from the decrease in corrosion resistance of specimen implanted at the dose 1×10^{17} ions/cm^2. These results suggest that the TiN formation on the implanted region is sufficient around the dose 7×10^{16} ions/cm^2, and above this dose the carbonitride precipitation may take place. The formation of carbonitrides during the high dose of irradiation has already been reported[21]. The stability of the passive film was determined by potentiotransient technique. The current density of the implanted specimens particularly, at the dose of 7×10^{16} ions/cm^2 showed very low value indicating the stability of the passive film even at high potential (1.5V). This stability of the passive film may be attributed to the formation of oxynitrides in the passive film in addition to TiO$_2$. The presence of oxynitrides was further confirmed by the appearance of small peak in the Ti 2p high resolution spectra at 457.5 eV along with Ti^{4+} 2p$_{3/2}$ and 2p$_{1/2}$ peaks. According to Chyou et al[23], the presence of oxynitrides was confirmed by the surface enrichment of titanium and nitrogen and a corresponding decrease in the intensity of oxygen. Similar observation found in this work indicated that titanium and nitrogen concentration was higher in the implanted-passivated specimen compared to the as-implanted specimen. The oxygen enrichment is high in the unimplanted passivated specimen which mainly consisted of TiO$_2$ and the intensity was comparatively low for implanted-passivated specimens. This additionally confirmed the presence of oxynitrides.

4 CONCLUSIONS

From the above results we can conclude that:

1. Nitrogen ion implantation on Titanium (CP) showed an improvement in the electrochemical behaviour of the passive film.
2. The shift of the OCP in the noble direction above the doses 1×10^{16} ions/cm^2 indicated the formation of a stable passive film.
3. Cyclic polarization studies showed that the passive current density decreases with increasing dose rates of implantation in comparison to the unimplanted specimens indicating least dissolution of passive film.
4. The high dose of implantation was found to be detrimental from the electrochemical point of view, and an optimum dose of 7×10^{16} ions/cm^2 is recommended.
5. XPS results showed that the formation of nonstiochometric titanium nitride (TiN$_{1+x}$ in the implanted layer. The surface enrichment of titanium and nitrogen during passivation of the implanted specimen indicates the formation of oxynitrides along with TiO$_2$ in the passive film.

Acknowledgements

The authors are grateful to Dr. Kanwar Knshan, Head, Materials Science Division , Dr. R K Dayal, Head, ACSS and Shri. J B Gnanamoorthy, Head, Metallurgy Division, Indira Gandhi Centre for Atomic Research, Kalpakkam, for their keen interest in the above work. Thanks are also due to IUC-DAEF, Indore. for the financial assistance.

References

1. H. Zitter and H. Plenk Jr, *J. Biomed. Mater. Res.*, 1987, **21**, 881.
2. S. Nasser, P. Campbell and H. C. Amstutz, Conference on 'Metal and their alloys in orthopaedic surgery', Gottington, 5–7 October 1987.
3. G. Meachim and D. F. Williams, *J. Biomed. Mater. Res.*, 1973, **7**, 555.
4. D. F. Williams, and G. Meachim, *J. Biomed. Mater. Res. svmp.*, 1974, **5**, 1.
5. P. G. Laing, A. B. Ferguson and E. S. Hodge, *J. Bone. Joint. Surg.*, 1959, **41A**, 737.
6. A. B. Ferguson, Y. Akahashl, P. G. Laing and E. S. Hodge, *J. Bone. Joint. Surg.*, 1962, **44A**, 323.
7. A. B. Ferguson, Y. Akahashl, P. G. Laing and E. S. Hodge, *J. Bone. Joint. Surg.*, 1962, **44A**, 317.
8. G. J. Williams, R. A. Palman, J. Colard and P. Ducheyne, *Analysis*, 1984, **12**, 443.
9. P. Ducheyne, G. Willems, M. Martens and J. Helsen, *J. Biomed. Mater. Res.*, 1984, **18**, 293.
10. J. A. Davidson and G. Schwartz, *J. Biomed. Mater. Res.*, 1987, **21**, 261.
11. W. Rostoker and J. Galante *J. Biomed. Mater. Res.*, 1976, **10**, 303.
12. D. A. Miller, R. D. Ainsworth, J. H. Durnbleton, D. Page, E. H. Mille and Chishen, *Wear*, 1974, **285**, 207.
13. R. A. Buchanan, E. D. Rigney and J. M. Williams, *J. Biomed. Mater. Res.*, 1987, **21**, 355.
14. E. D. Peterson, B. M. Hillberry and D. A. Heck, *J. Biomed. Mater. Res.*, 1988, **22**, 887.
15. R. Martinelia, S. Giovanardi, G. Paiombarini, M. Corchia, P. Delogu, R. Glorgi and C. Tosello, *Nucl. Instr. Meth. Phys. Res.*, 1987, **B19**, 236.
16. Pengxum Yan and Sl-Ze-Yang, *Thin Solid Films*, 1993, **282**, 204.
17. J Yu, Z H Zhao and L X Li, *Corr. Sci.*, 1993, **35**, p587.
18. E. Leitao, C. Sa, R. A. Silva, M. A. Barbosa and H. Ali, *Corr. Sci.*, 1995, **37**, 1861.
19. T. Sundararajan, U. Kamachi Mudall, S. Rajeswari, M Subbalyan, K. G. M. Nair and K. Krishan Presented at 'Discussion Meeting on Surface Science and Engineering ', Kalpakkam, January 8–10, 1996.
20. J. E. Elder, R. Thamburaj and P. C. Patnalk, *Surf. Eng.*, 1989, **5**, 55.
21. K. Hohmuth and B. Rauschenbacli *Mater. Sci. Eng.*, 1985, **69**, 489.
22. J. F. McAleer and L. M. Peter, *J. Electrochem. Soc.*, 1982, **129**, 1252.
23. M. A. Abdul Rahim, *J. Appl. Electrochem.*, 1995, **25**, 881.
24. J. Lausmaa, B. Kasemo and H. Mattson, *Appl. Surf Sci.*, 1990, **44**, 133.
25. S. D. Chyou, H. C. Shih and T. T. Chen, *Corr. Sci.*, 1993, **35**, 337.
26. K. S. Robinson and B. M. A. Sherwood, *Surj. Interf. Anal.* 1984, **63**, 261.
27. M. Miyagi, Y. Sato, T. Mizundo and S Sawada, in: Proc. 4th Int. Conf. 'TITANIUM 80', Kyoto, 1980, Eds. H. Kimura and O. Izumi (The Metallurgical society of AIME, 1980) p. 2867.
28. M. Shlrkhanzadeh, *J. Mater. Sci. Mater. Med.*, 1995, **6**, 206.
29. J. Schreckenbach, F. Schlotting, D. Dietrich, A. Hoffmann and G. Marx, *J. Mater. Sci. Lett.*, 1994, **14**, 1344.

3.1.3

Pitting Corrosion Studies on Nitrogen Implanted 316L SS for Biomedical Applications

M. Subbaiyan[1], K. M. Veerabadran[1], N. S. Thampi[2], Kanwar Krishnan[2], U. Kamachi Mudali[2] and R. K. Dayal[2]

[1]DEPARTMENT OF ANALYTICAL CHEMISTRY, UNIVERSITY OF MADRAS, GUINDY CAMPUS, MADRAS, INDIA

[2]METALLURGY AND MATERIALS GROUP, INDIRA GANDHI CENTRE FOR ATOMIC RESEARCH, KALPAKKAM, INDIA

1 INTRODUCTION

Fixation of fractures and correction of defects in the skeletal structure of the human body using devices of metals and alloys had been in practice for a long time.[1] However, in recent times ceramics, composites and polymers have found extensive use[2]. Despite this recent interest and development of polymeric and ceramic materials, metallic implants are preferred over these due to their inherent mechanical properties and ease of fabrication. Currently the surgical implants are usually made of any one of the three materials: austenitic stainless steels, cobalt-chromium alloys and titanium and its alloys[3]. Among all these materials titanium and its alloys are the most corrosion resistant[4]. However, the main disadvantages are their high cost, formation of metal oxide debris not related to wear and special welding procedures required for joining[5]. The austenitic stainless steels especially type 316L SS is the most popular material because of its relatively low cost, ease of fabrication and reasonable corrosion resistance[6]. However, 316L SS is prone to localised attack, releasing iron, chromium and nickel ions in the body fluid environment[7]. These leached out chromium and nickel ions are carcinogenic[3]. Hence high corrosion resistance is demanded for implants to be biocompatibile. Studies on failed and retrieved 316L SS implants revealed significant localized corrosion attack[8]. In every implant failure, the patient undergoes the trauma of repeated surgeries in addition to severe pain experienced during retrieval of the device. Moreover, its removel can cause great expense and hardship to the patient. Hence it is necessary to keep the failure incidences to a minimum by improving the service-life of the devices to overcome the above problems.

Surface modification of the metallic implant devices is one of the methods of choice to improve the corrosion and wear resistance of the materials. In the last two decades, ion implantation is considered to be a versatile technique to modify the surfaces for improving the wear and corrosion resistance[9-11]. Ashworth[9] found that 6% of ion implanted chromium on iron gives rise to corrosion resistance equal to 6% chromium in the bulk alloy. Buchanan et al[12] observed that nitrogen ion implanted Ti6Al4V alloy shared a 100-fold increase in wear resistance than the unimplanted Ti6Al4V samples. Nair et al[13] reported that nitrogen ion implanted 304SS obtains three fold corrosion resistance at the dose of 1×10^{17} ions/cm^2 compared to the plain sample in 3.5% NaCl solution at room temperature. Of the several ions reported for implantation, nitrogen ion implantation is proved to be the most suitable for biomedical applications[14].

Hence in the present study, the nitrogen ion implantation of the widely used 316L SS was undertaken to evaluate the optimum dose required for better performance of the material in physiological saline solution.

2 EXPERIMENTAL

2.1 Ion Implantation

Type 316L SS specimens were cut to 1 cm x 1 cm x 0.3 cm size, wet ground with SiC papers upto 600 grit followed by polishing with diamond paste of grain size 5 and 0.5 micron prior to implantation. The polished specimens were ultrasonically cleaned by soap solution, double distilled water and degreased in acetone. Nitrogen ion implantation was carried out at room temperature at 50 keV with TAM SAMES 150 kV accelerator using Iolar nitrogen (99.99%) gas, at doses 1×10^{15}, 1×10^{16} and 1×10^{17} ions/cm^2 under a vacuum (1×10^{-7} Torr).

2.2 Electrode Preparation

Nitrogen ion implanted 316L SS specimens were connected to the copper rod through silver paste to provide electrical contact. Barring the implanted region, the other parts of the electrode were masked with PTFE film, rinsed with double distilled water, acetone and then air dried.

2.3 Physiological Saline Solution

Ringer's solution (NaCl 9.0 g/l, $CaCl_2$ 0.24 g/l, KCl 0.43 g/l, $NaHCO_3$ 0.20 g/l in double distilled water) simulating the body fluid composition was used for the electrochemical studies. The pH of the solution was adjusted to 7.4 and the temperature was maintained at 37.4 ± 1°C using a water bath. The solution was continously purged with nitrogen during measurements.

2.4 OCP-Time Measurement

The implanted and unimplanted specimens were immersed in the Ringer's solution and their potentials were measured as a function time using Saturated Calomel Electrode (SCE) as the reference till the potential reached a steady state with respect to each of the specimens.

2.5 Cyclic Polarization Studies

Cyclic polarization study was carried out using Wenking ST 72 potentiostat. The electrochemical cell consisted of one litre flask, fitted with working electrode, platinum foil counter electrode and the reference electrode (SCE) with provision for nitrogen purging. The working electrode was allowed to remain in the Ringer's solution until a constant potential was reached. The working electrode was impressed with starting potential of 200 mV below the OCP and scanned towards the positive direction at 10mV/min till the current density reached a value of 1000 μA/cm^2 after passing through the breakdown potential E_b. Then the sweep direction was reversed to obtain the dynamic polarization hysteresis from which the pit protection potential E_p was arrived at.

2.6 Accelerated Leaching Studies

The accelerated leaching studies were carried out to understand the stability of the passive film and the amount of metal ions leached out from the passive film. The specimens were immersed in the Ringer's solution and allowed to attain the steady state to form the passive film. The implanted and unimplanted samples were imposed with respective breakdown potentials. In the case of plain 316L SS, a potential of 20 mV above the E_b was also imposed and their current densities were monitored as a function of time. The amount of metal ions leached out from the specimens was analysed by atomic absorption spectrometry (AAS).

2.7 Morphology of the Surface

The unimplanted and implanted specimens were observed under optical microscope before and after polarization studies to investigate the effect of ion implantation on pitting corrosion.

3 RESULTS AND DISCUSSION

3.1 Open Circuit Potential-Time Measurements

The open circuit potentials of the implanted and unimplanted samples measured as a function of time are given in Figure 1. The OCP of the implanted samples shifted towards the noble direction except for the dose rate 1×10^{15} ions/cm^2 compared to the plain sample (vide Table 1). At a dose rate of 1×10^{16} ions/cm^2, the OCP attained a constant value of -60 mV after a lapse of 40 minutes. Similarly, the sample implanted at a dose rate of 1×10^{17} ions/cm^2 attained a constant value of -23 mV. However, a dose rate of 1×10^{15} ions/cm^2 showed an OCP of -170 mV and the OCP shifted towards the active direction indicating an increase in the number of anodic sites. The dose rate of 1×10^{17} ions/cm^2 was found to be desirable which not only showed a shift towards the noble direction but also attained the steady state potential within a few minutes. These facts indicate that the formation of stable passive film is favoured at higher doses.

3.2 Cyclic Polarization Studies

The potentiodynamic anodic cyclic polarization curves for the unimplanted and the implanted samples are shown in Figure 2 and the relevant data in Table 1. The pitting potential E_b for the unimplanted sample was found to be $+140$ mV and the protection potential E_p -170 mV. The E_b values for the implanted samples at a dose rate of 1×10^{15}, 1×10^{16}, 1×10^{17} ions/cm^2 were

Table 1 *Summary of electrochemical data*

Doses (ion/cm^2)	Open Circuit Potential (mV)	Breakdown Potential E_b (mV)	Pit Protection Potential E_p (mV)
Plain	-64	$+140$	-170
1×10^{15}	-170	$+220$	-150
1×10^{16}	-60	$+340$	$+40$
1×10^{17}	-23	$+480$	$+70$

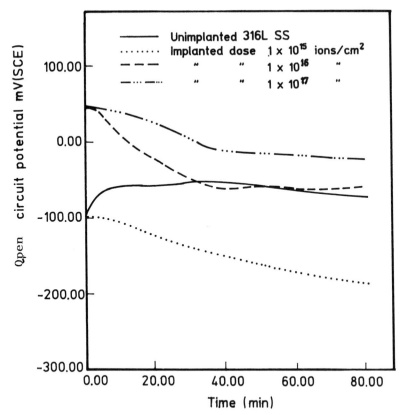

Figure 1 *Variation of OCP with respect to doses as a function of time*

found to be +220, +340, +480 mV respectively. The corresponding E_p values were observed to be -150, +40 and +70 mV. The shift in the E_b values of the implanted samples towards the noble direction compared to that of the unimplanted sample indicates improved corrosion resistance of the implanted specimens. A dose rate of 1×10^{17} ions/cm² which showed a maximum shift of E_b was chosen for the present study. The pit protection potential E_p determined from the reverse scan for the above samples also indicates a shift towards more noble direction compared to that of the unimplanted sample. This implies that the repassivating tendency of the implanted sample is enhanced.

The enrichment of CrN in the passive layer of the implanted samples would have impeded the release of metal ions through the passive film and accounts for the increased E_b values and greater repassivating tendency which can be inferred from the higher E_p values. Clayton et al[15] pointed out that the tendency of repassivation in implanted samples is facilitated by the fact that carbides and other inclusions tend to become partially redissolved under implantation and a homogeneous passive film is formed.

3.3 Accelerated Leaching Studies

The plot of variation of current density of the unimplanted and implanted samples at the respective E_b values as a function of time is represented in Figure 3. Implantation of the

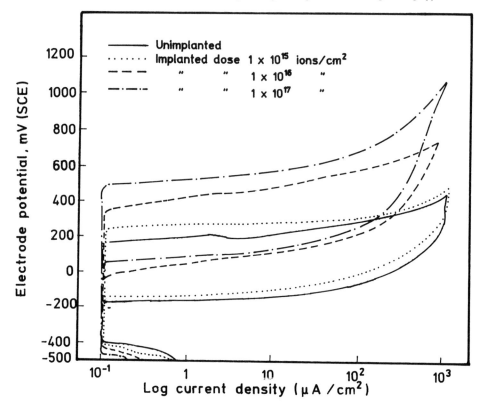

Figure 2 *Cyclic polarization curves for unimplanted and implanted specimens*

specimens tends to lower the current density with increasing dose rate compared to the unimplanted specimen. The current density of the specimen implanted at 1×10^{17} ions/cm^2 at an impressed potential of +500 mV compared to +160mV of unimplanted sample, remained close to zero for about 45 minutes. The solutions of the accelerated leaching studies were analysed for the leached out metal ions. The concentration profile of iron, nickel and chromium leached out from the unimplanted and implanted samples at the impressed potentials +140 mV, +220 mV, +340 mV and +480 mV respectively are illustrated in Figure 4.

It can be seen from the figure that a significant amount of metal ions namely 1.7 µg/ml of iron, 0.38 µg/ml of nickel and 0.7 µg/ml of chromium were leached out from 316L SS at an impressed potential of +160 mV (Eb + 20 mV). On the other hand the concentration of iron and nickel ions leached out from implanted samples (dose rate of 1×10^{15}, 1×10^{16} and 1×10^{17} ions/cm^2) at their respective E_b values which are much higher than the impressed potential of plain steel were found to be 0.22, 0.33 and 0.28 µg/ml of iron and 0.16, 0.16 and 0.14 µg/ml of nickel respectively. When the passive film breaks down, the base metal ion and the alloying elements nickel and chromium get dissolved. However, the amount of chromium leached out from all the implanted samples was 0.1 µg/ml, whereas the amount of chromium leached out from 316L SS at an impressed potential of 160 mV was 0.7 µg/ml. This can be

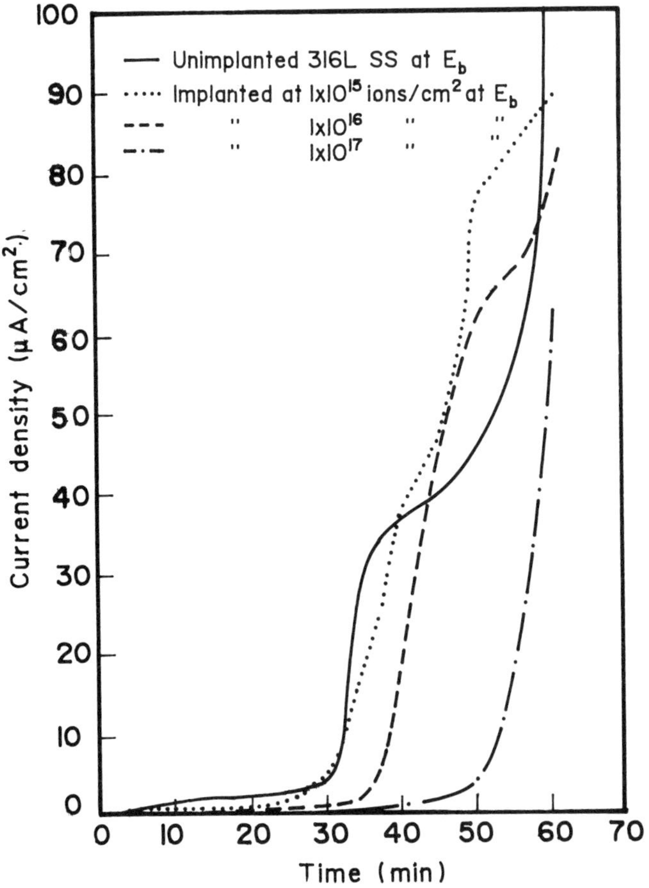

Unimplanted 316L SS at E_b
Implanted at 1x10^{15} ions/cm^2 at E_b
" 1x10^{16} " "
" 1x10^{17} " "
Current density (μA/cm^2)
Time (min)
0
10
20
30
40
50
60
70
80
90
100

Figure 3 *Variation of current density as a function of time of the specimens at their respective E_b values*

explained from the fact that in nitrogen ion implanted samples, the stable passive film which incorporates chromium nitride and iron nitride inhibits the dissolution of chromium. The amount of iron and nickel leached out are very low at a dose rate of 1×10^{17} ions/cm^2. This is in conformity with the reports of Kothari et al[16] wherein the improved corrosion and wear resistance of nitrogen ion implantation on 304 SS is attributed to the formation of $Fe_{2+x}N$ and chromium nitride. The incorporation of CrN in the passive film was also confirmed by studies conducted by Moore et al.[17]

3.4 Surface Morphology

Optical Micrographs of unimplanted 316L SS samples are presented in Figure 5, Figure 5a depicts the surface nature of the polished surface. Figure 5b shows the pit morphological

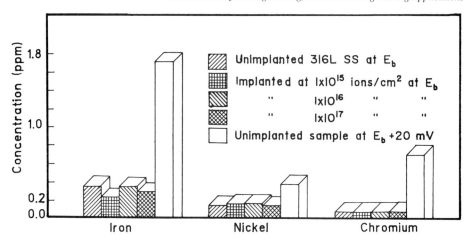

Figure 4 *Leaching pattern of the unimplanted and implanted specimens at their respective E_b values*

behaviour of the unimplanted specimen. The large pits account for the delay in the repassivation process.

Figure 6 depicts the optical micrographs of the nitrogen ion implanted (dose rate of 1×10^{15} ions/cm²) specimens and the corrosion pattern of the implanted surfaces at different doses. From the micrography it can be seen that implanted specimens exhibit very few and

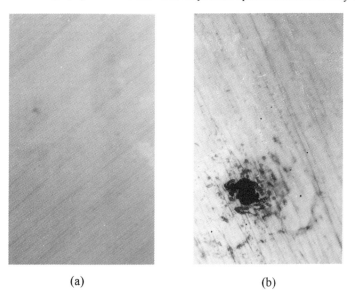

(a)   (b)

Figure 5 *Optical micrograph of 316L SS of (a) polished sample (b) corrosion pattern of above at E_b*

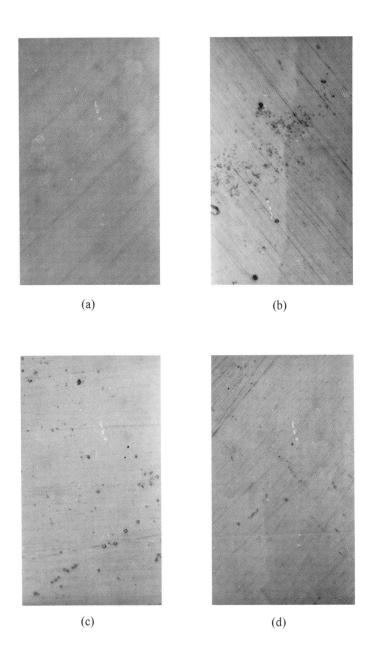

(a) (b)

(c) (d)

Figure 6 *Optical micrograph of implanted sample (a) sample implanted at 1×10^{15} ions cm^{-2}, (b)–(d) corrosion patterns of implanted sample at doses 1×10^{15}, 1×10^{16}, 1×10^{17} ions cm^{-2} respectively at their corresponding E_b values*

very small pits compared to unimplanted specimen. A dose rate of 1×10^{17} ions/cm² indicates immunity towards localized attack. The higher pit protection potentials observed with the implanted specimens confirm that the initated pits are easily repassivated.

4 CONCLUSIONS

- The OCP tends to shift towards the noble direction with nitrogen ion implantation.
- Pitting and Pit-protection potentials of nitrogen implanted 316L SS are found to be more noble, compared to the unimplanted 316L SS.
- The higher pitting and pit- protection potentials indicate the beneficial effect of nitrogen ion implantation in improving the pitting resistance of 316L SS.
- The repassivation of actively growing pits is considerably increased in the nitrogen ion implanted 316L SS.
- The accelerated leaching study indicated that the release of iron, chromium and nickel from the nitrogen ion implanted 316L SS was considerably less compared to type 316L SS plain sample (unimplanted)
- Nitrogen ion implanted surface enhances the pitting corrosion resistance of the sample.

Acknowlegements

The authors are grateful to Dr. K.G.M. Nair, Scientific Officer, Materials Science Division and Shri.J.B.Gnanamoorthy, Head, Metallurgy Division, Indira Gandhi Centre for Atomic Research, Kalpakkam, for their keen interest in the above work. Thanks are also due to UGC, New Delhi, for the financial assistance.

References

1.   R. J.Hadad, S. D.Cook and K. A.Thomas, *J. Bone. Joint. Surg.*, 1987, **69A**, 1459.
2.   A. F. Tencer, *J.E.E. Eng in Med and Bioc.*, 1989, **40**.
3.   D. F. Willams, *J. Mat. Sci.*, 1987, **22**, 3421.
4.   N. Bruneel and J. A. Helson, *J.Biomed. Mater. Res.*, 1988, **22**, 203.
5.   L. H. Boulton and A. J.Betts, *Br.Corros.J.*, 1991, **26**, 287.
6.   Tzyy-Ping Cheng, Wen-Ta Tsai and Ju-Tung Lee, *J. Mater. Sci.*, 1990, **25**, 936.
7.   K. Nielsen, *Br. Corros. J.* 1987, **22**, 272.
8.   M. Sivakumar and S. Rajeswari, *J. Mater. Sci. Lett.*, 1992, **11**, 1039.
9.   V. Ashworth, *Corros. Sci.*, 1976, **16**, 661.
10. L. G. Sudensen, *Corros. Sci.*, 1980, **20**, 63.
11. H. Ferber, *Corros. Sci.*, 1980, **20**, 117.
12. R. A. Buchanan, E. D. Rigney and J. M. Willams, *J. Biomed. Mater. Res.*, 1987, **21**, 355.
13. M. R.Nair, S.Venkatraman, D. C. Kothari, K. B. Lal and R. Raman, *Nucl. Instr. and Meth.*, 1988, **B-34**, 53.
14. A. Chen, and K. Sridharan, *J. Surf. Coat. Technol.* 1991, **50**, 1.
15. C. R. Clayton, K. G. K. Doss, H. Herman, S. Prasad, Y. F. Wang, J. K. Hirvonen and G. K. Hubler, Proceedings of a symposium held as part of the Annual Meeting of the

Materials Research Society, Cambridge, Mass, 1979, November 30, 65.
16.   D. C. Kothari, M. R. Nair, A. A. Rangwala, K. B. Lal, P. D. Prabhawalkar and P. M. Raole, *Nucl. Instr and Meth.*, 1985, **B7/8**, 235.
17.   R. C. Moore, G. L. Grobe and J. A. Gardella, *J. Vac. Sci. Technol.*, 1991, **A9**, 1323.

Section 3.2 Aerospace

3.2.1

Characterization of High Temperature Metal Matrix Composite Coatings

A. R. Marder

ENERGY RESEARCH CENTRE AND DEPARTMENT OF MATERIALS SCIENCE AND
ENGINEERING, LEHIGH UNIVERSITY, BETHLEHEM, PA 18015, USA

1 INTRODUCTION

Surface engineering has been defined as, "manipulating the properties of materials through the deposition of films and coatings"[1] and the use of composite materials as coatings has added another opportunity for improving the properties of structural components. Protective coatings, coatings that can withstand high temperature corrosion and erosion, have found acceptance in a large number of industries including: electrical utilities, engine manufacturers, pulp and paper production and chemical processing. Thermal spray processing methods, e.g. air plasma spray (APS), vacuum plasma spray (VPS) and high velocity oxyfuel (HVOF), have been utilized to produce composite coatings of metal carbides in a corrosion resistant matrix for improved high temperature properties. More recently, electrochemical deposition processes have been applied to the production of metal matrix composite (MMC) functionally graded coatings. It is the purpose of this paper to report on the microstructural and properties characterization of metal matrix composite coatings.

2 THERMAL SPRAY COMPOSITE COATINGS

2.1 Microstructural Characterization

Thermal spray coatings are formed by melting materials in particulate form or wire feed stock and accelerating the molten or partially molten droplets toward a substrate. Once these molten droplets strike the substrate they expand out and form a splat[2]. As additional particles impact the specimen, splats will eventually interact and form a continuous coating, Figure 1. Splat to substrate, as well as splat to splat bonding, is primarily mechanical and can be weak, providing little resistance to pullout during cutting, grinding and polishing of the coating for metallographic preparation. In addition, multiphase MMC coatings, such as cermets, can have enhanced pullout due to the different sizes and densities of the as-sprayed powders. Consequently, careful preparation techniques must be used to prevent pullout which can cause erroneous measurement of porosity and volume fraction of phases.

Blann[3,4] has reported on the effect of preparation procedures on the porosity level of various thermal spray coatings and Bluni, et al.[5] have used quantitative image analysis techniques to evaluate the structural features of thermal spray coatings. The addition of fluorescent dye to

Figure 1 *Typical splat morphology found in thermal spray coatings*

the epoxy mounting material[6], enables porosity to be more easily identified and a microstructural feature, the shortest splat boundary path from the coating surface to the substrate, has been correlated to the high temperature corrosion properties of spray coatings[7,8]. Recent research in our laboratory has made use of light optical microscopy (LOM), electron probe microanalysis (EPMA), X-ray diffraction (XRD) and quantitative image analysis (QIA), to characterize the microstructure of HVOF composite coatings of varying amounts of chromium carbide in an FeCrAlY matrix.

The typical microstructure of the HVOF coating consisted of large white semi-circular FeCrAlY alpha grains within a region of white and grey lenticular splats, Figure 2. EDS/WDS elemental dot mapping showed that the white and grey splats consisted of iron aluminum oxides, chromium oxides and chromium carbides. A quantitative analysis of each of the oxide, carbide and alpha phases was conducted by first measuring the total oxide (same grey level) in the coating by QIA. The remaining white phases, consisting of chromium carbide and alpha phase, were distinguished by overetching the alpha phase and counting the remaining carbide

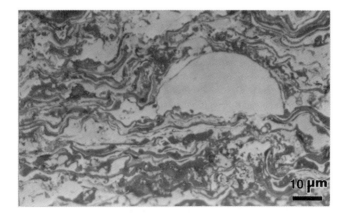

Figure 2 *As sprayed microstructure of HVOF coating with 80% chromium carbide and 20% FeCrAlY*

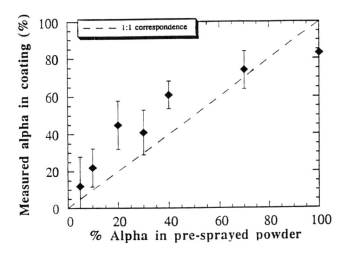

Figure 3 *Measured alpha phase versus percent alpha in the pre-spryed powder*

phase. Figure 3 shows the measured alpha phase versus the percent alpha in the pre-sprayed powder for the range of coating compositions studied. The dotted line shows a 1:1 correspondence of percent alpha in the coating to percent alpha in the pre-sprayed powder. The alpha has good spray efficiency as the amount of alpha in the coating is actually slightly higher than that in the pre-sprayed powder.

The measured chromium carbide phase versus the nominal chromium carbide in the presprayed powder is shown in Figure 4. The spray efficiency of the carbide is poor, e.g. only 29% of the coating is chromium carbide when the pre-sprayed powder contained 95% chromium carbide. The chromium carbide may not have been fully molten when sprayed (MP ~1900°C compared to alpha MP~1500°C), therefore some of the carbide could have deflected

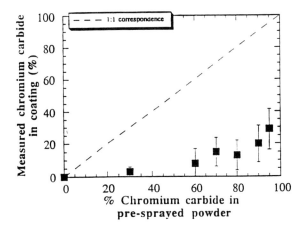

Figure 4 *Measured chromium carbide versus percent chromium carbide in the pre-sprayed powder*

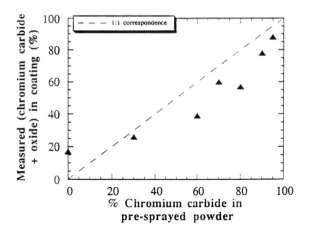

Figure 5 *Measured (chromium carbide plus oxide) phases versus percent chromium carbide in the pre-sprayed coating*

from the coating surface. Coating thickness was found to decrease from approximately 400 μm in the 0% chromium carbide alloy to approximately 50μm in the 95% chromium carbide alloy. In addition, some of the chromium carbide could have also been reduced to oxide during spraying, accounting for the lower chromium carbide in the coating. It was found that measured oxide increased with increasing chromium carbide in the presprayed powder and that when total hard phase, i.e. carbide plus oxide, is plotted versus chromium carbide in the pre-sprayed powder the results approach the ideal 1:1 (Figure 5).

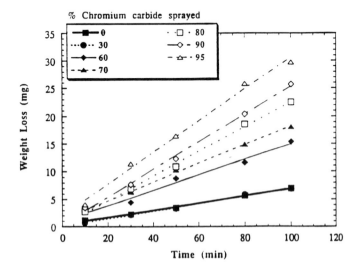

Figure 6 *Erosion test results at a 90° impingement angle*

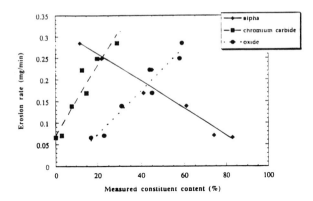

Figure 7 *Erosion rate as a function of microstructural constituent*

2.2 Erosion Resistance

Erosion tests were conducted at 30° and 90° impingement angles. Figure 6 shows the 90° impingement data and Figure 7 is a plot of erosion rate as a function of alpha, chromium carbide and oxide content measured in the coating. Erosion rate increases with increasing carbide and oxide content and the coatings with the highest alpha content have the best erosion resistance. The chromium carbide and oxide phases are assumed to play a similar rôle in the coating since they are both relatively hard and brittle in relation to the soft ductile alpha phase. Figure 8 shows the erosion rate versus total oxide and carbide content at a $90°$ impingement

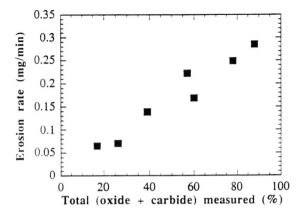

Figure 8 *Erosion rate versus total (oxide plus carbide) content at 90° impingement angle*

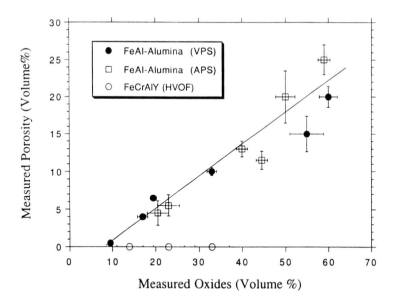

Figure 9 *The effect of measured oxide on porosity in APS, VPS and HVOF Fe-Al type coatings*

angle. These results show that an increase in steady state erosion rate occurs with increasing total hard phase (carbide plus oxide).

A thermal spray processing study on Fe-Al type alloys with deliberate additions of Al_2O_3, compared APS, VPS and HVOF. Figure 9 shows that for plasma sprayed coatings, porosity

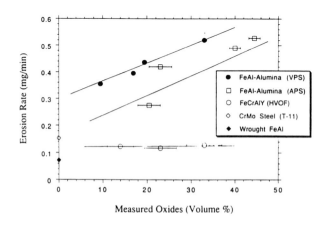

Figure 10 *The effect of measured oxide on the erosion resistance of APS, VPS and HVOF Fe-Al type coatings*

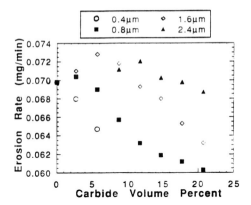

Figure 11 *The effect of volume percent carbide on erosion rate at various carbide sizes*

increases with measured oxide in the coating; however, the HVOF process significantly reduces porosity over the VPS and APS processes. Similarly, the HVOF processed alloys have the best erosion resistance, Figure 10, being superior to both APS and VPS over the range of measured oxides in the coating. In fact, the HVOF sprayed coatings with Al_2O_3 are even superior to the wrought substrate Cr-Mo steels, but not as good as the wrought Fe-Al alloy. These results also show that decreasing hard phase Al_2O_3 for every process increases erosion resistance. Therefore, in preparing a composite coating containing hard and soft constituents, it is important that each constituent is uniformly distributed. Spraying parameter control is essential in limiting the amount of oxide formed in the MMC coating and decreasing the hard phases (chromium carbide plus oxides) to below 20 volume percent will improve the erosion resistance of these composite coatings.

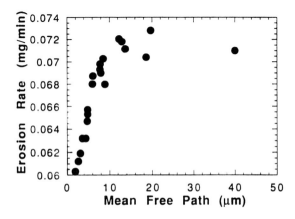

Figure 12 *Effect of carbide mean free path on erosion rate*

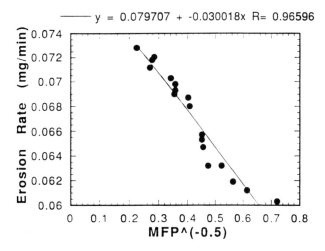

— y = 0.079707 + -0.030018x R= 0.96596

Figure 13 *The effect of the inverse square root of the carbide mean free path
on erosion rate*

2.3 Microstructural Modelling of Erosion Behaviour

The iron-spheroidized carbide system is an ideal model system to determine the role of second phase hard particles on the erosion resistance of a ductile metal matrix composite. In this system it is relatively easy to control and change the carbide size and spacing. A series of iron-carbon alloys was quenched and tempered to develop an experimental matrix of particle size from 0.4μm to 2.4μm and amount of carbide particles from 0 to ~22 volume percent. These specimens were erosion tested at a 90° impingement angle and the steady state erosion rate was determined. Figure 11 shows that as the carbide volume percent increases up to 22%, the erosion rate decreases. Although it was shown that the erosion resistance of thermally sprayed MMC coatings improved with decreasing volume percent of hard phase (carbide plus

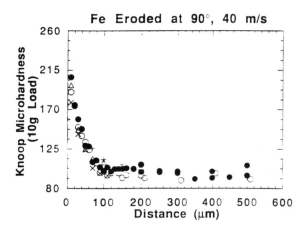

Figure 14 *The plastic zone size in an iron sample eroded at 90°*

Figure 15 *The etched cross-section of a pure nickel deposit at current densities*
of 15 A/dm²

oxide), Figure 10, an optimum volume percent second phase may be required for the best erosion resistance. Figure 11 also shows that the erosion rate decreases with decreasing particle size, therefore another microstructural feature that combines volume percent and particle size may be necessary to characterize erosion rate.

The mechanical properties of spheroidized steels are well documented and have been found to depend on volume fraction, size and shape of the carbide particles. Ductility appears to depend on volume fraction carbide[9,10], and yield strength[11], flow stress, ultimate tensile strength

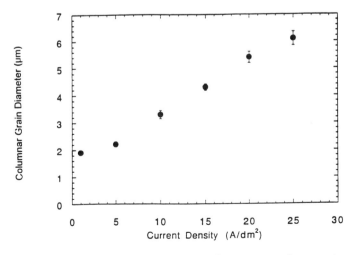

Figure 16 *The effect of current density on columnar grain diameter in pure Ni deposits*

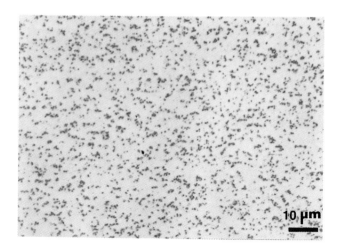

Figure 17 *Typical light optical micrograph of a nickel-alumina electro-composite*

and failure strength[12] all are dependent on the inverse square root or Hall-Petch type relationship with the particle spacing. Plotting erosion rate versus mean free path between carbide particles (MFP), Figure 12, shows that as the mean free path increases, the erosion rate also increases up to an apparent saturation value at a mean free path in excess of approximately IS 4m. A Hall-Petch plot, Figure 13, shows excellent correspondence between decreasing erosion rate and increasing the inverse square root of the MFP.

It has been shown that the plastic zone size below the erosion surface increases with increasing erosion resistance (lower erosion rate)[13] and similar results have been found for some of theses alloys, Figure 14. Although this would suggest that work hardening or increasing the strain hardening coefficient would be an important parameter in increasing erosion

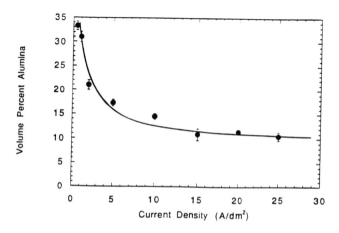

Figure 18 *Volume percent alumina as a function of current density*

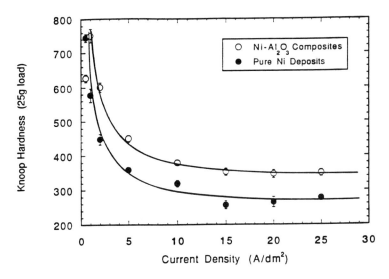

Figure 19 *Hardness as a function of current density*

resistance, Liu[14] has found that increasing the carbide mean free path in the same alloys increases the strain hardening coefficient. However, according to Figure 12, an increase in mean free path will increase the erosion rate, opposite to the effect expected. Liu[14] also found that yield strength and the strength coefficient decrease with carbide MFP. Thus it appears that the mechanical properties' relationships controlling erosion resistance are complex, but may relate to the ability of the material to absorb energy or a toughness parameter as proposed by Levin, et al.[13]

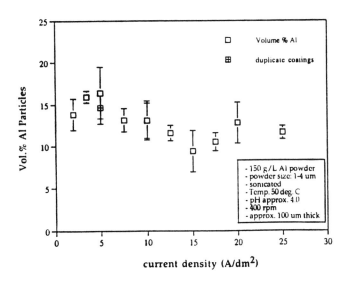

Figure 20 *The effect of current density on the volume percent Al particles retained in the coating*

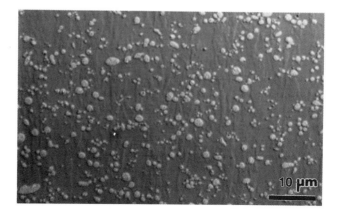

Figure 21 *Microstructure of the as-plated Ni-Al coating*

(a)

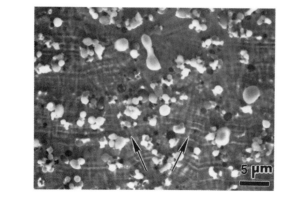

(b)

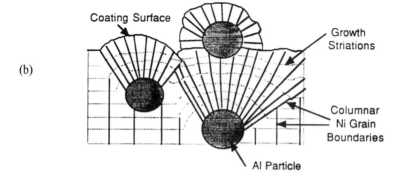

Figure 22 *The relationship between Al particles acid growth striations in the coating (a) SEM image and (b) schematic*

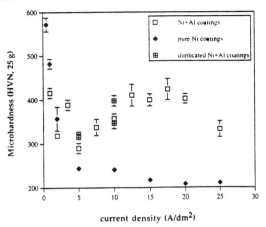

Figure 23 *The effect of current density on the hardness of composite coatings*

3 ELECTRODEPOSITED COMPOSITE COATINGS

The fabrication of coatings by electrochemical methods has the advantage of versatility and the ability to coat complex shapes at low cost. In addition, inert particles added to the plating bath will co-deposit with the electroplated metal creating a metal matrix composite coating. The volume fraction of codeposited particulates depends upon a number of plating parameters including: electrolytic bath chemistry, current density, particle size, shape, amount and composition. The effect of these parameters on Ni electrodeposited coatings in a sulfamate bath with Al and Al_2O_3 particulates has been investigated[15,16].

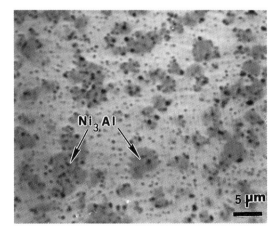

Figure 24 *BSE photomicrograph of the Ni-Al coating heat treated at 635°C for 10 hours*

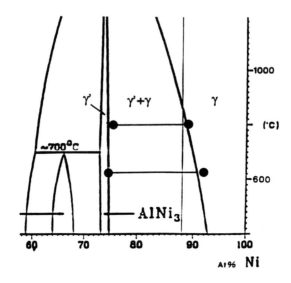

Figure 25 *The Nickel Aluminum phase diagram*

3.1 Microstructural Characterization of Electrodeposited Nickel Composite Coatings

3.1.1 Nickel-Alumina Composites. The characteristic light optical microstructure of an etched cross-section of a Ni electrodeposited coating is seen in Figure 15. The structure is made up of apparent long columnar grains that are perpendicular to the interface. The columnar grain thickness, measured parallel to the substrate interface, increases with current density, Figure 16, in agreement with Saleem, et al.[17] Introduction of a-Al_2O_3 Particulates to the bath

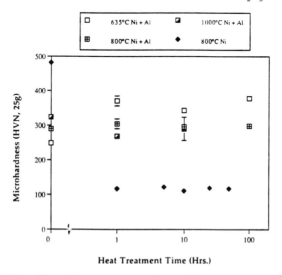

Heat Treatment Time (Hrs.)

Figure 26 *The effect of heat treatment time and temperature on the hardness of electrodeposited MMC alloys*

Table 1 *Summary of the measurements for stepped FGM composites*

Field	Sigle layer thickness (μm)	Stepped thickness (μm)	Single layer vol %	Stepped vol %	Single layer HKN	Stepped HKN
1	40	39.2 ± 2.8	0	0	278	285 ± 7
2	40	43 ± 3.2	15	14.6 ± 0.7	391	411 ± 13
3	40	46 ± 5.1	30	33.1 ± 1.2	740	722 ± 21

produced a relatively uniform distribution of the ceramic particles within the matrix, Figure 17 and Figure 18 show that the amount of alumina incorporated in the composite decreases with current density. On etching the microstructure, the etchant preferentially attacked the interface between the ceramic particle and the matrix, making it difficult to observe the grain structure. As expected, interparticle spacing decreased with increasing volume percent alumina. The hardness of the electrodeposits increases with current density, Figure 19, and as expected, the alumina composites have a higher hardness than the pure Ni deposits.

3.1.2 Nickel-Aluminum Composites. The effect of current density on the volume percent aluminum found in the electrodeposited coatings shows that the amount of aluminium varied little with current density, Figure 20. A light optical micrograph, Figure 21, demonstrates the uniform distribution of the aluminum particles in the nickel matrix. Contrasting the microstructure between a nickel deposited coating and a Ni plus Al deposited coating at a current density of 15 A/dm², shows that the addition of Al particles to the Ni matrix refines the columnar morphology of the coating. This effect may be due to the nucleation of small nickel grains onto the surface of the Al particles. In the vicinity of Al particles, especially larger ones, coating growth direction changes and striations normally found in electrodeposited coatings, are no longer parallel to the substrate interface, Figure 22a. This results in a surface mound above the large particle, which is developed in Figure 22b. A significant hardness

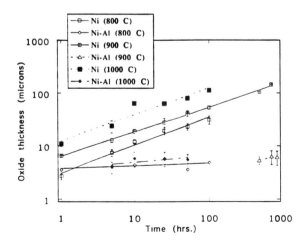

Figure 27 *Oxidation of Ni and Ni-Al composite coatings*

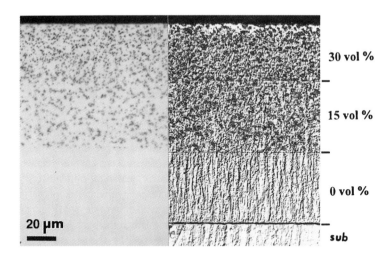

Figure 28 *Light optical micrograph of the nickel-alumina FGM stepped coating. Right side is etched*

increase is found for these coatings in contrast to the Ni and Ni/alumina coatings, Figure 23, due to the structural refinement from the Al particles.

The heat treatment of the Ni/Al coatings results in the formation of Ni_3Al, Figure 24, the amount being dependent upon the temperature as seen in the Ni-Al phase diagram, Figure 25. The hardness results for three different diffusion treatments is given in Figure 26, where the

Figure 29 *SEM photomicrograph of an etched Ni-alumina stepped composite showing the continuous columnar structure*

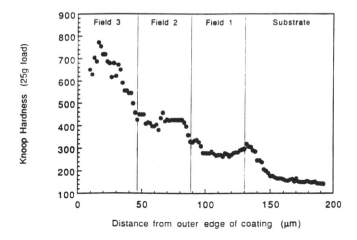

Figure 30 *Characteristic hardness profile of an FGM stepped composite*

results of the heat treatment of a Ni coating are included for comparison. Whereas the hardness of the Ni coating decreased sharply with time as a result of recrystallization and grain growth of the as-plated columnar structure, the Ni-Al coatings did not exhibit the same behaviour. For a given heat treatment, the hardness of the Ni-Al coatings remained constant for extended heat treating times, suggesting the formation of a very stable microstructure. Oxidation studies of the Ni-Al composite coatings indicate a significant improvement over as-plated Ni due to the availability of Al in the coating to produce a protective Al_2O_3 scale on the surface.

3.2 Characterization of FGM Coatings

New design of engineering components has led to the introduction of coatings with gradients of chemical composition, phase distribution or microstructures[18]. Through the manipulation of the processing variables, discretely graded microstructures of a coating show a volume percent Al_2O_3 in a Ni electrodeposited matrix were produced. Table 1 shows a summary of the measurements for the stepped composite. The processing parameters were chosen so that two extremes (0 and 30 volume percent), plus a median composition (15 volume percent), would be represented in the coating. Therefore, it would be easier to determine from bulk layers, if the microstructures of the individual layers were modified during fabrication.

A characteristic micrograph of the unetched, as-plated sample can be seen in Figure 28, left. The structure consists of an inner pure nickel layer and two electrodeposited composite coatings, with increasing volume percent alumina. No visible interface can be observed between the individual deposited layers. In the etched microstructure, Figure 28, right, the interfaces can be seen to be linear with no apparent defects. It appears that the columnar structure is continuous from one layer to the next as shown in Figure 29, however, because of the difficulty in etching the higher volume percent Al_2O_3 deposit, no observations can be made of the interface between the two composite layers. Figure 30 shows the characteristic hardness profile of the discretely graded structure. The hardness of the single electroplates and the corresponding layer within the stepped structure are in good agreement, Table 1, which suggests that the microstructure and properties of the individual layers have not been altered in producing the FGM stepped composites.

4 SUMMARY

Protective coatings, i.e. coatings that can withstand high temperature corrosion and erosion, often rely on composites with hard phase oxides and/or intermetallics, in a ductile matrix to obtain the necessary properties. Two processing methods of obtaining composite coatings are thermal spraying and electrodeposition. Thermal spray processing methods such as air plasma spray (APS), vacuum plasma spray (VPS) and high velocity oxyfuel (HVOF), often are used to produce spray composites of a metal carbide in a corrosion resistant matrix. However, in order to obtain optimum properties, microstructural characterization is needed to determine the relationships between spray powder and coating composition, volume percent hard phase and erosion resistance, etc. Often times model systems can be used to evaluate the composite microstructural features that control properties, as in the Fe-Fe$_3$C system that shows interparticle spacing to be the controlling parameter for improved erosion resistance. Electrodeposition is a well known process that has recently been used to build composite coatings. Different types of coatings can be constructed, depending upon the deposited particulate used, e.g. Al or Al$_2$O$_3$. Intermetallic coatings for high temperature oxidation have been produced by electroplating and heat treating a Ni matrix with Al particulates and stepped composite FGMs have been produced with varying percentages of Al$_2$O$_3$ hard phase. Characterization of the electroplating parameters and microstructure are essential for the production and application of these composite coatings.

Acknowledgments

The author wishes to thank his colleagues and students for there major contributions to this paper: Prol. K. Barinak, Dr. S.Bluni, Dr. B. Smith, A.O. Benscoter, J.N. DuPont, S.W. Banovic, D.F. Susan, B.S. Schorr, K.J.Stein, B. Levin, C.M. Petronis and Prof. R. Chaim of the Technion, Israel.

References

1. E. D. Hondros, 'Surface Engineering', ed. by P.K. Datta and J.S. Gray, The Royal Society of Chemistry, Cambridge, 1993, 1, 1.
2. H. Herman, *Scientific American*, 1988, 9, 112.
3. G. A. Blann, 'Proc. of the Fourth National Thermal Spray Conference', ed. by T.F. Bemecki, ASM International, 1991, p. 175.
4. G. A. Blann, 'Proc. of the International Thermal Spray Conference', 1992, p. 959.
5. S. T. Bluni, K. Goggins, B. Smith and A.R. Marder, 'Metallography: Past, Present, and Future (75th Anniversary Volume)', ASTM STP 1165, ed. by G.F. Vander Voort, F. J. Warmuth, S.M. Purdy and A. Szirmae, ASTM, Philadelphia, 1993, p. 254.
6. S. T. Bluni and A.R. Marder,'Proc. of the Fourth National Thermal Spray Conference', ed. by T F Bemecki, ASM International, 1991, p. 89.
7. S. T. Bluni and A.R. Marder, 'Surface Engineering', ed. by P.K. Datta and J.S. Gray, The Royal Society of Chemistry, Cambridge, 1993, Vol. 2, p. 220.

8.   S. T. Bluni and A R Marder, *Corrosion*, 1996, **52**, 213.
9.   B. I. Edelson and W M Baldwin, Jr., *Trans.* ASM, 1962, **55**, 230.
10. C. T. Lui and J. Gurland, *Trans.* ASM, 1968, **61**, 156.
11. L. Anand and J. Gurland, *Met Trans.*, 1976, **7A**, 191.
12. C. T. Liu and J. Gurland, *Trans.* AIME, 1968, **242**, 1535.
13. B. Levin, J. N. DuPont and A.R. Marder, *Wear*, 1995, **181-183**, 810.
14. C. T. Liu, PhD Dissertation,Division of Engineering, Brown University, 1967.
15. S. W Banovic, C. M. Petronis, K. Barmak and A R Marder, Proc. of the Symposium on High Temperature Coatings 11, TMS, 1996, Anaheim, CA.
16. K. Barmak, S. W. Banovic, C. M. Petronis, D. G. Puerta, D. F. Susan and A. R. Marder, 'Proceedings on Microscopy of Composite Materials III', 1996.
17. M. Saleem, P. A. Brook and J W Cuthbertson, *Electrochemica Acta*, 1967, **12**, 553.
18. B. Ilschner and N. Cheradi, 'Proc. 3rd International Symposium on Structural and Functional Gradient Materials', ed. by B Ilschner and Cheradi, Presses Polytechnique et Universitaires Romandes, 1995, p. V-Ix.

3.2.2

Simulation of Failure Mechanisms in Al$_2$O$_3$/Al Metal Matrix Composites and the Effect of Fibre/Matrix Interface Bond Strength Using Finite Element Modelling Techniques

D. Griffin, A. Daadbin, J. S. Burnell-Gray and P. K. Datta

SURFACE ENGINEERING RESEARCH GROUP, SCHOOL OF ENGINEERING, UNIVERSITY OF NORTHUMBRIA AT NEWCASTLE, UK

1 INTRODUCTION

Because of their high specific strength and stiffness, together with their high temperature capabilities, fibre reinforced metal matrix composites (MMCs) are becoming increasingly utilised. Well designed MMCs combine the light weight and ductility of the matrix material, with the high specific strength and stiffness of the reinforcement.

One of the main difficulties in analysing the failure mechanisms in composites is the inapplicability of the rule of mixtures, i.e.

$$G_{Ic} \text{ composite} \neq G_{Ic} \text{ matrix } V_m + G_{Ic} \text{ Fibre } V_f$$

where, G_{Ic} = critical strain energy release rate, V_m and V_f = volume fraction of matrix and fibres respectively. The difficulty is due to the fact that composite failure is made up of a series of energy absorbing mechanisms – debonding, delamination, fibre bridging and matrix

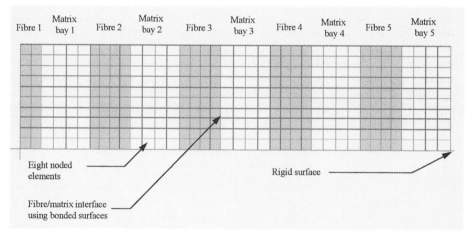

Figure 1 *Finite element mesh showing four and a half fibres situated in five matrix bays*

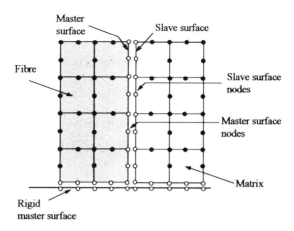

Figure 2 *Elements and nodes making up the master/slave surfaces for the fibre/matrix interface*

microcracking. Additionally, the interface between fibre and matrix plays a crucial rôle in the load transfer characteristics of the composite as a whole[1-3]. Strengthening by fibre reinforcement depends on the process of load transfer, and is limited by the shear strength of the interface. Some of the difficulties can be overcome by adopting the approach of fracture analysis using finite element analysis (FEA).

2 MODELLING PROCEDURE

A two-dimensional, axisymmetric quarter model of a continuous fibre Al_2O_3/Al MMC was constructed in *ABAQUS*. From fractographic analysis and from manufacturer's data, the fibre diameter and volume fraction were determined to be 10μm and 45% respectively. The composite modelled was 25μm in length and 39μm in width and consisted of four and a half fibres situated in five matrix bays (Figure 1). Table 1 shows the material properties applied to the fibres and matrix bays. The differing failure mechanisms incorporated into the model used the "contact surface", "bond surface" and "debond" options available in *ABAQUS* together with a "fracture criterion" suboption. Critical nodal stresses in front of the crack tip(s) were used to determine the particular time, and type, of failure. Mode I cracking for both fibre and matrix, fibre bridging and fibre/matrix debonding were all incorporated into the model.

The *ABAQUS* debond options, all use a master/slave surface concept – as indicated in Figure 2. In the models discussed here element faces making up the fibre side of the fibre/matrix interface were chosen to constitute the master surfaces and element faces comprising

Table 1 *Material properties of fibres and matrix bays*

Property	Al_2O_3	Al
Young's modulus E (GPa)	210	68.9
Poisson's ratio v	0.3	0.33

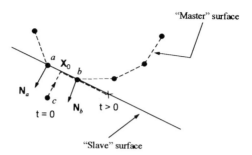

Figure 3 *Schematic showing debonding algorithm in ABAQUS*

the matrix side were selected as the slave surfaces. For the fibre/matrix interface, the fibre side surface was chosen as the master and the matrix side surface as the slave – i.e. the stiffer element set forming the surface was chosen as the master [4]. A rigid surface made up the *r*-axis and the composite was bonded to this surface in the normal direction only, thus allowing for mode I failure, but also allowing for Poisson contraction of the composite.

Three debond options are available in *ABAQUS* – finite sliding, small sliding and infinitesimal sliding. The finite sliding formulation allows sliding of a finite amplitude, and arbitrary rotation, of the two surfaces in contact. In the small sliding formulation, the nodes between the master and slave surface elements should not slide more than one element length and the rotation and deformation of the master surface should not cause the local tangent planes to become a poor representation of the master surface (see Figure 3). Finally the infinitesimal sliding algorithm is the same as that of the small sliding and in addition ignores non-linear geometric effects.

All three types of sliding algorithm were incorporated in test models to determine the most appropriate algorithm:

i.   Finite Sliding Algorithm. Because of the way the finite sliding algorithm works, *ABAQUS* changes the geometry of the model to correctly place nodes along the contact surfaces. Although no undesirable effects should be introduced by these changes in geometry, in practice, for *all* the test models, the algorithm was found to be unstable and gave unreliable stress distribution results compared to the results from small- and infinitesimal-sliding algorithm models. Also this algorithm proved extremely computationally expensive and model runs were very lengthy.

ii. Small Sliding Algorithm. This algorithm proved much more stable than the finite sliding algorithm. Compared to identical conventional models of composites, i.e. without any

Table 2 *Failure properties of composite*

	First run UTS (MPa)		Second run UTS (MPa)	
	σ_c (MPa)	τ_c (MPa)	σ_c (MPa)	τ_c (MPa)
Fibre	1775	887.5	1775	887.5
Matrix	190	95	190	95
Interface	190	95	982.5	491.25

bonded surfaces, the stress distributions were found to be identical.

iii. Infinitesimal Sliding Algorithm. The infinitesimal sliding algorithm gave the best results with algorithm stability to near total failure of the composite and the most economic model run time.

Since the infinitesimal sliding algorithm gave the best results in the test models, this was chosen to be used for the rest of the modelling work. Figure 3 shows a schematic describing how the small- and infinitesimal-sliding algorithms work: using initial nodal coordinates, unit normal vectors are first computed for all nodes on the master surface, e.g. N_a and N_b. These unit normal vectors are used to define a smooth varying normal vector $N(x)$, at any point, x, on the master surface. The algorithm then determines which master nodes will interact with slave node c for the entire analysis. A point on the surface, X_0, is computed for slave node c such that the vector formed by the slave node and X_0 coincides with $N(X_0)$. The example in Figure 3 assumes that X_0 is on the element face with end nodes a and b. A potential contact condition between node c and the line perpendicular to $N(X_0)$ will be enforced. At any time t > 0, node c is constrained not to penetrate this line. The load transfer always occurs between node c and nodes a and b.

The nodal failure stresses for the fibre/matrix interface were assigned the UTS values of the matrix[4]. Mode I failure of the fibre was set at the UTS of the fibre and similarly mode I failure of the matrix was set at matrix UTS. The failure criterion was established as follows:

$$\sqrt{\left(\frac{\hat{\sigma}_n}{\sigma^f}\right)^2 + \left(\frac{\tau_1}{\tau_1^f}\right)^2 + \left(\frac{\tau_2}{\tau_2^f}\right)^2} = f, \ \hat{\sigma}_n = \max(\sigma_n, 0)$$

where σ_n is the normal component of stress carried across the interface, τ_1 and τ_2 are the shear stress components in the interface, and σ^f, τ_1^f, τ_2^f are the normal and shear failure stresses. A crack tip node debonds when the fracture criterion, f, reaches the value 1.0 within a given tolerance: $1 - f_{tol} \leq f \leq 1 + f_{tol}$. f_{tol} was set at 0.05 for the duration of the modelling[4]. Also the shear stresses τ_1^f and τ_2^f were set at 50% σ^f.

The model was run twice, once with the fibre/matrix interface set at the matrix UTS, the second time the interface set at the average fibre/matrix UTS (see Table 2).

A total tensile strain of 1% was applied to the composite. The left hand side of the model was constrained radially. A crack was seeded, in the right hand side matrix bay, by excluding the end node from the node list making up the bonded surfaces (Figure 4). The strain was applied incrementally and the type and progress of the failure mechanisms were observed for each increment. The model run was complete at either 1% applied strain or total composite failure.

3 RESULTS AND DISCUSSION

Crack propagation started along the matrix as soon as the critical nodal stress reached the matrix UTS (~ 0.2% strain). At 0.25% strain, mode I failure caused the crack to reach the first fibre (Figure 5). The crack was blunted, but started to propagate up the fibre/matrix interface. At 0.29% strain, cracking began in the second matrix bay. Thus the crack bridged

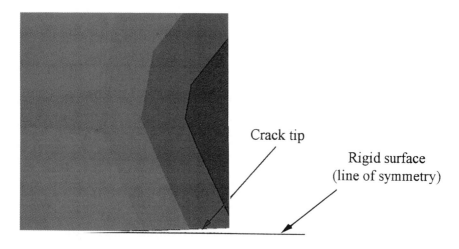

Figure 4 *Seed edge crack on right hand side of matrix bay*

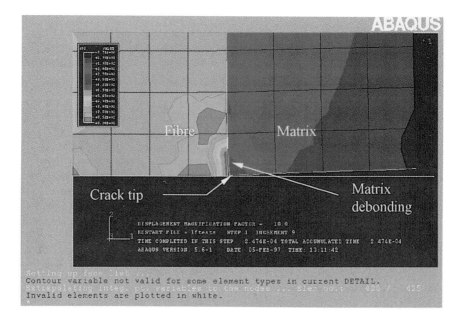

Figure 5 *Crack tip reaches first fibre*

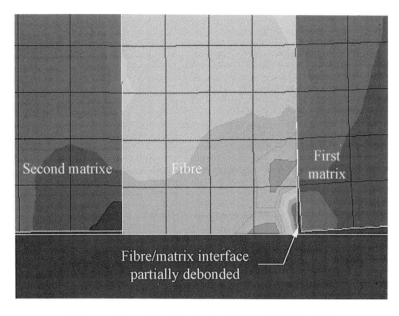

Figure 6 *First fibre bridged and partially debonded*

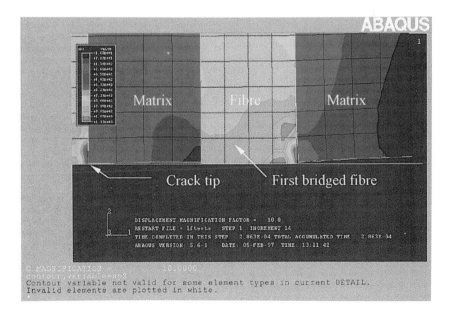

Figure 7 *Total failure of matrix bays 1 and 2*

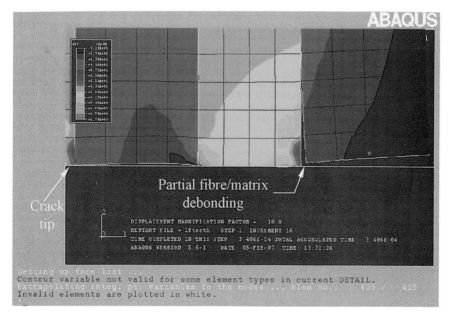

Figure 8 *Crack reaches second fibre*

the first fibre (Figure 6). As the crack continued through the second matrix bay, delamination became evident on both sides of the first fibre. The crack continued to the next fibre, and at 0.4% strain the second matrix bay failed completely in mode I (Figure 7). This failure mechanism

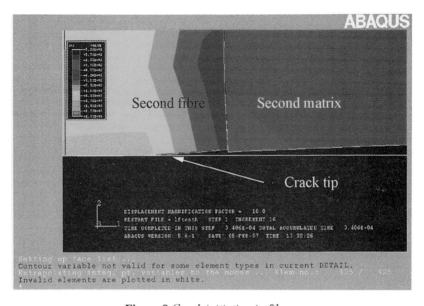

Figure 9 *Crack initiation in fibres*

was repeated until the crack reached the last fibre and all the matrix bays had failed in mode I – with fibres now taking all the strain. The model run was terminated before the fibres started to fail.

The second model run showed a slightly different failure mechanism. Although, as in the previous model run, crack propagation started along the matrix as soon as the critical nodal stress reached the matrix UTS, the first fibre was bridged at a higher strain of 0.37% furthermore, the second matrix bay did not completely fail until the total strain was at 0.48% (Figure 8). As can be seen from Figure 9, crack propagation was also occurring in mode I in the fibres too, indicating that the stress in front of the crack tip at the fibre/matrix interface was sufficiently high to start propagation into the fibre. This is perhaps not unsurprising, when the fibre/matrix bond strength was set at the fibre UTS.

4 CONCLUSION

The dominant mode of failure for both model runs was mode I failure of the matrix followed by fibre bridging, with partial fibre debonding. For the higher bond strength of the interface the composite failed at a higher strain and cracks began to initiate in the fibres due to the high bond strength at the fibre/matrix interface. Although at the present time the *ABAQUS* contact surface algorithm is not sufficiently stable to simulate complete failure of the composite, the results presented here suggest that a high fibre/matrix bond strength encourages crack propagation into the fibres, thus lowering the total tensile load the composite can withstand before failure.

There have been many papers written outlining analytical and semi-analytic methods to predict composite failure[5,6]. Begley and McMeeking[5] used a complex system of semi-analytic integral equations, to predict the occurrence or not of fibre bridging. However, while their technique required significantly less computational effort than the FE approach described here, the amount of man hours required to create and implement the solution was at least the same as the time required to create the FE model, and did not address any other failure mechanisms.

References

1. P. K. Datta, K. N. Strafford and J. S. Burnell-Gray (editors), 'Surface Engineering Practice: Processes, Fundamentals and Applications in Corrosion and Wear', Ellis Horwood, 1990, p. 397.
2. D. Griffin, A. Daadbin, P. K. Datta and J. S. Burnell Gray, Proceedings of the 10th Irish Materials Forum Conference IMF10, Key Engineering Materials, Vol. 99–100, September 1994.
3. N. Melantis, C. Galiotis, P. L. Tetlow and C. K. L. Davies, *Composites.*, 1993, **24**, 459.
4. Abaqus user's manual, V5.5, Vol 1, 1995.
5. M. R. Begley and R. M. McMeeking, *Materials Science and Engineering*, 1995, **A200**, 12.
6. J. Aboudi, *Composites Science and Technology*, 1988, **33**, 79.

3.2.3

High Temperature Air Oxidation Performance of Modified Aluminide Coatings on a Nickel-Based Superalloy

W. Y. Chan, J. S. Burnell-Gray and P. K. Datta

SURFACE ENGINEERING RESEARCH GROUP, SCHOOL OF ENGINEERING, UNIVERSITY OF NORTHUMBRIA AT NEWCASTLE, UK

1 INTRODUCTION

The modern generation of high temperature alloys/coatings depends on the formation of protective oxide for long-term protection. Aluminum and/or chromium are often added to nickel-base superalloys to provide good oxidation resistance at elevated temperatures. Such scale must be stable, slow-growing, compact, and remain adherent under thermal cycling conditions. Improvement of the resistance of nickel-base superalloys can be achieved by the use of aluminide diffusion coatings[1-11].

The increase in operating temperatures of industrial energy systems and gas turbines, often coupled with the decrease in fuel quality being employed, has led to the extensive use of coatings capable of providing improved service life. Aluminide diffusion coatings, commonly deposited by pack cementation process, were first used to improve the oxidation resistance of turbine blades which were inadequate due to breakdown of the protective surface oxides and co-diffusion with the substrate materials; these coatings were superseded by platinum-aluminide coatings[6-12]. It has been shown that the high oxidation performance of platinum- or rhodium-modified diffusion aluminide coatings is superior to those of the unmodified coatings. The protective capability of the aluminide coatings is conferred by the ability of the β-NiAl layer to provide aluminum to maintain an Al_2O_3-based scale at the surface, and its stability governs the effectiveness of the coating[13-14].

The influence of precious and reactive elements on the high temperature performance of advanced coatings is of particular interest. Precious element additions (platinum or rhodium) on Ni-8Cr-6Al alloy have been found to be beneficial for oxide adherence[9]. Such an improvement in oxidation resistance appeared to be associated with the ability of Al_2O_3 to reform after scale spallation, and promote oxide adherence. However, detailed mechanisms by which platinum or rhodium improve oxide adherence were not elucidated, but they were expected to be quite different from which are operative where oxygen-active elements or oxide dispersions are employed.

The oxidation studies conducted in a variety of alumina-forming alloys have shown that spallation of oxide scale is the major scale life-limiting factor and that the adherence of protective oxide scales has to be improved via additions of reactive elements, for example, Y, Hf and Zr. It has been proposed that there is a fixed temperature range over which reactive elements are beneficial[15-19]. The use of minor additions of reactive elements, including the rare earth metals and their oxides to improve scale adhesion is well documented[20-27]. It is well recognised that minor additions (1%) of rare earth and other active elements (e.g. Ti, Hf, Zr or Si) have a number of beneficial effects, viz: (1) enhanced selective oxidation of Cr or Al; (2) a shortening of the transient oxidation period; (3) a reduction in the growth rate of Cr_2O_3 and Al_2O_3 scales: (4) a reduction in the extent of void formation (coalescence of vacancies) at the alloy/scale interface; and (5) improved scale adherence.

Tsipas[25] studied the effect of Hf addition on the cyclic oxidation behaviour of Ni_3Al in an air environment. The weight change data showed that the stoichiometric Ni_3Al and Al-rich Ni_3Al suffered severe spallation during cyclic oxidation test while the alloys containing 2 at% Hf exhibited weight increases and no spallation during cyclic oxidation. NiO was identified as the predominant oxide phase on both the retained and spalled scales in all alloys studies. $NiAl_2O_4$, Al_2O_3 and HfO_2 were reported to be present in the Hf-containing alloys. Hf-protrusions from the scale into the alloy substrate in the Hf-containing alloys were attributed to the enhanced adhesion of the scale to the substrate. In addition, lack of spalling, buckling, ridging, or scale detachment noted in Hf-added alloys was attributed to a reduction in the void formation at the alloy substrate, possibly due to vacancy elimination at preferential sites such as HfO_2.

The objectives of the present work were to determine:

(a) the kinetics of oxidation of the nickel-aluminide, platinum-modified, and platinum-rhodium-aluminide coatings;
(b) establish the composition and microstructure of the oxide scales; and
(c) assess the influence of platinum- and platinum-rhodium-modified additions on the thermal stability of these advanced coatings.

2 EXPERIMENTAL PROCEDURE

The nominal composition of the directionally solidified MAR M002 alloy substrate used in the present investigation is presented in Table 1. The materials used in the current studies were supplied by Rolls-Royce, and Johnson Matthey. The platinum modified coatings were prepared by Chromalloy (UK) Limited. Rod-shaped (25mm in length and 6mm diameter) specimens were coated with a platinum aluminide of the RT22LT-type which nominally contains 35–55% Pt. First, a layer of platinum was electroplated on the alloy surface. Following a diffusion treatment, the surface was aluminised in a high activity pack. After coating, all specimens were then diffusion heat treated for 1h at 1100°C in argon. Finally, all the specimens were thermally aged for 16h at 870°C to precipitate the strengthening gamma prime phase. Following heat treatment the test specimens were washed in distilled water and then ultrasonically cleaned in acetone, followed by hot air drying prior to testing.

To determine the effect of temperature on the structure and composition of the coatings, the as-received and as-coated specimens were exposed to laboratory air at 1000 and 1150°C for up to 580h. The corrosion resistance of the modified aluminide coatings was evaluated by

Table 1 *Nominal chemical composition of MAR M002 substrate material (wt.%)*

Cr	Al	Ti	Co	W	Ta	Mo	Hf	Fe	Zr	B	C	Ni
9	5.5	1.5	10	10	2.5	0.5	1.25	0.5*	0.055	0.015	0.015	Bal

*denotes maximum

using a cyclic oxidation test method. Cyclic oxidation tests were conducted at 50h intervals. The kinetic data were obtained by means of a discontinuous thermo-gravimetric method. The values of the weight of the specimens were obtained prior to discontinuous oxidation test and further weight changes were recorded after various periods of exposure.

Detailed microstructural investigations were conducted using optical and electron microscopy. Quantitative EDX analyses of elements were performed at an accelerated voltage (20kV) using a spot size of 100nm. The results were quantified using the ZAF-4 programme of the Link Analytical EDX system connected to a Hitachi 2400 scanning electron microscope. To identify all phase constitutions X-ray diffraction was carried out using Cu-K$_\alpha$ radiation on a Siemens D-5000 diffractometer. Phase identification was accomplished with the aid of the Joint Committee Powder Diffraction Standards (JCPDS) file[28].

3 RESULTS AND DISCUSSION

Figure 1 illustrates comparative morphologies of the transverse sections of coating microstructures associated with the straight aluminide, platinum-modified, and platinum-rhodium-modified aluminide coatings on MAR M002 alloy substrate. It is expected that the structure and coating compositon will reflect a difference in the scaling performance of each coating system.

3.1 Microstructure of Straight Aluminide Coating

The morphological features associated with the straight aluminide coating in the heat treated condition can be seen in Figure 1a . The overall coating thickness is about 57 μm, consisting of an outer layer about 47 μm in thickness, containing fine precipitates, mostly Cr particles in a matrix of hypostoichiometric β-NiAl and an inner interdiffusion zone (~10 μm) consisted of alternating lamina of NiAl phase containing substrate elements and expected to be β-phase, and a (Ni+Co)-Cr-W type sigma phase. In addition, Hf-, Ta- and Ti-rich particles possibly associated with MC-type carbides were observed in the interdiffusiom zone (15 μm in thickness).

Figure 2 shows the elemental concentration profiles of Ni, Al and W across the simple aluminide coating into the substrate. It can be seen that the concentration of Ni is approximately 18% higher than Al across the outer layer indicating that it consisted mainly of hypostoichiometric β-NiAl (nickel-rich) and its long term cyclic oxidation performance is inferior to β-NiAl phases containing higher levels of aluminum as observed in the Pt- and/or Pt-Rh-modified aluminide coatings.

3.2 Microstructure of Platinum Aluminide Coating

The initial coating microstructure of the platinum aluminide coating (~ 59 μm in thickness) was found to consist of two layers as shown in Figure 1b. Most of the platinum was concentrated in the outer layer (~26 μm in thickness) as illustrated in Figure 3. X-ray diffraction data revealed that this layer consisted of a mixture of PtAl$_2$ and β-NiAl. In addition to the main three elemental constituents (nickel, platinum and aluminum), relatively small concentrations of chromium and cobalt were detected. The outer coating layer was relatively free of tungsten and other refractory elements. XRD analysis (Table 2) indicated that PtAl$_2$ was present as a

Table 2 *X-ray diffraction analysis (α-2θ) of straight aluminide, Pt- modified, and Pt-Rh-modified aluminide coatings after cyclic oxidation test in 1 atm(1.0 x 10⁵ Pa) at 1000 and 1150°C*

Coatings	Temperature (°C)	Products
Straight aluminide	1000	NiAl, NiAl$_2$O$_4$, Al$_2$O$_3$, TiO$_2$, NiO
Straight aluminide	1150	NiAl, NiAl$_2$O$_4$, NiO, TiO$_2$, Al$_2$O$_3$
Pt- aluminide	1000	NiAl, PtAl$_2$, TiO$_2$, Al$_2$O$_3$
Pt-aluminide	1150	NiAl, PtAl$_2$, TiO$_2$, Al$_2$O$_3$
Pt-Rh aluminide	1000	NiAl, PtAl$_2$, ZrO$_2$, Al$_2$O$_3$, TiO$_2$
Pt-Rh aluminide	1150	NiAl, PtAl$_2$, ZrO$_2$, Al$_2$O$_3$, TiO$_2$, NiO

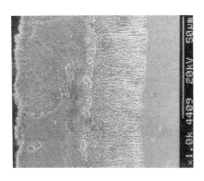
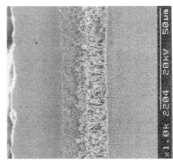

(a) Straight aluminide   **(b) Pt-modified aluminide**

(c) Pt-Rh-modified aluminide

Figure 1 *Scanning electron micrographs showing the morphological features associated with the cross-section of the straight aluminide, Pt-modified, and Pt-Rh-modified aluminide coatings on MAR M002 substrates*

Table 3 *Parabolic rate constants (k_p) for the platinum-, platinum-rhodium-modified aluminide coatings after cyclic oxidation test up to 580h at 1000 and 1150°C*

Coatings	Temperature °C	k_p (g^2 cm^{-4} s^{-1})
Pt-aluminide	1000	3.42×10^{-15}
Pt-Rh aluminide	1000	3.56×10^{-15}
Pt- aluminide	1150	4.28×10^{-14}
Pt-Rh aluminide	1150	3.56×10^{-14}

secondary phase in a matrix of hyperstoichiometric β-NiAl (aluminum-rich) common in RT22LT-type coatings. Pt enriched β-(Ni,Pt)Al phase was also detected in the β-phase containing outer coating.

Detailed EDX analysis (Figure 3) revealed that the inner coating layer (~20 μm thick) was relatively free of platinum and the inner coating layer consisted of mainly hypostoichiometric β-NiAl (nickel-rich) with other elements, such as cobalt, chromium, titanium and tungsten in solid solution. This inner hypostoichiometric β-NiAl is likely to have formed by interdiffusion, i.e. outward diffusion of nickel, and inward diffusion of aluminum. Fine precipitates were also observed in the inner coating layer, EDX analysis shown to be consisted of Cr rich phases. Detailed EDX analysis (Figure 4) showed that the interdiffusion zone (~15 μm in thickness) to be consisted of blocky particles of sigma phase dispersed in a matrix of hypostoichiometric-NiAl, containing (Ni,Co)-Cr-W. Hafnium-rich, and Ta rich MC type carbide particles were also detected in this region, indicating that hafnium could have also also diffused into the coating.

3.1 Microstructure of Platinum-Rhodium Aluminide

Figure 1c illustrates the morphological features associated with Pt-Rh-modified aluminide coating in the heat treated condition. A two layered structure was observed. It is interesting

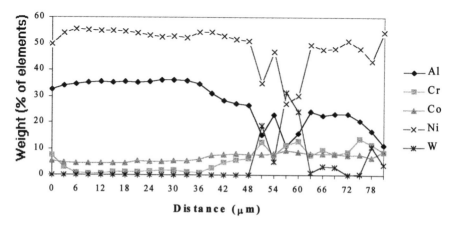

Figure 2 *Elemental profile across the straight aluminide coating from the surface to substrate on MAR M002 alloy substrate*

to note that the platinum concentration decreased gradually from the surface to a thickness at about 25 μm from the surface in the hyperstoichiometric β-NiAl (aluminium-rich) outer coating layer as revealed in Figure 5. X-ray diffraction analysis showed that this outer layer (~25 μm thick) consisted of a mixture of PtAl$_2$, in a matrix of β-NiAl. Elemental analysis revealed that a rhodium-rich second layer, about 20 μm in thickness was also located at ~12 μm to 33 μm from the surface of the coating, and the concentration of rhodium gradually decreased towards the inner layer. The inner layer of platinum-rhodium aluminide exhibited a very similar microstructure and composition to the platinum-aluminide coating as described earlier; detailed EDX data showed that this inner coating layer consisted of mainly hypostoichiometric β-NiAl (nickel-rich) matrix in a solid solution such as cobalt, chromium, titanium, tantalum and tungsten. Very fine Cr-rich particles were also observed in this inner coating layer. SEM micrographs such as Figure 6, provided a detailed analysis of the interdiffusion zone (~15 μm in thickness) which exhibited a laminated structure of blocky sigma phase particles containing (Ni+Co)-Cr-W in a matrix of hypostoichiometric β-NiAl. Occasionally, Hf, Ta and Ti-rich MC-type carbide particles were identified in this interdiffusion layer. It is most likely that this coating layer was formed by the outward diffusion of nickel and the inward diffusion of aluminum.

3.2 Oxidation Kinetics of Modified Aluminide Coatings

Cyclic oxidation tests were performed at 1000 and 1150°C to evaluate the scale spallation resistance of the protective α-Al$_2$O$_3$ scale. The general oxidation kinetics of the straight aluminide, Pt-, Pt-Rh-modified aluminide coatings on MAR M002 alloy substrate at 1000 and 1150°C are presented in Figures 7-9. In all of these tests the platinum and the platinum-rhodium modified coatings performed better than the straight aluminide coating. In particular, protective alumina scales were maintained for longer periods of time on the Pt- and platinum-rhodium-modified coatings compared to that on the straight aluminide coating as revealed in Figures 7–9 and their respective scaling rates are listed in Table 3. The oxidation behaviour of the modified aluminide coatings exhibited protective reaction kinetics and followed a parabolic rate law. It is clearly revealed that the platinum-rhodium-modified coating displayed smaller weight increases than the platinum aluminide coatings for cyclic oxidation tests conducted for prolonged exposures up to 580h at 1150°C. The Pt-Rh-aluminide showed a very low parabolic

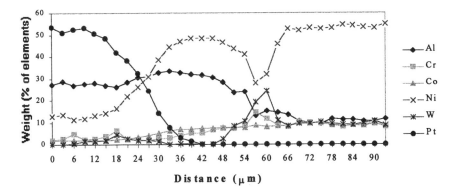

Figure 3 *Elemental profile across the Pt-modified aluminide coating from surface to substrate on MAR M002 alloy substrate*

rate constant values of 3.56×10^{-14} and $4.28 \times 10^{-14} g^2 cm^{-4} s^{-1}$ respectively attributed to the development of dense, protective and slow growing α-Al_2O_3. Very low oxidation scaling kinetics were also observed for the Pt-, and Pt-Rh-modified coatings exhibiting oxidation rates (k_p) values of 3.42×10^{-15} and $3.56 \times 10^{-15} g^2 cm^{-4} s^{-1}$ respectively, following cyclic oxidation test at 1000°C (Table 3). It is interesting to note that the straight aluminide also exhibited parabolic kinetics after exposure up to 265h at 1000°C. It is most likely that the protective capability of the aluminide coatings is conferred by the ability of the β-NiAl layer in the modified aluminide coatings to provide aluminium to maintain an Al_2O_3-based scale at the surface. The β-NiAl phase has the best oxidation resistance of the nickel aluminides[13], and therefore its stability will govern the effectiveness of the coating[14]. The addition of Pt, and Pt-Rh to the aluminide coating provided further improvement in the thermal stability of the coatings as described in the following section.

3.3 Morphology of the Scale Formed on the Straight and Modified Aluminide Coatings After Cyclic Oxidation

Detailed analysis of the weight change versus time data for the three coating systems after oxidation test performed at 1000 and 1150°C showed that the platinum-rhodium coating

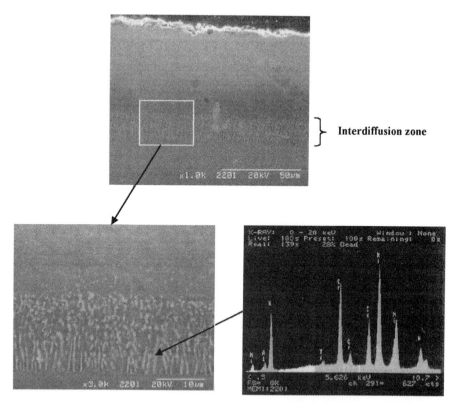

Figure 4 *EDX analysis of "white" particles at the interdiffusion zone of Pt-modified aluminide coating on MAR M002 alloy substrate*

exhibited the best oxidation resistance at the two temperatures as shown in Figures 7-9 and detailed in Table 3. The platinum- and platinum-rhodium-modified aluminide coatings also exhibited significant improvement in scale stability over the straight aluminide coating. The internal oxidation corrosion resistance of the straight aluminide coating was also found to be superior to the uncoated alloy which exhibited internal oxidation as revealed in Figures 10 and 11.

The coating microstructures beneath the scale were very different as revealed in the optical micrographs (Figure 11) for the three coating systems following 50h oxidation at 1150°C. The simple aluminide had formed an appreciable amount of gamma prime, and in some areas, next to the scale. Figure 11a shows the transverse section of the scale formed on the straight aluminide coating after 50h oxidation at 1150°C. The scale is composed mainly of NiO, $NiAl_2O_4$, and Al_2O_3.

The platinum aluminide and the Pt-Rh-modified coatings on the other hand displayed a continuous layer of β-NiAl phase adjacent to the scale with gamma prime appearing at the grain boundaries as revealed in Figure 11. It is also indicated that the oxidation rates during the early stage of oxidation were found to decrease with decreasing alloy grain size as depicted in Figure 11. This may be attributed to the increased grain-boundary diffusion of solute (Al) through the coating to the oxide/coating interface – leading to the selective oxidation of Al. Albeit the influence of the alloy grain size was less important once a protective Al_2O_3 scale was formed; the oxidation rate was then controlled by diffusion through the oxide.

Figures 12 and 16 show the types of degradation for which Pt and Pt-Rh additions are effective means to extend coating lives and describe the mechanisms by which Pt- and Pt-Rh-modified aluminide produces the beneficial effects on the oxidation performance at elevated temperatures. The addition of Pt and Pt-Rh was demonstrated to have significant influence on the transient and steady state oxidation. Metastable Al_2O_3 phases, such as δ-Al_2O_3 and θ-Al_2O_3 are commonly formed[29–30], as revealed in Figure 15. The metastable Al_2O_3 phases

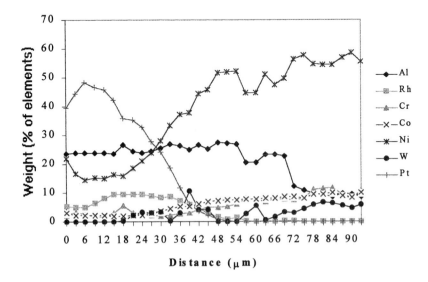

Figure 5 *Elemental profile across the Pt-Rh-modified aluminide*

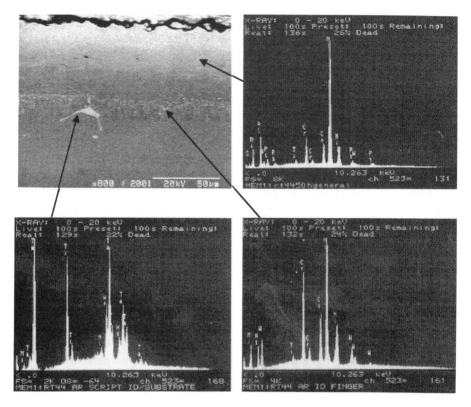

Figure 6 *EDX analysis of "white" particles at the interdiffusion zone of Pt-Rh-modified aluminide coating on MAR M002 alloy substrate*

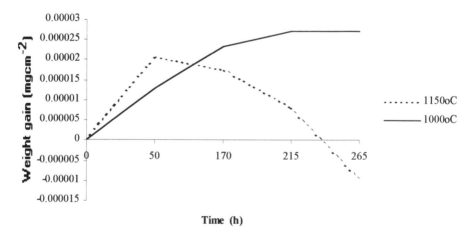

Figure 7 *Weight gain versus time for straight aluminide coatings after cyclic oxidation test at 1000 and 1150ºC for exposure up to 265h*

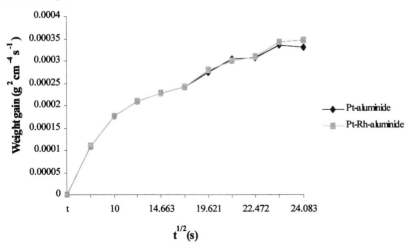

Figure 8 *Weight gain versus square root of time (s) for Pt-, and Pt-Rh- modified aluminide coatings after cyclic oxidation test at 1000ºC for exposure up to 580h*

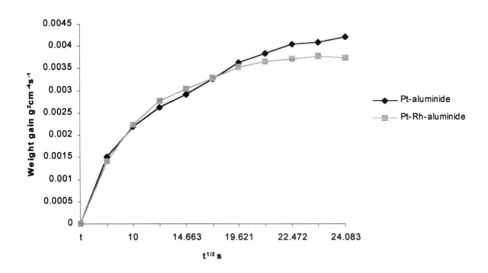

Figure 9 *Weight gain versus square root of time (s) for Pt-, and Pt-Rh-modified aluminide coatings after cyclic oxidation test at 1150ºC for exposure up to 580h*

transform to α-Al_2O_3 after prolonged exposure at high temperature beginning at the gas/oxide interface and moving inward to the scale/coating interface[29-30]. θ-Al_2O_3, a cubic structure with a cation-vacancy network, grows by outward diffusion[19] and α-Al_2O_3 grows by inward diffusion[31]. The results of SEM analysis combined with morphological studies of oxides at 1000 and 1150°C evidently showed that the changes are due to the phase transformation of θ to α-Al_2O_3[32-33] as represented in Figure 15. The incorporation of Hf, Ti, Pt and Rh in the oxide scale accelerates the θ to α-Al_2O_3 transformation and hence quickly establishes a dense, protective scale. The tendency of scale spallation at the scale/coating interface is caused by

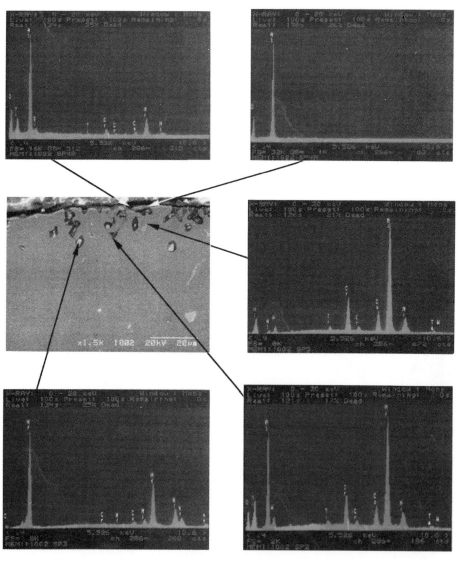

Figure 10 *Scanning electron micrograph and EDX images showing the transverse section of MAR M002 alloy after cyclic oxidation test at 1150°C for 50h*

Straight aluminide coating on MAR M002   **Pt-modified a luminide coating on MAR M002**

⊢——⊣
10µm

Pt-Rh modified aluminide coating on MAR M002

Figure 11 *Photo-micrographs in cross-sections of the straight aluminide, Pt-aluminide and Pt-Rh-modified aluminide coatings after cyclic oxidation test at 1150°C for 50h*

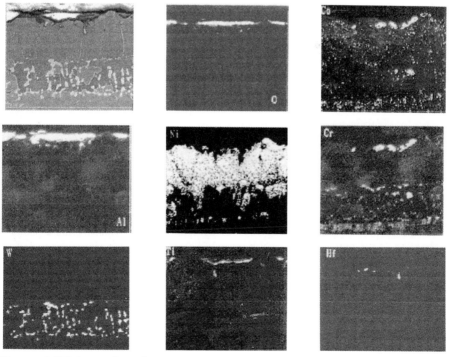

Figure 12 *SEM secondary electron and X-ray images of elements associated with straight aluminide coating after cyclic oxidation test at 1150°C for 265h*

the lateral growth of α-Al_2O_3. The adherence of the alumina scale to the coating is improved by doping the α-Al_2O_3 with Hf and/or Zr. The oxide grows by outward diffusion of Al^{3+} cations in the bulk and at grain boundaries and inward diffusion of oxygen at the grain boundaries. This leads to oxide formation within the oxide layer at the gain boundaries, causing a certain contribution to lateral growth. This lateral growth induces stresses in the scale, which could lead to separation of oxide from the coating. However on the Pt- and Pt-Rh-modified aluminide coatings, the α-Al_2O_3 protective scale was doped with Hf and Zr, and a well-adherent layer developed and grew at a reduced rate. This is consistent with the hypothesis put forward earlier to account for the reactive element effect[15,18]. Current study also illustrated that the reactive elements, such as Hf and Zr, improved the alumina scale adherence by the formation of pegs at the scale/coating interface Figures 16 and 17.

The presence of platinum at the outer layer excludes refractory transition elements, such as tungsten from the outer coating layer. Platinum also promotes the selective oxidation of aluminum[34] as represented in Figure 13 and 14. The addition of rhodium together with Zr reduced the aluminum content required to form a protective alumina scale. The Ti, Hf, Zr and Rh acted as oxygen getters and reduced both nickel oxide formation and internal oxidation. At a subsequent stage, the reactive element phases decrease the outward diffusion flux of aluminium to the oxide/coating interface. The analysis in Figures 16 and 17 shows that a healing Al_2O_3 layer was formed on the surface of the modified aluminide coating.

Based upon the above findings, it can be concluded that the Pt- and Pt-Rh-modified aluminide coatings exhibited better thermal stability than the straight aluminide and the uncoated MAR M002 alloy substrate, as revealed in Figures 10 and 11, at the temperatures investigated. The scale developed on Pt-Rh aluminide coating is more protective in comparison with that of the Pt aluminide and the straight aluminide coatings. This behaviour can be related to differences in coating composition, particularly the Pt-, and Pt-Rh rich-, layers as represented schematically in Figures 18 and 19.

During the initial stage of oxidation, a thin scale of metastable Al_2O_3 quickly formed[35-36], on the Pt- and Pt-Rh- aluminide coatings. The stability and adherence of the alumina scale is greatly enhanced as a result of the presence of $PtAl_2$-, hafnium- and zirconium-rich phases in the scale/coating interface and at grain boundaries. The presence of oxide ridges provided evidence for aluminum transport through oxide grain boundaries[37], arising from a countercurrent of aluminum and oxygen. This leads to outward growth of the oxide ridges at the scale/coating interface. During exposure at 1000 and 1150°C the structure of the outer coating changed from a mixture of $PtAl_2$ and NiAl into mixture of NiAl and Ni_3Al. Significant reduction in platinum concentration as shown in Figures 13 and 14 could be attributed to thermodynamic instability of $PtAl_2$ and inward diffusion of platinum. Clearly, in the as-deposited condition the interdiffusion zone was virtually free of platinum, but after 50 h oxidation at 1000 and 1150 °C a noticeable amount ($\sim$ 10 wt.% Pt) was observed. As the exposure increased the composition of the outer coating layer approached that of the inner coating layer due to interdiffusion. Although the structural stabilities of the Pt- and Pt-Rh-modified coatings are very similar as revealed in Figures 3 and 5. The Pt-Rh-modified coating developed a more protective scale in comparison with the Pt aluminide coating- this can be attributed to the synergistic effect of the Rh-rich secondary diffusion barrier which effectively reduced the outward diffusion of Al from the β-phase coating and Ni from the substrate and enhanced the coating stability over a longer oxidation test duration.

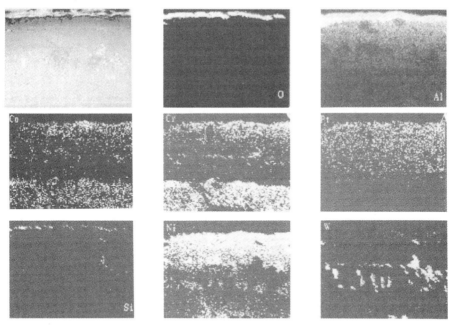

Figure 13 *SEM secondary electron and X-ray images of elements associated with Pt-modified aluminide coating after cyclic oxidation test at 1150°C for 50h*

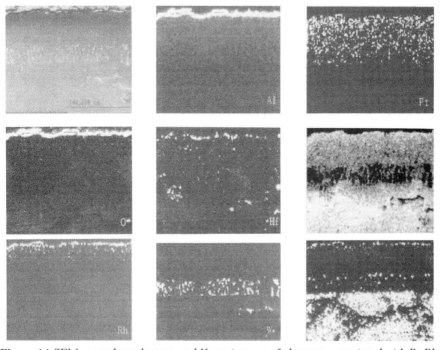

Figure 14 *SEM secondary electron and X-ray images of elements associated with Pt-Rh-modified aluminide coating after cyclic oxidation test at 1150°C for 50h*

(a) Pt-modified aluminide

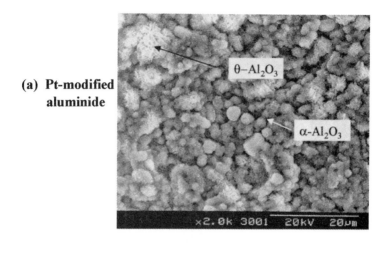

(b) Pt-Rh-modified aluminide

(c) Pt-Rh-modified aluminide

Figure 15 *SEM secondary electron images of the morphology of alumina scale formed after oxidation test for 50h at 1150°C on ; (a) Pt-modified aluminide, and (b), (c) Pt-Rh-modified aluminide coatings*

4. CONCLUSIONS

The following conclusions may be drawn concerning the oxidation behaviour of the straight aluminide, Pt-,and Pt-Rh-modified aluminide coatings on MAR M002 alloy substrate in 1 atm $(1 \times 10^5 \, \text{Pa})$ at 1000°C (1273K) and 1150°C (1423K) for exposure up to 580h.

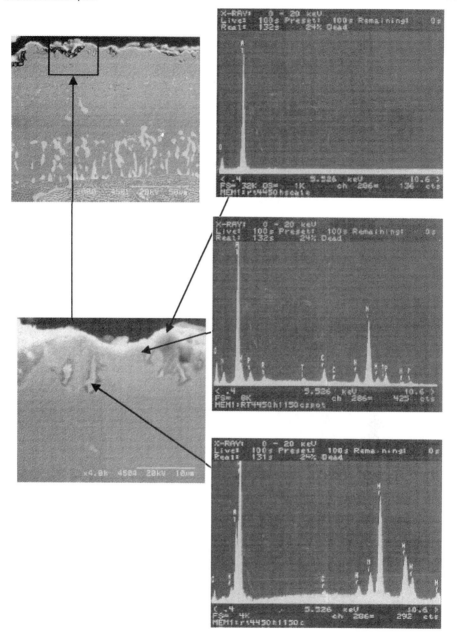

Figure 16 *Scanning electron micrographs and EDX analyses on Pt-Rh-modified aluminide coating after cyclic oxidation test for 50h at 1150°C*

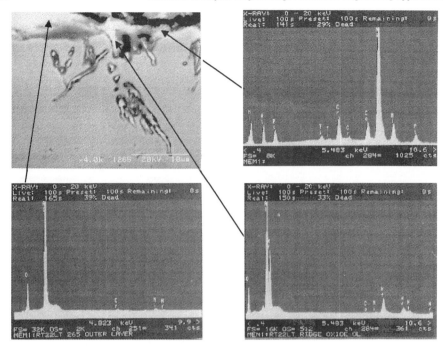

Figure 17 *Scanning electron micrographs and EDX analyses on Pt-modified aluminide coating after cyclic oxidation test for 265h at 1150°C*

1. The nickel aluminide coating showed satisfactory oxidation protection on MAR M002 alloy substrate at 1000°C, Pt and/or Pt-Rh additions are necessary to obtain protection at higher temperatures.
2. The platinum- and platinum-rhodium-modified-aluminide coatings exhibited protective cyclic oxidation resistance between 1000 and 1150°C.
3. The straight aluminide, Pt-aluminide and the Pt-Rh-modified aluminide coatings exhibited very different microstructural features.
4. The straight aluminide sustained more internal oxidation attack and β-NiAl phase depletion than platinum-, and platinum-rhodium-modified coatings after oxidation at both temperatures.
5. The overall coating integrity observed in both platinum- and platinum-rhodium-modified coatings remained protective after prolonged exposure up to 580h at 1000 and 1150°C. The protective mechanism is apparently based on the formation of adherent, dense α-Al_2O_3 scales and the distribution of Pt, and/or Pt-Rh phases throughout the β-NiAl layer.
6. The observance of Hf, Rh, Zr and Ti particles at the scale/coating interface indicated that the scale is formed at least partially by an outward growth mechanism.

Acknowledgement

The financial support from EPSRC under Contract No.GR/K02367 is acknowledged. The

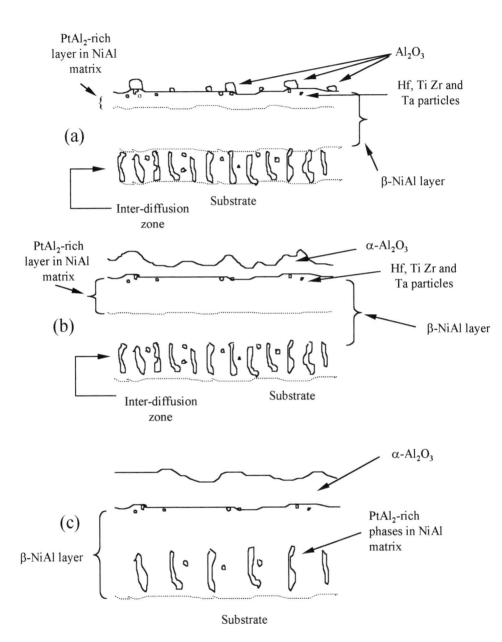

Figure 18 *Schematic drawing showing the reaction mechanism for Pt-modified aluminide coating on MAR M002 after cyclic oxidation test , (a) early stage, (b) transient stage and (c) early steady state*

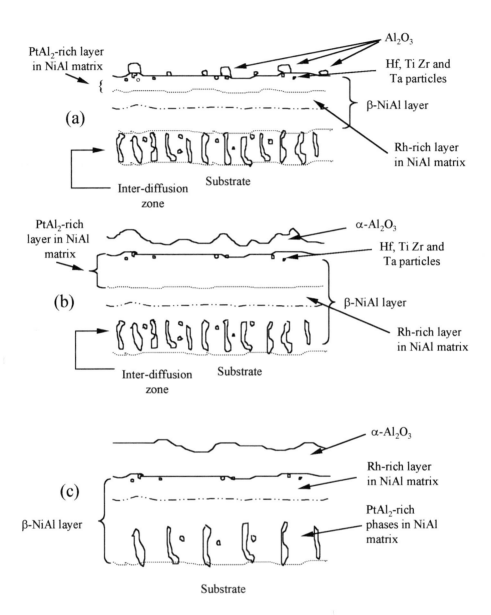

Figure 19 *Schematic drawing showing the reaction mechanism for Pt-Rh-modifed coating on MAR M002 after cyclic oxidation test, (a) early stage, (b) transient stage and (c) early steady state*

technical collaboration, provision of materials and coatings from Rolls-Royce plc, Johnson Matthey and Chromalloy (UK) Ltd and their useful discussions are also gratefully acknowledged.

References

1. W. Goward, D. H. Boone and .C S. G. Giggins, Transactions ASM, 1967 **60**, 228.
2. A. T. Cape, US Patent 3, 107175, 15 October 1963.
3. H. H. Todd, US Patent 3, 494748, 10 February, 1970.
4. G. Lehnert, British Patent, 1210026, 1970.
5. D. A Joseph , US Patent No. 3,495,748, 1970.
6. G. Lehnert and H. W. Meihardt, Electrodeposition Surface Treatment, 1972, **1**, 189.
7. K. Bungard, U S PatentNo. 33,677,789, 1972.
8. R. J. Stueber and S. J. Klach, US Patent,No.551,615, 1975.
9. E. J. Felten and .F S. Pettit, Oxid Metals, 1976, **10**, 189.
10. E. F. Felten, ibid, 1976, **10**, 23
11. R. G Wing and I R McGill, *Plat. Metals Review*, 1981, **3**, 94.
12. T. N. Rhys-Jones, In ' Materials Development in Turbo-Machinery Design' Eds D M R .Taplin, J F Knott and M. H. Lewis, Institute of Metals, London and Dublin , 1989, 218.
13. N. R. Lindblad, *Oxidation of Metals*, 1969, **1**, 143.
14. M. M. .P Janssen, G. D. Rieck, *Trans Met Soc.* 1967, *AIME*, **239**, 1372.
15. D. P. Whittle and J. Stringer, *Phil. Trans. Royal Soc. London* 1980, **Ser.A295**, 309.
16. H. Hindam and D. P Whittle, *Oxid. Met.*, 1982 **18**, 245.
17. J Bennett, in *Proc. Conf. on 'High Temperature Corrosion'* ed. R A Rapp, Houston TX, NACE-6, 1983, 145.
18. J. Stringer, *Mat. Sci. Eng.*, 1989 **A120**, 129.
19. R. Prescott and M. J. Graham, *Oxid. Met.*, 1992 **38**, 233.
20. F. S.Pettit, *Trans Met Soc AIME*, 1967, **239**, 1296.
21. K. N. Strafford and P. J. Hunt, Werkstoffe u Korr, 1979, **30**, 41.
22. K. N. Strafford, *High Temperature Technology*, 1983, **1**, 307.
23. K. N. Strafford and P J Hunt, *'Materials and Coatings to Resist High Temperature Corrosion'* Eds, D. R. Holmes and A Rahmel, Chapter 3, 1978.
24. K. N. Strafford, W. Y. Chan, and J. F. Norton, *In Proceeding of 8th International Cong on Metallic Corrosion, 8th ICMC*, 1981, **3**, 1576.
25. D. N. Tsipas, *Trans Japan Inst. Metals, Supplement*, 1983, 569.
26. K. N. Strafford, P. K. Datta, A. F. Hampton, A. Starr and W. Y. Chan, *Corrosion Science*, 1989, **29**, 755.
27. K. N. Strafford, P. K. Datta and J. S. Gray, *'Advances in Surface Engineering; Processes, Fundamentals and Applications in Corrosion and Wear '*, Eds, K. N. Strafford, P. K. Datta and J. S. Gray, Chapter 3.11, 1990, p. 397.
28. Joint Committee on Powder Diffraction Standards, International Centre for Diffraction Data, Swartmore, PA, 1995.
29. J. Doychak and M Ruhle, *Oxid. Met.*, 1989, **31**, 431.
30. J. Doychak, J L Smialek, and T. E. Mitchell, Metallurgical Transaction A, 1989, **20A**, 499.

31.　J. Jedlinsky and Borchardt, *Oxid Met*, 1991, **36**, 318.
32.　G. C. Rybichi and J. Smialek, *Oxid. Met*, 1989, **31**, 275.
33.　J. Jedlinsky, Oxid. Met, 1993, **39**, 55.
34.　M. R. Jackson and J. R. Raiden, *Met Trans.*, 1977, **8A**, 697.
35.　F. A. Golightly, G. .C Wood and F. H. Stott, *Oxidation of Metal*, 1980, **14**, 217.
36.　E. W. A. Young and J. H. W de Wit, *Oxidation of Metal*, 1986, **26**, 351.
37.　H. Hindam and W W Smeltzer, J. *Electrochem. Soc.*, 1980, **127**, 1630.

3.2.4

Taguchi Optimization of an LPPS MCrAlY Coating for use as a Bond Coat for Thermal Barrier Coatings

R. E. Jones[1], J. Cawley[1], D. S. Rickerby[2] and J. Green[3]

[1]SHEFFIELD HALLAM UNIVERSITY, SHEFFIELD, S.YORKS, UK

[2]ROLLS-ROYCE PLC, DERBY, UK

[3]CHROMALLOY UK LTD, DERBYSHIRE, UK

1 INTRODUCTION

The development of modern aero gas turbine engines is reliant on the ability to further increase the temperature within the "hot-section". This allows both performance benefits and lower NO_x emissions. The superalloy materials used to manufacture components within the high pressure (HP) turbine stage already require considerable protection from the environment. This protection is provided by oxidation/corrosion protective coatings such as aluminides, platinum aluminides or MCrAlY overlay coatings. The current drive towards higher flame temperatures, however, has lead to the introduction of thermal barrier coatings (TBCs) . Insulating ceramic thermal barrier coatings have the potential to provide a significant temperature drop between the air stream and the metallic surface of the component with only limited concessions to efficiency. Partially yttria stabilised zirconia (PYSZ) has been adopted as the industry standard for TBCs due to its very low thermomechanical stability over the range from ambient temperature up to 1400°C. TBCs are usually produced in one of two ways, either by air plasma spraying (APS) or by electron beam physical vapour deposition (EB-PVD). One over-riding aspect with regard to the adoption of TBC technology is the underlying "bondcoat" which establishes a thermally grown oxide (TGO) at the bondcoat/ ceramic interface, maintaining TBC adhesion. This paper reviews the various bondcoat technologies currently being considered for use with EB-PVD TBC's and goes on to describe the optimisation of low pressure plasma sprayed overlays and the improvements which result in terms of oxide chemistry and TBC performance.

2 PRODUCTION PROCESSES FOR EB-PVD BOND COATINGS

MCrAlY coatings, where M represents a metal usually Fe, Co or Ni or mixtures thereof, are two phase structures of β-'M' aluminides in a matrix of a γ-phase. Under high temperature oxidising conditions the β-phase supplies the aluminium required to form a continuous α-alumina scale. As the scale grows the β-phase becomes destabilised through the loss of aluminium and degrades to γ-phase via γ'. This process produces a β-phase depleted zone towards the outer portion of the coating; the zone extends with increasing temperature or prolonged exposure. The aluminium activity within this zone can be shown to follow a typical diffusion profile. When the aluminium activity falls below a critical level, the growth of a pure alumina scale cannot be sustained and mixed oxide spinels form. Areas of localised β-depletion, through porosity or finishing defects can also lead to spinel formation. It has been shown that spinel formation weakens the interfacial bonding between the alumina and the underlying metal, this provides an easy path for stress-generated cracks to run, causing debonding of the scale. When MCrAlY coatings are used in the rôle of bondcoat for TBCs the spinel formation and subsequent scale spallation leads to TBC loss. It is therefore essential to discourage any process which could potentially result in this occurring. It is reported that the onset of oxide spallation is delayed by the addition of active elements such as yttrium. Yttrium migrates to the grain boundaries and the oxygen-rich surface/interface. Once incorporated in the scale the yttrium is reported to limit the grain size, this aids oxide plasticity. Yttrium sitting on the grain boundaries near the interface encourages inward oxygen transport leading to yttrium-rich alumina peg formation. It has been proposed that pegs help to mechanically key the oxide to the underlying coating and hence increasing the scale adhesion. It is evident that the performance of an MCrAlY coating is dependent on the quality of the production process in order to ensure yttrium is mobile within the coating and that internal oxidation or porosity does not occur.

There are many processes currently being used to produce MCrAlY-type coatings. The majority of these are either thermal spraying or EB-PVD techniques.

2.1 Thermal Spraying

Thermal spraying techniques produce coatings by propelling superheated particles from a gun nozzle towards the substrate. On contact with the relatively cold component surface, the particles plastically deform into splat shapes and condense. The splats combine to form a continuous mechanically keyed coating structure. The subsequent highly stressed coating is then heat treated and the splats diffuse together to form an homogeneous structure. Early systems provided the heat to the particles through the burning of fuels, the subsequent gaseous expansion propelled the particles, introduced into the flame as a powder, towards the target. Development of these early systems has lead to plasma spraying techniques where the heat input is provided from a plasma. Additional propellant is introduced in order to increase the overall energy of the impinging particles, with a resultant increase in coating cleanliness and integrity. With the high temperatures experienced within a plasma exclusion of oxygen is essential in order to produce metallic coatings. An oxygen-free environment can be achieved in several ways, for example by flooding a sealed chamber with an inert gas such as argon or by coating within an evacuated chamber. These techniques are known as argon shrouded plasma (ASP) and low pressure plasma spraying (LPPS) respectively. Both systems can produce high quality metallic coatings, the LPPS technique, without the braking effect of an atmosphere, has additional benefits of extended nozzle-to-workpiece distance and higher energy impinging particles. Both techniques require considerable finishing procedures due to the high degree of surface roughness.

2.2 EB-PVD

In EB-PVD a pre-alloyed ingot is evaporated in a low pressure chamber of an inert gas. The steered electron beam is focused onto the top of the ingot causing it to melt and evaporate. The vapour, propelled by the energy released during the evaporation process, condenses on everything within line-of-sight of the melt pool. The pre-heated substrate provides some mobility to the arriving species allowing less than favourable lattice sights to be occupied, resulting in the production of a fully dense coating. The atomistic nature of EB-PVD produces a very smooth surface which requires only limited finishing. The process is limited, however, by the compositions that can be produced. The different vapour pressures of constituent elements of complex multicomponent coatings mean that deposition rates differ for different elements. This results in an inability of this technology to produce some coating compositions. Combined with the high initial equipment costs this has meant that the concentration of interest has centred around plasma spraying techniques. The EB-PVD process, however, produces coatings of a significantly superior quality to those available through plasma spraying, see Figure 1.

2.3 The Rôle of Yttrium

The importance of free yttrium metal within MCrAlY coatings has already been mentioned and as yttrium has a high affinity for oxygen it can readily form oxides during production. Once saturated with oxygen, yttrium no longer retains the chemical driver that causes diffusion to oxygen rich environments. If yttria (Y_2O_3) forms then the benefits demonstrated from yttrium additions no longer persist. Wavelength dispersive spectroscopy (WDS) mapping for yttrium and oxygen can be used to highlight whether a coating has free or combined yttrium within it and indicate the distribution of that phase. Free yttrium would be redistributed after isothermal soaking at elevated temperatures, whilst yttrium combined with oxygen would remain static. Comparison of WDS maps before and after isothermal soaking would reveal any elemental movement. As is evident the ASP sample, when analysed, had yttrium throughout the coating profile, this was however extensively combined with oxygen. After soaking, therefore, only a limited amount of yttrium diffused.

The LPPS sample showed yttrium throughout the coating, mostly segregated to the splat boundaries, but not significantly associated with oxygen. Subsequent to the isothermal soak the yttrium in the LPPS sample remained distributed through the coating, this suggested that it was, to some degree, combined with oxygen. In comparison the EB–PVD sample had yttrium concentrations at the substrate/coating interface and the coating/TBC interface. After the soaking this segregation became even more distinct. As previously suggested there was

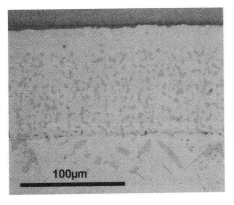 

EB-PVD MCrAlY   LPPS MCrAlY

Figure 1 *Processing rout comparison isothermally soaked for 100hrs at 1100°C.*

only a slight oxygen signal associated with the EB–PVD coating. The segregation in the as-processed state was deemed unsatisfactory because of the possibility of having to re-work a component during production and subsequently leaving it devoid of yttrium. The LPPS coating had yttrium fully distributed throughout the coating as for the ASP system; however, a smaller proportion of this was associated with oxygen. This suggests that a greater proportion of yttrium would be available to aid with oxide and TBC bonding. This analysis, coupled with the compositional freedom associated with the LPPS process, suggested that it offered the greatest potential as a viable production process to produce MCrAlY bondcoats for TBC's.

2.4 Surface Finish

The LPPS process produces surface finishes that are considerably rougher than is aerodynamically favourable for aero engine components. Coatings, therefore, require a finishing procedure prior to use. Two methods have been examined that result in a smoother surface finish after a post spraying diffusion heat treatment.

1. Zirconia based ceramic bead peening at an intensity of 10A
2. Barrel polishing with CE7 media.

The former process smoothed the coating by plastically deforming the peaks of the MCrAlY laterally across the surface, pushing them into intimate contact with the bulk of the coating. The subsequent light barrel polish and vacuum heat treatment resulted in a "sintered" structure with a surface finish value of: $R_a = 80\mu$ inch. This procedure was however unsatisfactory as the surface roughness achieved was not deemed sufficiently smooth. There was also concern over the potential damage to thin wall sections on some components. Figure 2a shows the typical "10A" structure after isothermal soaking in still air. Surface features termed cold shuts are evident; these can lead to localised premature spinel formation, due to aluminium depletion as the aluminium content is used more rapidly in these regions to form pockets of alumina. Figure 2b shows the typical finish produced by polishing only. The change in polishing media, from that used with the peening process, combined with the increase in polishing time resulted in an improvement to the surface finish: $R_a = 20\mu$ inch. Comparison of the soaked microstructures from Figure 2 shows the polished-only sample displayed almost twice the depth of β-phase depletion when compared to the peened example. Following the previous argument, the onset of spinel formation would begin earlier in the polished-only sample. WDS mapping of the peened and of the polished-only samples revealed that a greater proportion of the yttrium, in the latter system, was associated with oxygen. It was proposed that the peening and heat treatment operation was closing and sealing fissures and porosity in the outer portion of the coating. The polishing process did not bring internal interfaces into intimate contact and hence the post-polishing diffusion heat treatment did not sinter the coating; this left internal pathways for oxygen ingress, resulting in internal oxidation. Deep etching of the two aged systems, in 10% bromine in methanol, revealed convoluted internal oxide paths linking back to the surface of the coating in the polished-only sample, these were not in evidence in the peened version. The indication of porosity suggested the possibility of improving the as-sprayed quality of the LPPS MCrAlY. By optimising the spraying parameters a more dense coating could be produced. This would lead to a greater degree of free yttrium within the coating and therefore a more adherent and longer lasting oxide scale would result delaying the onset of spinel formation. The direct result of this would be an improved bondcoat for EB-PVD TBCs.

3 OPTIMIZATION OF THE LPPS PROCESSING ROUTE

Due to the large number of variables associated with the LPPS process, full factorial statistical process control (SPC) design would be impractical. For example a full factorial design of a system with seven variables set to two levels (high and low) would require 2^7 (128) experiments to obtain an optimised system set up. This number can be significantly reduced by using fractional factorial design. When carefully chosen, fractional factorial designed experiments can yield almost the same amount of information as a full design. Several design strategies utilise fractional factorial experiments, notably the Taguchi Method.

3.1 Background to the Taguchi Method

The Taguchi Method of fractional factorial design is based around a specially constructed set of orthogonal arrays. An array consists of columns of variable factors (temperature, pressure etc.) and rows of trial or experiment numbers. When a two tier design, the most basic array, is considered, factors can be set at two levels (1 and 2). When considering two adjacent columns of an array, the two levels can combine in one of four ways: (1,1) (1,2) (2,1) and (2,2). When two adjacent columns of an array form these combinations the columns are said to be balanced or orthogonal. When this premise is extended to include any two columns of an array, it too is said to be balanced.   In order to use the Taguchi Method, the operator simply needs to identify all the variables within the system and fit them to the orthogonal array most suited to that particular process. In the ideal case all the variables are independent of each other, however, complications can arise if two variables interact. Variable interactions can be accommodated by careful array design, see below. Conventional fractional factorial design can result in different conclusions being drawn from the same experiment; the Taguchi Method overcomes this by ensuring experimental design conformity.

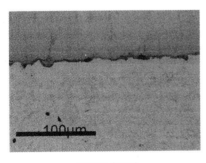 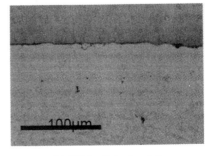

LPPS (10A)	LPPS (polished only)
100h @ 1100°C	100h @1100°C

Figure 2 *(a) shows typical "10A" structure after isothermal soaking in still air (b) shows the typical finish produced by polishing only*

3.2 Assessment of Results

The Taguchi Method requires that analysis of the results from each trial is consistent and must produce a numerical response. The response can be anything that can be quantified such as density or pressure, however, a more subjective response may be used if it, too, can be given a value, e.g. Quality on a scale of 1 to 10. The most straightforward way of assessing the results is from an average response graph. The average response for a particular factor is plotted against the relevant level. In a two level design this results in two points connected by a, presumed, linear relationship. In actuality a two level average response graph only shows which level produces the better result. By selection of the most appropriate levels, as indicated by the average response graph, the ideal parameters can be attained. These often lie outside the experience of the series of trials. To ensure the optimal parameters have been chosen, the resultant responses can be mathematically predicted. Comparison of the predicted and actual results give an indication of the completeness of the experiment design. The most likely reason for significant variation from the predicted result is from an interaction between two or more variables.

3.3 Interactions

Identifying those variables which are most likely to interact with each other is key to Taguchi design. Once aware of the possible interactions, the associated variables can be placed in positions in the array which enables the combined effect to be addressed. Each trial number consists of a set of parameters which together produce a result. Each result is affected to differing degrees by the combination of variables. By calculating how much each variable affects a response, when compared to the average, a graph can be plotted showing the effects of the individual parameters. The average response graph for each variable is affected by "aliased pairs". These represent potential interactions and are mostly insignificant when compared to the major contributing factor; however, if an interaction occurs which has a major effect it can seriously alter the interpretation of the results. Careful positioning of the variable factors in the original array design can help to identify interactions by ensuring the potentially significant aliased pair falls with a non-significant variable. The subsequent assessment of the average response graph would show a significant effect where none was expected. Further array design could then incorporate an interacting term.

It was decided that the Taguchi Method was the ideal tool with which to optimise the spraying parameters of the LPPS MCrAlY in order to produce the best available bondcoat from this technology.

4.0 EXPERIMENTAL METHOD

4.1 Phase One

The initial phase of the fractional factorial design was to identify all the variables involved in the process and possible interactions between them. With the aid of an experienced operator eleven variables were identified, see Figure 3, there was also the possibility of three of these interacting.

Chamber Pressure A   Powder Flow Rate B

Gun Current C   Argon Pressure D

Helium Pressure E   Nozzle Type F

Powder Fraction G   Spray Distance H

Sample Preheat I   Sample Rotation J

Carrier Gas Pressure K

Figure 3 *Process variables*

The initial stage of the design was aimed at being a scoping trial in order to identify a suitable operating window for the process, for this a two level array was deemed sufficient. In order to accommodate eleven variables at two levels the " $L_{12}(2^{11})$ " array was utilised. See Figure 4. Unfortunately this array does not allow for aliased pair or interaction analysis. Any effects from interactions are spread evenly across the array, however, if a non-significant factor produces an unexpectedly strong response it is likely to be as a result of an interaction from somewhere in the array. The two levels, indicated as 1's and 2's in the array, were chosen to obtain as greater spread as possible in the results. In the ideal situation the twelve trials required of the L_{12} array would be completed in a random order; however, in order to reduce

L_{12}	A	B	C	D	E	F	G	H	I	J	K
1	1	1	1	1	1	2	1	1	1	1	1
2	1	1	1	1	1	1	2	2	2	2	2
3	1	1	2	2	2	2	1	1	2	2	2
4	1	2	1	2	2	2	2	2	1	1	2
5	1	2	2	1	2	1	1	2	1	2	1
6	1	2	2	2	1	1	2	1	2	1	1
7	2	1	2	2	1	2	2	2	1	2	1
8	2	1	2	1	2	1	2	1	1	1	2
9	2	1	1	2	2	1	1	2	2	1	1
10	2	2	2	1	1	2	1	2	2	1	2
11	2	2	1	2	1	1	1	1	1	2	2
12	2	2	1	1	2	2	2	1	2	2	1

1 = low setting

2 = high setting

Figure 4 *An $L_{12}(2^{11})$ orthogonal array*

the "down time" of the production facility the experiments were ordered in such a way that limited the number of times the rig was opened to air. This was not considered detrimental to the experiment as the Taguchi Method is sufficiently robust to withstand this slight modification. The experiment was run twice in order to ensure reproducibility.

4.2 Responses

The MCrAlY optimisation required that the structure became more dense, with a greater resistance to β-phase depletion and preferably with reduced surface roughness.

An improvement in surface finish was assumed to follow an optimisation of the spraying parameters, as a reduced number of unmelted particles would be present in the structure. The three parameters chosen as the feedback responses for phase one of the experiment were, therefore, density, surface finish and β-phase depletion. The coatings were sprayed onto test bars nominally 110mm in length with a circular cross section of 8 mm. The density of the coatings was calculated from measurements of the weight gain after spraying and the measured volume of the coating, taken to be a hollow cylinder. As the errors in measuring the volume of the coating were relatively small, it was assumed that they did not affect the ranking of the testpiece results. Surface roughness measurements were made using a portable talisurf device, along the length of the test bars, with the stylus being positioned normal to the surface of the coating. The final response, β-phase depletion, was taken from two sections taken from the centre of the test bar. These were isothermally soaked for 25 hours at 1100°C in still air. Thirty two measurements were taken around the circumference of the test pieces in cross-section after metallographic preparation. The average of these readings was quoted as the β-depletion response for that trial number. The average of the responses for each trial, and it's repeat, was used to calculate and plot the average response graphs. From these the ideal parameters were obtained in order to optimise for density, β-depletion and surface roughness. As greater density was required, the level was chosen for variable that gave the highest value for the average response. Theoretically these would give, collectively, the optimised parameters in order to produce the most dense coating as possible. Surface roughness and β-depletion, however, were required to be as low as possible, the levels taken from the relevant response graphs were therefore those which produced a lower average response. If increased density, reduced surface finish and reduced β-depletion were associated, then similar optimised spraying parameters would be expected. Conformance tests pieces were produced with the predicted parameters for each of the three responses. These varied considerably from the predicted results when the responses were measured, this suggested there was an interaction. The strength of the response for a supposed non-significant variable also suggested an effect from an interaction.

4.3 Phase Two

The mismatch between the predicted and the actual results indicated that an interaction between two or more of the variables was likely. Assessment of the LPPS process implied a likely interaction between either:

1. the powder feed rate and the carrier gas pressure, or
2. the powder feed rate and the gun current.

This was supported by assessment of the most significant responses from the average response graphs from phase one. A second set of experimental conditions was designed varying the three parameters above and the next most influential variable factor. The remaining conditions were set at the level which was predicted to produce the best results for the β-depletion response. An L_8 (2^7) array was most suited to the second phase experiment, see Figure 5. β-depletion was adopted as the most decisive response, whilst surface roughness and coating density were ignored because of the potential inaccuracies of measurement. As a result from phase one, a second response was introduced. This was an arbitrary assessment of the LPPS rig to operate with the imposed spraying conditions. This was termed "sprayability" and was rated on a scale from one to ten by the rig operator, one being poor and ten being ideal. Several of the columns are used from this array to determine the interacting term and so do not contain spraying variables. The aliased pairs of concern were placed in the empty columns so that the interaction could be fully assessed.

	B	K	BK	C	BC	BI or KC	I
1	1	1	1	1	1	1	1
2	1	1	1	2	2	2	2
3	1	2	2	1	1	2	2
4	1	2	2	2	2	1	1
5	2	1	2	1	2	1	2
6	2	1	2	2	1	2	1
7	2	2	1	1	2	2	1
8	2	2	1	2	1	1	2

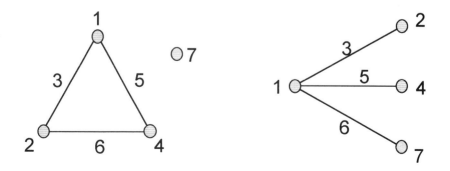

Figure 5 *An L_8 (2^7) array and associated line maps*

4.4 Phase Three

The second phase of the experiment exposed the interacting terms, this enabled the final matrix to be designed, taking into account the interaction. In order to gain the most information from the final array, whilst still keeping the processing time to a minimum, a mixed level experiment was chosen. The L_{16} array with the levels set at four and two, see Figure 6, allowed a greater degree of information to be obtained from two of the variables whilst the remainder were set at two levels as before. The L_{16} array allowed for the interacting terms to

	1	2	3	4	5	6	7	8	9	10	11	12	13	14	15	16
Q	1	2	2	1	2	1	1	2	2	1	1	2	1	2	2	1
P	1	2	2	1	2	1	1	2	1	2	2	1	2	1	1	2
K	1	2	2	1	1	2	2	1	2	1	1	2	2	1	1	2
I	1	2	2	1	1	2	2	1	1	2	2	1	1	2	2	1
E	1	2	1	2	2	1	2	1	2	1	2	1	1	2	1	2
D	1	2	1	2	2	1	2	1	2	1	2	2	1	2	1	1
A	1	2	1	2	1	2	1	2	2	1	2	1	2	1	2	1
N	1	2	1	2	1	2	1	2	1	2	1	2	1	2	1	2
Y	1	1	2	2	2	2	1	1	2	2	1	1	1	1	2	2
X	1	1	2	2	2	2	1	1	1	1	2	2	2	2	1	1
W	1	1	2	2	1	1	2	2	2	2	1	1	2	2	1	1
C	1	1	2	2	1	1	2	2	1	1	2	2	1	1	2	2
B	1	1	1	1	2	2	2	2	3	3	3	3	4	4	4	4

Figure 6 *An L_{16} 4/2 mixed level array.*

be accommodated for by again placing the aliased pairs in columns purposefully left vacant. The most influential variables, namely the interacting terms, were set at four levels allowing the relationship between them to be assessed. Some of the variables from phase one were deemed of minor significance and were therefore set at the optimum level obtained from the earlier trials, if they had no significant effect on coating performance the level was chosen for operating convenience. Figure 7 shows how the aliased pairs were fitted into the array. The resultant coatings were assessed, as before, for β-depletion and sprayability.

Figure 8 shows the average response graph for phase one for the β-phase depletion measurements. The variables to which the response was most sensitive shows up as having a very steep gradient, whist less critical variables have a much shallower angle. From phase one it is evident that the most significant variables were B, C, F, I and K. As β-depletion is required to be a minimum then the factor level which produced the lower average response was chosen for the optimum spraying parameter.

The predicted spraying conditions were therefore:

A2, B1, C2, D1, E1, F2, G1, H1, I2, J2 and K1.

Where 1 and 2 indicate the factor levels. The predicted result from these parameters was calculated and a confirmation run was completed in order to compare to them. When the results were compared they fell short of the prediction. This discrepancy was attributed to an interaction between two or more factors. The significance of factor I was unexpected as substrate pre-heat was most likely to effect coating-to-substrate adhesion and not coating performance. The most significant factors B, C and K were examined and it was suggested that these were the most likely candidates for the interaction; factor I was included to ensure that the strong response shown in phase one was not real. Figure 9 shows the average response graph for β-depletion for the second phase of the design. The most significant terms within this design were B, C and the "BC" interaction term. This highlighted that these were the interacting terms affecting the results from phase one. This interaction was subsequently included in the final array.

Figures 10 (a) and 10 (b) show the response graphs for β-depletion and sprayability respectively from the final experiment. The responses are more difficult to interpret because

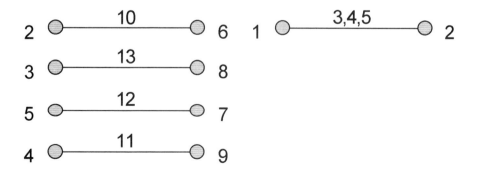

Figure 7 *Linear graphs for the $L_{16}(b)$ array*

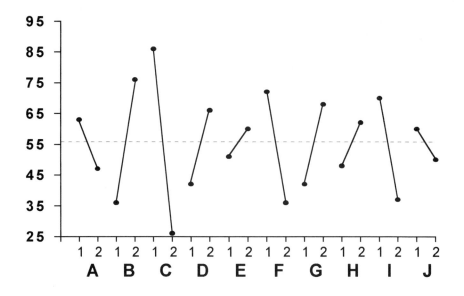

Figure 8 *Average response graph for phase one for the β-phase depletion measurements*

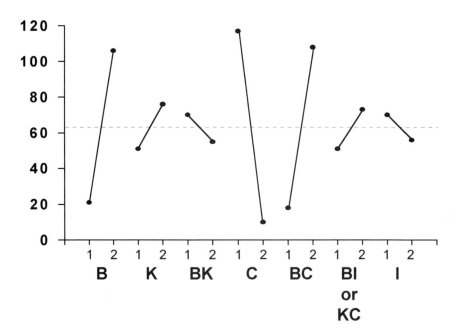

Figure 9 *Average response time graph for β-depletion for the second phase of the design*

of the four level responses for some factors and the interaction term. There are apparently more responses than factors in the final design as these allow for the calculation of the interaction and represent the aliased pairs and are therefore ignored during the selection of the optimal spraying parameters.

Optimising for β-depletion as before gave the final conditions, when combined with the fixed terms, as:

A1, B2, C2, D1, E2, I2 and K1.

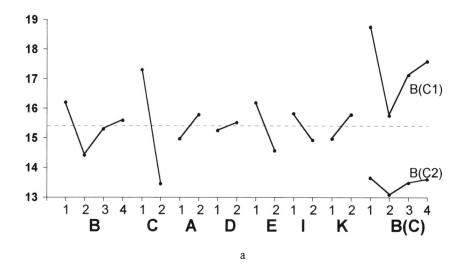

a

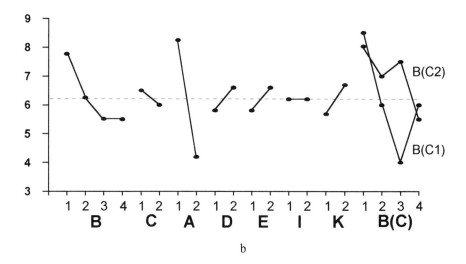

b

Figure 10 *(a) and (b) show the response graphs for β-depletion and sprayability respectively from the final experiment*

The way in which B and C interacted is demonstrated in the superimposed traces, where C2 produced the overall best β-depletion response. The shape of the response demonstrates that the relationship to B was not linear but dipped and rose again. The optimum level in this case was therefore the lowest result B2. When comparing these results to those optimised for sprayability, it can be seen that they differ slightly. Namely:

A1, B1, C1, D2, E2, I1 or 2 and K2

The difference between the less significant figures can be ignored at this stage because they only have a limited effect on the results. This leaves the significant factors A,B and C. A at level one benefits both responses. The ideal levels for B and C, however, oppose each other. When taken in isolation, factor C was not significant to the sprayability response but was very significant to β-depletion, where C2 was preferable. When examining the BC interaction term for β-depletion, C2 was again preferable when combined with B2. The effect of B was however minimised by the interaction and so any level was acceptable. The result of C2 on the interaction, with regards to sprayability was marked, B1 was most preferable. The compromise position, where a coating with good resistance to β-depletion was readily producible, resulted in the following set of spraying parameters:

A1, B2, C2, D2, E2, I2 and K2.

Figure 11 shows the percentage changes to the original parameters required to produce the optimised conditions.

The new spraying parameters resulted in a coating that could be readily sprayed, with no associated dripping or melting of the powder as seen during phase one of the trials, and had a much reduced β-depleted layer after isothermal soaking than the original polished-only samples.

Identified Process Variables	Percentage change to the Original Process Parameters
A - Chamber Pressure	0
B - Powder Flow Rate	-15%
C - Gun Current	0
D - Primary Gas Pressure	-20%
E - Secondary Gas Pressure	0
F - Nozzle Type	0
G - Powder Fraction	0
H - Spray Distance	-23%
I - Sample Preheat Temp	-50%(in time)
J - Sample Rotation	0
K - Carrier Gas Pressure	-23%

Figure 11 *Comparison of the origional and optimised parameters*

5 CONCLUSIONS

The appearance of the coatings produced with the optimised spraying parameters before and after isothermal soaking is significantly better than that of the earlier LPPS coatings. Figure 12 shows that the β-depletion after ageing is now comparable to that seen in the EB-PVD systems and the apparent density seems to be comparable to that of the peened LPPS coatings.

Supportive evidence that the optimisation process has reduced the amount of yttrium tied up with oxygen, can be seen in the near-surface region of the coating where, after ageing, yttrium-rich pegs are evident. These can only form if yttrium is free to diffuse during high temperature exposure.

With the new parameters there is no need for an intense peening operation after spraying and therefore the associated dangers of deforming thin walled components is removed.

Comparison of the adhesion of TBCs to the range of MCrAlY bondcoats mentioned is shown in Figure 13 this shows a significant improvement with the new parameters over the comparable, polished-only sample. The peened sample shows a slight benefit over the optimised sample, however further improvements may be possible, to the latter, with a modified peening process of a much lesser intensity. Future work will assess the potential of this process. All the evidence suggests that the use of the Taguchi Method as a tool for optimising the spraying parameters of an LPPS MCrAlY coating for use as a bondcoat for EB-PVD TBCs was beneficial to the life of the system; this cannot be confirmed, however without real testing in an aero gas turbine engine.

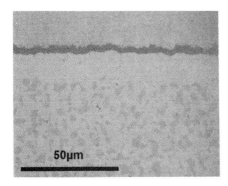

EB-PVD Coating
25hrs at 1100°C

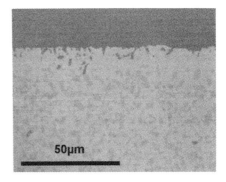

Optimised LPPS Coating
25hrs at 1100°C

Figure 12 *Comparison of optimised LPPS coating and EB-PVD equivalent*

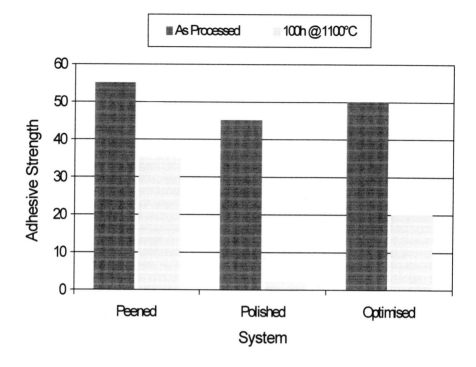

Figure 13 *Comparison of adhesive strengths of the original systems and the optimised system*

3.2.5

Joining of Ferritic ODS Alloys for High Temperature Application

M.W.Carroll and N. J. Wood

BRITISH GAS PLC, RESEARCH & TECHNOLOGY DIVISION, GAS RESEARCH CENTRE, ASHBY ROAD, LOUGHBOROUGH, UK

1 INTRODUCTION

To improve efficiency in modern power stations the use of a high temperature heat exchanger is required. For temperatures in excess of 900°C, ceramics have been considered, but they are brittle. The most promising materials are the Oxide Dispersion Strengthened (ODS) alloys.

ODS alloys are not processed via the casting route used for more conventional alloys, but are mechanically alloyed in a high energy ball mill. The oxide particles which help to give these alloys their high temperature strength, are very small and finely dispersed throughout the matrix. Any melting process, such as conventional welding, would disrupt the structure of the ODS alloy, causing it to lose its strength above 700–800°C. Hence, to make possible the fabrication of a heat exchanger for use at 900–1200°C, a solid state joining route is necessary. The difficulty of joining ODS alloys has been their Achilles heel for some time. However, there are currently three solid state routes under consideration by British Gas (BG), and the manufacturers of ODS alloys:

i.   Diffusion Bonding
ii. Vacuum Brazing
iii. Explosive Welding

Diffusion bonding has been investigated for relatively small components fabricated from sheet[1]. Vacuum brazing is also promising, but is limited by the size of available vacuum chambers. Hence the route investigated the most by BG is explosive welding.

Explosive welding of the ODS alloy Incoloy MA956 has been carried out by Jackson and Marsh[2]. Much of the work has been carried out on sheet, but also tubes, and tubes to tube plates were joined.

One of the factors expected to be life limiting for ferritic ODS alloys is the oxidation resistance of the α-alumina films which form the protective surface layer. A model which allows for the calculation of the time at which ferritic ODS alloys will show the catastrophic oxidation which leads to failure has been developed . However, the interdiffusion of elements between dissimilar metals at high temperature could potentially lead to shorter times to failure than those predicted by the Quadakkers model[3].

Hence in this work the ODS alloy ODM751 has been explosively welded to a cast reformer alloy and the effect of temperature on the joint has been investigated.

The object of this paper is to report on the interdiffusion occurring at 1150°C, and to suggest, based on our observations, steps that could be taken to improve the long term integrity

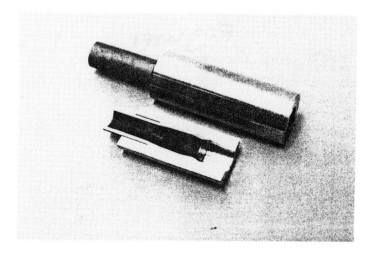

Figure 1 *Macrophotograph of a welded joint*

of ODS/reformer alloy joints. The expected life of power plant is about 30 years, and at least initially, it is hoped that the heat exchangers would operate for up to 20,000 hours without replacement.

2 EXPERIMENTAL

The welds investigated were made by explosively joining ODM751, manufactured by Dour Metal of Belgium, to a sand cast version of the reformer alloy Paralloy T57. These compositions are shown in Table 1. Note that the T57 contains only 0.1 wt% C, a relatively low level for such alloys. A macrophotograph of a successfully welded stub is shown in Figure 1.

2.1 Heat Treatments

One of the ODM751/T57 welds was cut longitudinally, and further sectioned into six, again longitudinally. The following heat treatments were carried out in air at 1150°C: 0, 50, 100, 250, 500 and 1000 hours.

Some of the welds which received the longer heat treatments were jacked apart by the effect of the oxidation at the sample edges. This finding resulted in testing a whole joint for 500 hours, at the same temperature.

Table 1 *Composition of ODM 751 and Paralloy T57 materials*

Mat.	C	Si	Mn	Ni	Cr	Mo	Nb	Al	Ti	Y	S	P	Cu
T57	0.11	1.0	0.4	33.7	23.4	0.06	0.8	-	-	-	0.01	0.02	0.05
751	0.03	-	-	-	16.0	1.5	-	5	0.5	0.3	0.01	0.02	-

The results from this test are also included in this paper and reveal an interesting comparison with the smaller samples.

Note that the heat treatments were carried out in still air. Exposure of the weld joints to a combusted gas atmosphere was considered, and Sigler[4] did find differences in the morphologies of α-alumina films formed in a synthetic exhaust gas atmosphere. However, recent studies at BG have shown that in terms of kinetics, there is no evidence that the corrosion rate of the ODS alloy would be accelerated[5], hence still air was used.

2.2 Low Power Microscopy

After exposure, the samples were examined visually, and by optical microscope at low magnification. It was clear that the edges of some of the samples had been attacked preferentially, despite having been polished to a 1200 SiC grit finish. The attack was usually concentrated at the ODM751/T57 interface edge. Some of the samples had even been jacked apart. X-ray diffraction (XRD) of the large scale pustules showed them to be an Fe-rich oxide. A macrophotograph of a badly corroded weld, after 1000 hours exposure, is shown in Figure 2.

The whole joint exposed for 500 hours was initially cut into two and polished. Visual examinations revealed fine nodules and suggested that enhanced corrosion occurred in the weld crevice.

2.3 Electron Microscopy and Microanalysis

The longitudinal samples were sectioned transversely to allow electron microprobe analysis across the ODM7521/T57 weld interface. The samples were mounted in cold setting resin prior to sectioning and were subsequently polished to a 1 μm diamond finish.

The complete weld was also sectioned and mounted. It was examined both longitudinally and transversely.

Figure 2 *Macrograph of a badly corroded ODM751/T57 sample after 1000 hours exposed at 1150°C in still air*

Figure 3 *Low magnification image of a sample after 100 hours. A diffusion zone between the two welded alloys is clearly visible. Cracking of the weld, initiated from the edge of the specimen, is also apparent*

2.3.1 Electron Microscopy. Electron microscopy confirmed that the weld joint between the two different alloys was subject to enhanced rates of attack, compared to the rest of the sample. Often the joint/edge of a sample was observed to corrode badly and from there the oxidation or nitridation spread preferentially along the joint. Thus, wherever the oxygen partial pressure was low enough, rapid nitridation could have occurred, as found by Bennett et al[6]. This 'crevice-line' type attack is visible in the 100 hour sample (Figures 3 and 4). These two micrographs also show a diffusion zone or interlayer formed at the weld, together with either a second phase, or porosity. The weld jacking was also very extensive on the 1000 hour sample and it was observed that a crack had grown right across the sample.

As expected, longer exposure times, in general, equated to more extensive attack. The exception to this was the 50 hour sample, which had cracked right through. Micrographs showing interesting features are presented in Figures 5 and 6.

In Figure 5 the 250 hour sample exhibits cracking, voidage and internal attack. The internal attack appears both as carbides and as acicular Widmanstätten-type platelets, within the T57.

Comparison of the joints at 0 and 100 hours, clearly shows that a diffusion zone (Figure 6)

Figure 4 *The edge of the welded sample from Figure 3, where internal attack and cracking coincide*

has developed even after a relatively short exposure to 1150°C. This zone was typically 150 microns in depth, stretching into the T57, not into the ODM751. After 500 hours, the diffusion zone had grown to the depth of 375 microns (Figure 7).

The large, angular dark coloured second phase particles visible in the diffusion zone (Figure 7) were analysed by energy dispersive X-ray analysis, and found only to contain aluminium. No other elements, including oxygen, were detected, suggesting that the particles were nitrides. The area immediately adjacent to these particles contained a high level of nickel, but there was no nickel in the particle.

The diffusion zone within the 100 hour sample also contained many smaller second phase particles, which were cross-shaped (Figure 8). These particles were all aligned in the same direction, probably along preferred crystallographic planes and are thought to be nickel aluminides. The surrounding matrix shows nickel depletion. This indicates that as the aluminium from the ODM 751 diffuses into the T57, the high chemical affinity of nickel for aluminium causes the former to diffuse to the aluminium, in order to form the thermodynamically stable NiAl phase.

The ODM contained a dispersion of small particles, both close to the original interface (~75 microns) and also 2mm away from the interface. The particles near the diffusion zone are much closer together. These particles are a similar size to the "cross shaped" nickel aluminides in the diffusion zone. However, their shape is different, being more even and rounded. The ODM also contained larger particles. One of these near the opposite edge of the specimen was analysed and found to be rich in aluminium. These were obviously oxides or nitrides, whereas the smaller particles were found to be titanium carbide or carbonitrides.

The ODS alloy close to the weld interface, contained particles of the same composition as those found in the bulk of the alloy, remote from the weld. They were analysed as being

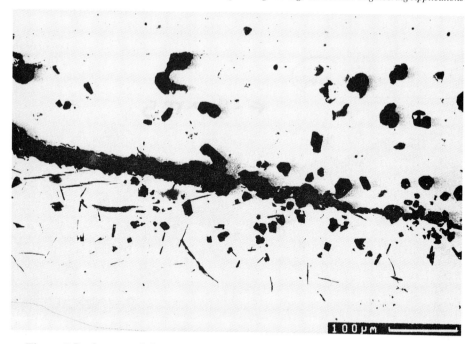

Figure 5 *Backscattered electron image of a sample after 250 hours exposure. Crack propagation is occurring within the diffusion zone in the T57*

principally iron, aluminium and chromium, with a small amount of nickel also present. These aluminium-rich particles are probably YAG, with the iron and chromium detected coming from the matrix.

The parent alloy T57 was also examined. A second phase was present, apparently delineating the grain boundaries in places. This light coloured phase (Figure 9) was clearly niobium carbides. These carbides are one of the primary strengthening mechanisms for T57 type alloys. A linescan of the area shown in Figure 9, i.e. the T57 adjacent to the diffusion zone/T57 interface, showed a nickel gradient within this area.

In general, only the 100 hour sample has been discussed in this section. However, except where otherwise indicated, these features were present on all the other samples, but to differing degrees.

Table 2 *Variation of depth of diffusion layer v time*

Sample/hours exposed	Depth/microns
0	0
100	150
250	300
500	375
1000	575
500 (complete weld)	700

Figure 6 *Backscattered electron image of a sample after 100 hours exposure. A diffusion zone approximately 150 μm wide can be seen within the T57*

The whole joint which had been heat treated for 500 hours was also examined in detail. It was hoped that this larger sample would have resisted attack, but this was not the case. This sample had also suffered accelerated attack as a result of crevice corrosion. The crevice edge has a thick surface oxide film. The internal attack has a darker appearance, suggesting it to be a nitride. The appearance was very similar to that of the 100 hour sample, shown in Figures 3 and 4. As with the other samples, most of the internal nitrides have formed within the diffusion zone. This zone was around 700 μm wide, which was thicker than the zone in the smaller section 500 hour specimen.

The internal nitrides in the diffusion zone were more prevalent close to the ends of the weld and in areas where the cracking had spread. The whole joint is best described schematically, Figure 10.

When the parent ODM of this sample was examined, an additional, interesting feature was noted. Optical microscopy of the ODM from the 500 hour sample showed large apparent voids deep within the parent alloy (Figure 11). The voids show a definite texture and are oriented parallel to the weld interface. Such massive voids are not usually found in ODM after such short exposures. They must be considered as Kirkendall porosity, due to aluminium diffusing into the T57. The orientation factor suggests that the aluminium has come from YAG or similar particles. This voidage could also limit the life of an ODM/T57 joint.

2.3.2 Composition Profiles. Line scans for the elements nickel, aluminium, chromium and iron were carried out across the interface of all the samples, so as to evaluate any overall mass migrations of elements between the two different alloys. Compositional profiles are shown in Figure 12, across uncracked materials, where possible.

From these data it can be seen that aluminium from the ODM diffuses into the T57 and Ni

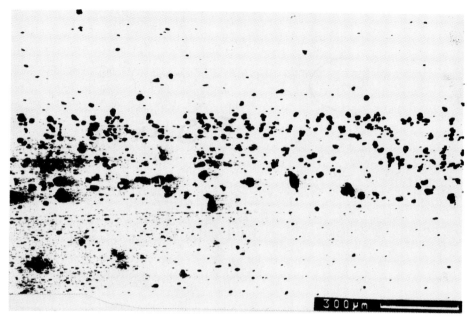

Figure 7 *Micrograph of a sample after 500 hours exposure, showing the diffusion zone to have grown to a width of over 300 μm. The large second phase particles are aluminium nitrides*

from the T57 diffuses into the ODM. It appeared that the Ni level in the ODM reached a peak of approximately 13 wt% at 200 μm from the interface after 500 hours. After 1000 hours the Ni content dropped to a uniform level of around 5 wt% in the ODM. Aluminium was not apparent in the T57 side of the interface until after 100 hours exposure and appeared to be reduced to lower levels after 1000 hours.

2.3.3 Diffusion Layer Depth v Time. Table 2 shows how the depth of the diffusion layer within the T57 varied.

The figures in Table 2 confirm a near-parabolic nature of the growth rate of this layer, and shows a more rapid growth rate in the complete weld, by a factor of two. This unexpected result confirms the value of testing the whole weld. It also suggests that a 2.7 mm thick tube of ODM will have lost a significant amount of aluminium to the diffusion layer. The amount of aluminium in the parent ODM will now have dropped from 4.5 wt% towards the critical level, about 1.4 wt%, below which breakaway corrosion occurs. The significance of the critical level value for aluminium in alumina forming ODS alloys is outlined in earlier BG work[7] . However, the loss of large amounts of aluminium could explain the large amount of voidage scan in Figure 11. The voidage is presumably a result of Kirkendall diffusion.

3 DISCUSSION

The exposure of explosively welded ODM751/Paralloy T57 joints to still air at 1150°C, for up to 1000 hours, has caused some loss of weld integrity. This weld degradation was evident as

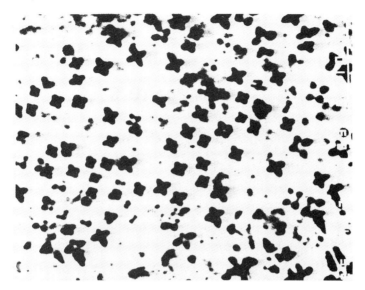

Figure 8 *Micrograph of a sample after 100 hours, showing very fine nickel-aluminide particles formed on preferred crystallographic directions*

cracking close to the weld interface. In all cases the crack initiated from the edge of the sample.

Weld degradation also appeared as interdiffusion of aluminium and nickel. Nickel ions had migrated from the T57 into the ODM and aluminium had diffused from the ODM into the T57. The nickel diffusion into the ODM was significant and after 1000 hours the whole sample showed a uniform level of around 3 wt%.

The result of this transfer of cations was a diffusion layer, stretching from the original T57/ODM interface into the T57, for hundreds of microns. The ODM did not contain any of this diffusion layer. The limited number of measurements carried out suggested the growth rate to be close to parabolic, i.e. controlled by diffusion.

The causes of the cracking of the welds after high temperature exposure can be attributed to a combination of this interdiffusion, and the crevice effect. Air is a mixed oxidant, as it contains both oxygen and nitrogen. Within a crevice, which is present on all tubular explosive welds, the access of oxygen is restricted. The surfaces of both alloys form oxides, but below certain crevice widths, nitrides form beneath the oxide. This is shown in Figure 4, where aluminium nitrides have formed beneath chromium oxide. The original ODM/T57 interface is visible at the bottom of the micrograph, indicating that the nitridation is within the diffusion zone. The nitridation appears to force the surrounding alloy apart causing microcracking in the manner suggested by Bennett et al[6].

It has been claimed by Aydin et al[8] that nitrogen atoms are virtually insoluble in oxides, hence nitrogen only transports into the underlying metal through pores, cracks and grain boundaries. A review of oxides and nitrides by Strafford[9] revealed that the solubility of nitrogen in bcc iron is very high, at 0.6 at%. Hence any nitrogen atoms getting through the cracks or pores in the surface chromic oxide will certainly have a significant solubility in the fcc Fe-Ni-Cr matrix of the T57. The most stable nitride of the major elements present in the diffusion zone, will be aluminium nitride ($\Delta G = -380$ kJmol^{-1}). These large cuboidal-shaped nitrides act as localised stress raisers within the diffusion zone. The volume change caused by nitridation

Figure 9 *Micrograph of the T57 alloy close to the weld interface. After 100 hours the grain boundaries still contain large amounts of niobium carbides*

within the growing 'nitride' crack – seen in Figure 4 – also causes stress. Hence cracks are able to jump from one nitride particle to the next very rapidly, through the areas of high stress concentration. Once the crack has grown to a significant depth, failure of the weld is imminent.

The interdiffusion of elements between the alloys seems to assist crack propagation thus accelerating failure.

In order to prolong the lives of explosively welded joints, three measures can be taken:

i.   Do not have welds in areas of high temperature. At present the component size and shape availability of ODS alloys does not make this solution possible.

ii. Slow down the interdiffusion between the two dissimilar alloys. A platinum, or platinum aluminide interlayer could increase the joint life.

iii. Prevent oxidant ingress, especially that of nitrogen, into the crevice. One way of doing this would be to in-fill the crevice with a suitable corrosion resistant filling. An ideal one would be a spray coating, of a FeCrAlY composition, thus ensuring that two of the three components in the system were compatible. The spray could be dressed, or polished to ensure there were no crevices at the in-fill edges.

4 CONCLUSIONS

1. Exposure of an explosively welded ODM751/Paralloy T57 joint to still air at 1150°C, for up to 1000 hours, has caused weld degradation. This degradation is caused primarily by

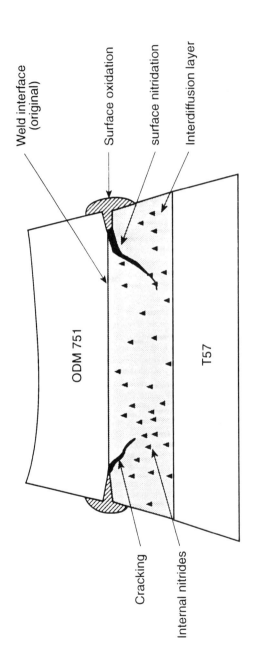

Figure 10 *A schematic representation of the corrosion of the sample exposed for 500 hours*

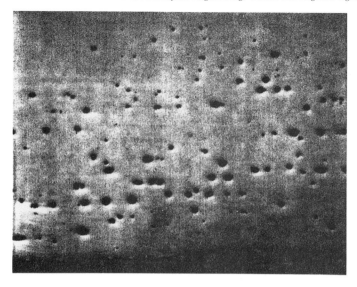

Figure 11 *Longitudinal section of the complete weld after 500 hours exposure. Large, oriented voids are apparent in the ODM7516*

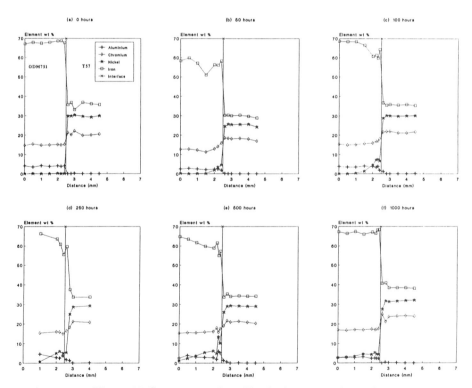

Figure 12 *Compositional profiles for heat treated samples*

diffusion of cations from one alloy into the other and by subsequent formation of large second phase particles.

2. Aluminium from the ODM751 diffuses into the T57. Nickel from the T57 diffuses into the ODM751. A large interdiffusion zone of up to 700μm in depth is formed. Also within this zone, small gamma phase NiAl intermetallic particles form.

3. Once nitrogen has entered the weld area, large blocky cuboids of aluminium nitride form. These cuboids act as localised stress raisers, to form an easy, active path for crack growth.

4 . The geometry of the samples exposed should be close to those experienced in plant, as the effect of crevices at high temperature is considerable. The testing of the small samples undoubtedly caused some acceleration in cracking.

Whole joints must be tested under pressure, with both crevice geometries, and sizes, similar to those likely to be present in service. Crevice size will affect factors such as rate of nitrogen pick-up, as would the nitrogen partial pressure.

5 FUTURE WORK

i. Longer term oxidation testing of whole joints. The effect of blocking up the crevice, and coating the chromia-forming material (T57) with a platinum aluminised or FeCrAlY coating should be considered.

ii. Joints need to be tested at pressure, as nitrogen ingress would be more rapid at higher pressures. Pressure testing would also assess the effect of nitridation on strength.

iii. Use coatings, such as platinum and/or aluminium to act as a diffusion barrier. Small changes in coating composition could markedly slow down interdiffusion.

Acknowledgements

To our many colleagues who helped with this work, especially J. Wonsowski, J. Booker, J. Jackson and P. Jackson.

References

1. M. Kopf, H. G. Mayer, G. Haufler, K. E. Stuttgart, T. Huber, N. Reheis and D. Sporer, 13th International Plansee Seminar, 896, 1993.

2. P. W. Jackson, N. P. Marsh, G. A. Hack and M. J. Shaw, *Welding and Metal Fabrication*, 1990, 84.

3. W. J. Quadakkers and K. Bongartz, *Werk und Korr*, 1994, **45**, 232.

4. D. R. Sigler, *Oxid. Met.*, 1991, **36**, 57.

5. N. Wood, Q. Mabbutt, J. Wonsowski and F. Starr, 13th International Plansee Seminar, 1993, 115.

6 . M. J. Bennett, J. A. Desport, C. F. Knight, J. B. Price and L. W. Graham, *Corr. Sci.*, 1993, **35**, 1159.

7. N. Wood and F. Starr, Microscopy of Oxidation 2, Eds. S B Newcomb and M. J. Bennett, 1993, p. 298.

8. J. Aydin, H. E. Buhler and R. Rahmel, *Werk. und Korr*, 1980, **31**, 675.

9. K. N. Strafford, *Corr. Sci*, 1979, **19**, 51.

Section 3.3 Automotive

3.3.1

Volume Production of Thermal Sprayed Coatings in the Automotive Industry

K. Harrison[1], A. R. Nicoll[2]

[1]SULZER METCO (UK) LTD

[2]SULZER METCO HOLDING AG. SWITZERLAND

1 INTRODUCTION

In the demanding automotive environment, the requirements for success in the area of surface enhancement can be identified as the ability to produce consistent, repeatable and the highest quality coatings in the most economical (low cost) manner.

Plasma spraying with automated equipment meets this challenging requirement. It also differs from other thermal spray processes such as conventional and high velocity flame spraying offering significant advantages and benefits for the manufacturer of coatings:

- higher spray rates and deposition efficiency,
- process gases are inexpensive and easily obtainable,
- gas consumption is lower,
- components are heated only minimally during spraying and less cooling is required,
- the plasma system is based on established technology and is an industrially proven and reliable process,
- wide range of spray guns available for many different applications,
- the process can be integrated into a spray cell offering full movement coverage and manipulation of the components,
- the cell can be integrated into a production line,
- the cell can be designed to meet all safety and environmental emission requirements.

The plasma spray process is automated whilst remaining operator-friendly and precisely controlled within a wide range of parameters thus producing consistent, reproducible coatings of the highest quality. This success is based on thoroughly tried and tested components and hence no start-up problems. Therefore, systems are reliable and cost-effective with minimal down-time and easy to maintain with low repair and maintenance costs.

In the following, the major key factors for the successful introduction of plasma spray coating applications in the automotive industry for industrial mass production are described. In particular, important human and environmental issues (concept and management) are discussed in the context of automotive applications.

2 PLASMA SPRAY PROCESS

The use of plasma sprayed coatings is based on a design philosophy which takes into account that the substrate is providing the mechanical properties and that the coating has to inhibit or enhance the surface phenomena.

The understanding of materials processing using plasma requires firstly that we segment all the processing steps in order to get an overview of the complexities involved[1]. This results in the five blocks representing:

1. surface Activation of the component,
2. powder to form the coating,
3. melting (Plasma system) of the powder,
4. coating Deposition (Manipulator) on the part,
5. coating Quality Control (Workshop or Laboratory) of the powder and coating.

All of the working parameters found within these segments are interactive and with the level of automation being used today are considered to be fully automated and reproducible. This level of reproducibility is based on the use of advanced technologies with the segments being matched to each other in terms of precision, accuracy and repeatability. The understanding of the way in which these segments interact results in coatings of consistent quality. Indeed, the question of quality in modern production schemes using plasma spraying has to take into consideration that quality relates directly to total quality and not just individual items such as the powder being used or the coating quality after deposition.

Compared with other coating methods, plasma spraying is unique in that the high temperatures (~10,000K) and specific energy densities achieved in thermal gas plasmas enable the melting of any material which has a stable molten phase and is available in powder form. Plasma spraying of materials such as ceramics, carbides, alloys and refractory metals, which have high melting points, has therefore become well established as a commercial process. Such coatings are increasingly used in many industries to impart specific surface properties such as corrosion resistance, thermal insulation, wear resistance and electrical insulation. These coatings are applied onto a variety of different types of component material such as steel, copper, aluminium, cast iron, nickel and cobalt base superalloys and more recently composite structures[2].

Powder is introduced into the thermal energy source i.e. a gas plasma stream in an air or low pressure/vacuum atmosphere and the particles become semi-molten, molten or remain as solid particles. The plasma is based on the ionization of gases such as argon, hydrogen, nitrogen and helium using an electric arc. Using grit blasting or a reversed transferred arc as surface activation processes we are able to provide the substrate with sufficient surface roughness that the accelerated, fused powder particles can impinge on the substrate surface, solidify and interlock.

It is the co-ordinated movement between the activated surface to be sprayed and the stream of molten particles which ensures the correct deposition of the coating in the required form and in the right place. Obviously, as a production process this has to be carried out in the most economical and reproducible manner according to each type of application.

The success of this process can be shown by the number of production and experimental coatings being deposited using plasma spraying. Figure 1 shows examples of components that are in production or being evaluated experimentally in the automotive industry.

3 PLASMA SPRAY MATERIALS

The selection of spray materials is connected with the functional surface requirement, type of surface exposure or attack. Examples are given in Table 1.

Table 1 *Examples of production and experimental coatings in the automobile industry*

Production coatings	Coating materials
Piston rings	Mo+NiCrSiB, ceramics, others
Lambda sensors	Spinel
Fuel injection nozzles	Molydenum
Experimental coatings	
Piston crowns	Thermal barrier coatings (TBC)
Engine bores	Anti-sliding wear
Alternator caps	Al_2O_3
Brake disks	HVOF-Cr_3C_2-NiCr
Synchron rings	Mo, Mo blends
Crankshafts	Repair
Valves	TBC, seat areas (VPS), anti-adherence on exhaust valves

Present applications involve coatings which meet precisely defined and continuously increasing requirements.

4 PLASMA SPRAY CONTROL

The purpose of the plasma gun is to provide constant thermal energy in the form of the plasma. Thus, the control of the plasma requires careful gas and arc regulation (feed back control) in order to ensure constant gas ionization and therefore, constant energy density at the point of powder injection and in this way controlled, reproducible melting. This also includes constant water temperature and flow rate of the cooling water circulating through the gun as this can affect the degree of ionization of the plasma and lifetime of wear parts in the plasma gun.

In designing an automated plasma system or simply a purchase of a stand-alone plasma system (i.e. plasma console, powder feeder, gun, power supply and heat exchanger), it is important to consider the ease of operation and whether the system will be operator friendly. Clearly, with an increasing level of plasma control the level of operator involvement in controlling the plasma process decreases. In addition, the system used in areas of R & D is equally important and experience demonstrates that the successful applications are those that use equipment in the R & D area where production parameters and movement patterns could be simulated and optimisation used, as a basis for the production equipment.

The operator probably is the biggest factor in the consideration of quality. All elements of the system are designed in such a manner that the operator can carry out his work without affecting the total quality of the sprayed component. This assumes that the equipment start-up and shut-down sequences are automated and that the gun wear parts have substantial lifetimes. Likewise changing a plasma gun (e.g. O.D. to I.D.) does not involve any changes in the plasma controller. A further consideration in the case of duplex and triplex coatings is that the gun has to be able to spray all the different powders without nozzle changes having to be made and for long periods of time.

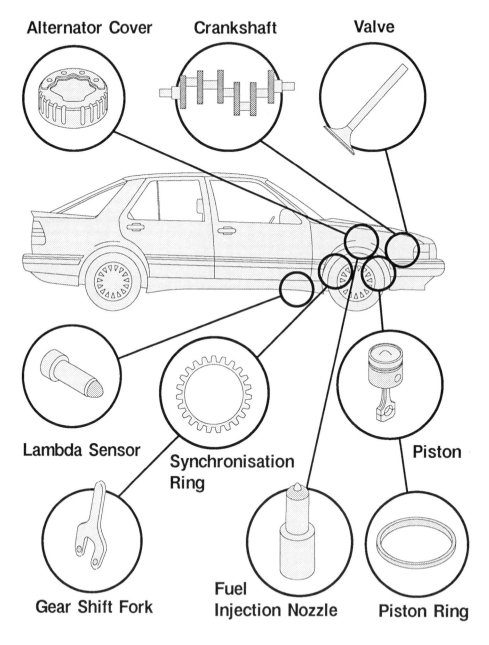

Figure 1 *Examples of production and prototype coatings on automotive components*

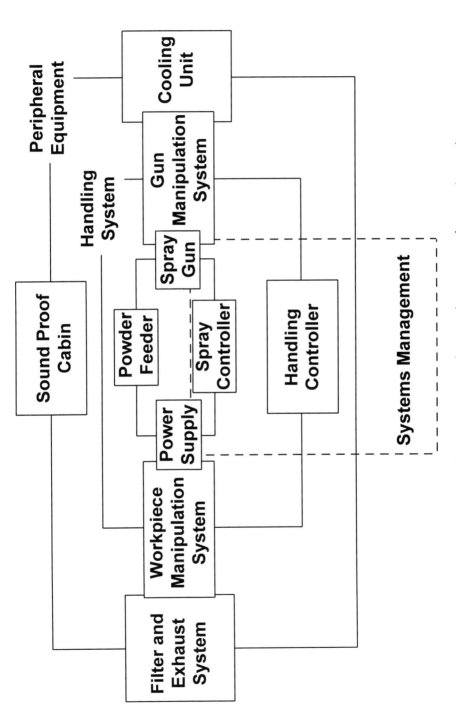

Figure 2 *Levels of plasma spray, manipulation and environmental systems integration*

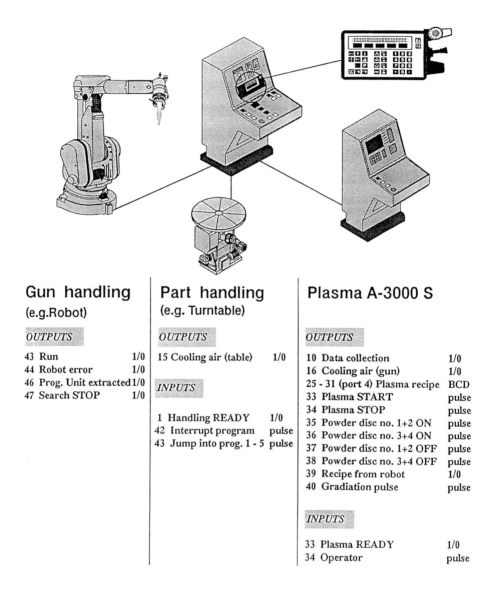

Gun handling
(e.g.Robot)

OUTPUTS

43 Run	1/0
44 Robot error	1/0
46 Prog. Unit extracted	1/0
47 Search STOP	1/0

Part handling
(e.g. Turntable)

OUTPUTS

15 Cooling air (table)	1/0

INPUTS

1 Handling READY	1/0
42 Interrupt program	pulse
43 Jump into prog. 1 - 5	pulse

Plasma A-3000 S

OUTPUTS

10 Data collection	1/0
16 Cooling air (gun)	1/0
25 - 31 (port 4) Plasma recipe	BCD
33 Plasma START	pulse
34 Plasma STOP	pulse
35 Powder disc no. 1+2 ON	pulse
36 Powder disc no. 3+4 ON	pulse
37 Powder disc no. 1+2 OFF	pulse
38 Powder disc no. 3+4 OFF	pulse
39 Recipe from robot	1/0
40 Gradiation pulse	pulse

INPUTS

33 Plasma READY	1/0
34 Operator	pulse

Figure 3 *Typical example of spray cell management bringing together the elements of plasma generation, plasma gun movement and component controls*

Parameter reproducibility covering gas flows and electrical requirements from location-to-location is obtained through the use of single source system calibration. This ensures that all systems are the same and that working parameters can be easily transferred and in use,

(a)

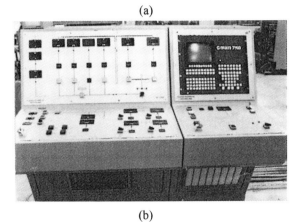

(b)

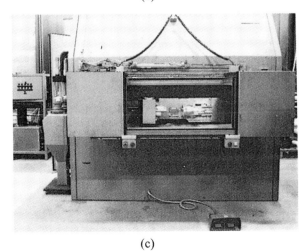

(c)

Figure 4 *Plasma spray piston ring facility showing the (a) grit blast unit, (b) control console and (c) the mandrel loading bay (Courtesy of Sulzer Metco AG, Switzerland)*

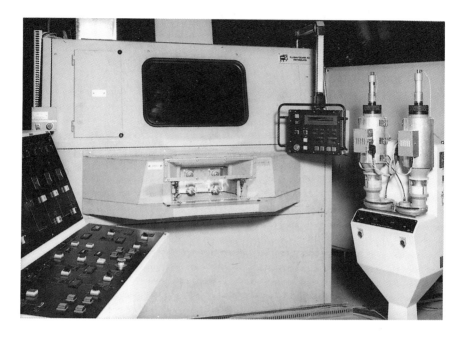

Figure 5 *Exhaust gas sensor and an automatic handling system for the plasma spraying of spinel powder (Courtesy of Bosch, Germany and Sulzer Metco, Switzerland)*

produce surface layers meeting the required specification independent of the operators and the location.

The economics of the plasma process are affected by several factors. They include deposit efficiency, target efficiency, gas consumption, energy and heat losses. An evaluation of maintenance costs is also crucial. What is the effective operating life of the various components? What is the system up-time compared to the down-time? In order to avoid scrap parts it is important to have a consistent, repeatable coating. Therefore, a system that controls gun start-up, ignition, steady state and shut-down is critical. A proper powder feed mechanism will ensure a consistent, repeatable feed rate thus optimizing the utilization of the powder[4,5].

In addition to the generation of the plasma for the melting of the powder, consideration has to be given to the manipulation of the components to be sprayed and the environment[6]. An overview is given in Figure 2. The levels of integration show the plasma core components, the plasma gun and workpiece manipulation together with the environmental systems i.e. the filter and exhaust system, the closed-loop water cooling unit and the sound proof cabin.

In summary, designing a fully automated plasma system requires analysis of the types of parts to be coated, types of coatings, lot sizes, masking, part cooling, part handling, gun manipulation, ease of programming and data acquisition etc.

5 ORIGINAL COMPONENT MANUFACTURE

In basic terms, automation of all aspects of the plasma spray and part manipulation process will ensure consistently acceptable quality each time. Processing costs and quality assurance

Table 2 *Typical plasma spray parameters for Mo + 30% NiCrSiB parameters*

Parameter	Ar-H$_2$	N$_2$-H$_2$
Argon (l/min)	47	-
Hydrogen (l/min)	10	4
Nitrogen (l/min)	-	26
Current (amps)	650	-
Powder Injector (ø mm)	1.8	1.8
Injector angle (°)	90	95
Injector distance (mm)	6	8
Powder gas (l/min)	2.5 Ar	2.6 N$_2$
Feed rate (g/min)	~ 47	~ 42
Spray distance (mm)	135	110
Powder Efficiency (%)	~ 67	~ 63
Coating Properties		
Bond strength (MPa)	~ 88	~ 49
Hardness (Hv100)	~ 557	~ 608

Figure 6 *Aluminium alternator cap coated with pure aluminium oxide*

Figure 7 *Inside of a spray gun cabin*

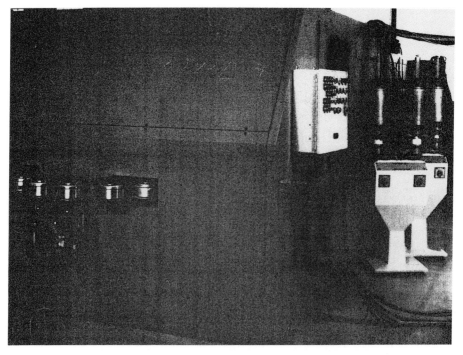

Figure 8 *Outside of spray cabin with powder feeder for the two guns shown on the right of the picture*

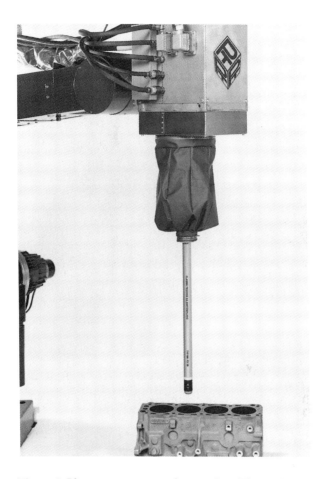

Figure 9 *Plasma gun rotating device – RotaPlasma® -500*

costs can be reduced by selecting and managing the optimum spray movement. The overall cost of the process must include an evaluation of the economics of the plasma process itself. When selecting the equipment required to satisfy the needs of a spray application, one should think in terms of a work cell. This consists of various pieces of equipment totally integrated under the supervision of one control. An example is shown in Figure 3 for spray cell management. This kind of system places the plasma gun on a robot and the component on a manipulator. The underlying requirement is that the system elements can provide the necessary electronic signals and have accurate feedback control. The production system is encapsulated in a sound proof cabin fitted with air inlets and dust emission controls. The powder feeder and system management processes are mounted outside the cabin where they can be easily supervised by the operator. This becomes all the more important when the system is required to run up to 24 hours a day and 6 days a week[7]. Areas requiring maintenance or subjected to wear have to be given easy access which requires a certain amount of system manufacturing experience during the design of the system. The exact design of the work cell will depend on

(a)

(b)

Figure 10 *Thermal barrier coatings on (a) valves and (b) piston crowns*

the types of parts to be coated, types of coatings, lot sizes and whether off-line programming and data acquisition are required.

The productivity of this kind of system depends very much on the operator to change the component after each spray cycle. As this involves stopping and starting of the plasma process, the next development is to automate the handling of the parts. In the plasma spraying of piston rings, the rings are compressed (gap is closed) onto a mandrel, which is then grit blasted, plasma sprayed and ground. Transfer of the mandrel is usually by the operator whilst the grit blast, spraying and grinding operations are automatic and supervised by the operator. An example of a typical plasma spray unit is shown in Figure 4. The coating is usually sprayed into a groove in the ring-inlaid, half-inlaid or full-face periphery[3] using argon-hydrogen or nitrogen-hydrogen spray parameters (see Table 2).

It has been estimated that between 30 and 40% of all piston rings manufactured in Europe for automotive applications (approx. 220 million) are thermally sprayed (wire arc and plasma spraying). The major advantage of using plasma is the ability to select a specific powder composition according to the application that can be based on metals, alloys, carbides or ceramics[3] in this way the optimal coating on the piston ring can be found in the wear system with wear occurring on the top ring of the piston or on the liner.

This type of complete system tooling ensures that the spray system can be installed in a factory production line without disturbing existing tooling or other operators. Both sound and dust emissions are controlled and the operators are not subjected to factors that can affect working conditions.

In addition to ensuring the systems provide reproducibility, they also have to be capable of handling large volumes of components for long periods of time. Components themselves can be a limiting processing factor based on shape/accessibility considerations. Thus, the machine design for plasma processing has to provide a high level of flexibility covering gun - substrate movements and the ease of gun accessibility to the complex surfaces that have to be coated. Figure 5 shows a dedicated machine for the spraying of a lambda sensor, necessary for the control of the exhaust gas emissions from the engine. These components are manufactured from zirconia and have an electrode on the inside and outside surfaces. A plasma sprayed spinel coating was developed to protect this electrode during operation from the exhaust gases, but also at the same time to provide gas access to the sensor. An advantage of the plasma process is the ability to spray porous coatings as well as dense coatings thus providing the possibility to adopt a specific porosity to a particular type of engine. To deal with the high part volumes involved, systems were developed with multiple guns. Figure 5 shows the operator interface of this type of dedicated machine which has been used to manufacture more than 50 million sensors since 1976 which demonstrates the high reproducibility of the plasma process.

Ford published information for the first time at the National Thermal Spray Conference in 1992 at Orlando, USA on the system developed to spray alternator caps. The aluminium alternator cap, Figure 6, is coated with pure aluminium oxide. The purpose of the coating is to provide electrical insulation for the power electronics that are mounted directly on the coated cap.

Originally, the specification called for a coating thickness of 160mm ($\pm$ 20 μm) with a spray time of <12 seconds, and an allowed downtime of per day of 20 minutes every 17 hour working period (4 million parts per year). This called for the selection of special parameters giving the wear parts in the spray gun extended lifetimes.

Figure 7 shows the inside of the spray cabin where two plasma guns are working continuously spraying parts in about 7–8 seconds. The parts are plated on their mounts, two to each section for spraying, by a "pick-and-place" robot (see Figure 8) with parts being supplied on a conveyor system.

The parts enter from the left and are moved by the part-transfer system to the gun. Directly under the parts being sprayed is the exhaust hood to ensure that the power overspray is collected efficiently.

Figure 8 shows the outside of the spray cabin with the robot and conveyor not shown. The powder feeder for the two guns is shown on the right of the picture.

In order to validate the system in Switzerland, it was actually run with parts at the required production rates. The transfer of parts was carried out by teams of people as the robot and conveyor system were supplied directly in the USA.

6 EXPERIMENTAL COATINGS

The field of machine tooling has to take into account the requirement of applying coatings to external and internal surfaces. A recent development in the area of internal surface spraying is the rotation of the plasma gun while the component remains stationary. This provides the opportunity of spraying the cylinders of engine blocks without having to rotate the part. This would allow the plasma technology to be integrated directly into existing engine block fabrication lines. This gun rotating device is shown in Figure 9.

At the present time, the cylinder bore is:
- galvanic; and
- contains nickel.

Obviously, with the present environmental concerns, there is a certain drive to replace the galvanic type processes because of the problems involved with emissions and the disposal of contaminating liquids.

The coating contains nickel and this is also seen as a major factor because the wear debris removed from the cylinder wall during operation becomes entrained in the exhaust and is released into the environment as nickel contamination.

Selection of the appropriate powder composition and spray parameters results in coatings that have the necessary wear properties to replace the cylinder liner used in aluminium engine blocks.

Due to the fact that the plasma process uses powder then the choice of material can be directed specifically to compositions that:
- are not based on strategic materials such as cobalt or nickel,
- in the form of a coating provide the required wear and friction characteristics and, with the correct selection of spray parameters, can have a designed porosity content which in combination with the oil provides enhanced lubrication characteristics. The result is that the user can look towards low priced materials that are easily obtainable, e.g. iron based or ceramic materials.

The use of TBC on piston crowns, heads and valves has been reported in the literature[8]. Recently, progress has been reported[9] which reduces the exhaust gas emission. Success with piston crowns requires the ability to be able to accurately deposit the coating material (robot/manipulator) without creating excessive thermal strains (effect of cooling) which will cause failure in service and thickness control so that the parts do not need to be ground after spraying. Many of the advantages of using TBC's have not been realised in the past due to the pressure of cost regarding secondary factors such as masking materials, grinding, powders (alternatives to zirconia) and the belief that the process is not sufficiently stable for long-term volume production (Figure 10).

The atmosphere surrounding the plasma can play a significant part in determining the properties of the deposit. The plasma interacts with the surrounding environment and, besides lowering the plasma temperature, the entrainment of gas can also affect deposit characteristics. Protective coatings can contain elements which are oxygen sensitive and readily oxidise when sprayed in air. The oxidation of such elements can be reduced or totally eliminated by using the plasma in an inert atmosphere i.e. by spraying in a chamber at a low residual gas level. In this case, the spraying facility consists of a plasma-spraying system (plasma gun, powder feed and power source etc. with a plasma gun and a workpiece manipulator) mounted in a vacuum chamber. In competition to plasma transferred arc (PTA) welding, vacuum plasma spraying has been evaluated for the deposition of stellite powders in the seating areas of valves. The advantage of VPS is the ability to preheat the valve in vacuum and then apply the layer at a constant temperature thus avoiding the build-up of thermal strains in the layer.

7 ENVIRONMENTAL AND SAFETY CONCERNS

In carrying out the production process described above, it is important to protect the operator (and existing equipment) from the thermal spray environment – dust, noise and ultra-violet emissions. It is important to consider the efficiency of the dust collection system for the overspray and its disposal (as well as the disposal of the plastic powder cans). Operators working with robotic systems typically increase their spray time dramatically.

A typical manually operated plasma coating sequence might be as follows:

i.   inspect, clean, mask and grit blast the part,
ii. check masking and insert in the coating fixture,
iii. install the part in the spray unit, start the exhaust system, start plasma torch and if necessary preheat part,
iv. introduce powder and allow plasma conditions to stabilise,
v.   manually deposit the coating (with cooling if necessary),
vi. shut down plasma torch, cool part, and remove,
vii. post-coating surface processing (with heat treatment if necessary).

In a production process, parts are grit-blasted automatically in a machine with grit control systems (see masking and surface activation). Grit-blasted parts are loaded continuously into a plasma spray unit in which the spray process is operating continuously as well. Savings are generated primarily because of the transfer of tasks from the human operator to the robot. As we have seen above, the robot in the spray cell now carries and manipulates the spray gun, relieving the operator to perform other tasks. The *human* operator is still responsible for the performance of the spray robot, because a poorly programmed robot will continue to do so until corrected by a correctly trained operator or a specialist from a systems supplier.

Because robots and manipulators are advancing as a new technology, manufacturing personnel must be trained on their proper use and installation. The installation sequence can be broken down into a step-by-step process which will result in accurate establishment of objectives, proper selection of handling equipment and thermal spray system vendor, orientation of the existing labour force and a technically sound, successful application of the technology.

Having the operator outside of the spray cabin and robot/manipulator systems doing the spraying means that spray parameters such as workpiece stand-off, spray angle and relative movement become highly reproducible and efficient, thus increasing the total quality level. In systems management control as shown above the movement is controlled from a single source and the spray process can be slaved into the movement control. In this way the operator is only concerned with loading the component, checking the powder feeder, setting the parameters for spraying and coating deposition and checking that the coating meets certain standards such as thickness or hardness etc. and can supervise a complete centre, e.g. preparation, spraying, finishing and QC.

8 OUTLOOK

Plasma spraying is undoubtedly at the beginning of a new technological expansion, which will lead to an even stronger industrial integration of this process in the coming years.

In the examples given above, coatings are produced using plasma spray processing equipment, machine tooling, manipulators and robotic systems that are considered to be individual, isolated work stations. In the future, integration of both the grit blast operation and the plasma spraying of coatings will be necessary in a single processing area. Here, both these operations are connected to a host computer which, parallel to other host computers controlling other mechanical machine tool operations can communicate to each other and with the main frame computer. This will allow complete integration of plasma processing in factories using computer integrated manufacturing.

References

1. A. R. Nicoll, 'Protective Coatings and their Processing – Thermal spray', Tech. Rep. Publ. No 86002E, Plasma-Technik AG, Wohlen, Switzerland, 1985.
2. A. R. Nicoll, Production Thermal Spray Equipment and Quality Control Considerations, Plasma-Technik AG, Rigackerstrasse 21, 5610 Wohlen, Switzerland, 1986
3. U. Buran, Mader, H. C.and Morsbach, M., Goetze AG, Germany, Technical Paper K 35 .
4. G. Wuest, S. Keller, A. R. Nicoll, and J. Donelly, J. *Vac. Sci Tech. A*, **1985**, 3, 2464.
5. A. R. Nicoll, Retesting Plasma Spray Powders, Plasma-Technik AG, Rigackerstrasse 21, 5610 Wohlen/Switzerland, 1986.
6. K. D. Borbeck, Robotics and Manipulators for Automated Plasma Spraying and Vacuum Spraying, Proc. 10th Int. Thermal Spray Conf., DVS-Berichtsband 80, Deutscher Verlag für Schweisstechnik (DVS) GmbH, Düsseldorf, p. 99.
7. T. Peterman, Thermal Spray: International Advances in Coatings Technology, Proc. ITSC Orlando USA, 1992, (309), publ. by ASM Int. USA.
8. W. F. Calosso and A. R. Nicoll, Process Requirements for plasma sprayed coatings for internal combustion engine components; Energy Sources Technology Conference, Feb.15-20, 1987, ASME paper 87-ICE-15.
9. No smoking on the buses, The Economist June 12th 1993 (90).
10. A. R. Nicoll, H. Gruner, R. Prince and G. Wuest, *Surface Engineering* 1985, 1, 59.

3.3.2
Development of coatings for lubricant-free ball bearings and of an adapted test bench for the evaluation of the coated ball bearings

O. Knotek, E. Lugscheider, F. Löffler and M. Möller

MATERIALS SCIENCE INSTITUTE. AACHEN UNIVERSITY OF TECHNOLOGY. AACHEN, GERMANY

1 INTRODUCTION

To fit bearings for different stresses mostly the geometry has been varied. This led to bearings with balls, needles, tons, etc. for roller bodies and adapted bearing rings. But in some applications it is necessary to change not only the form but also the material.

One example for this is the exchange of conventional bearings by lubricant-free bearings in food machines, where a contamination with the lubricant is not permitted. Nowadays the degree of design is high to ensure that no lubricant can be lost. Conventional bearings are still used because bearings that ensure a sufficient lifetime during a lubricant-free run are expensive. These bearings are mostly made of bulk ceramic materials.

In this work the development of a coated hybrid ball bearing is described. The hybrid ball bearing consists of steel bearing rings (100 Cr 6, tempered), ceramic balls (silicon nitride) and plastic cages (polyamide). For a lubricant-free running the bearing rings are deposited with hard coatings. The deposition is done with the magnetron-sputter-ion-plating (MSIP) technique. In the first stage the coatings are optimized by coating specimens with simple geometry. After this the optimized coatings are deposited on the bearing rings.

For examination of the coated ball bearings in lubricant-free applications a test bench has been developed. In this test bench the ball bearings are loaded with defined stresses and tested for their lifetime.

2 PRE-EXPERIMENTAL EXAMINATION OF SPECIMENS WITH SIMPLE GEOMETRY

Before coating the bearing rings the deposition was optimized by coating specimens of simple geometry. The material consists of tempered 100 Cr 6 with a hardness of 60–63 HRC. Five coating systems were examined:

Ti-B, Si-C, Si-N, Zr-O, AL-O

To ensure no reduction of hardness the temperature of the rings during coating had to be less than the tempering temperature of 180°C. A reduced hardness causes a lack of coating support. Therefore a low temperature physical vapour deposition (PVD) process was used.

The MSIP coating plant allows a wide range of usable targets. The sputter plasma can be created by a DC or a high frequency (HF) power station. This allows the sputtering of conductive and non-conductive target materials. With the help of an additional heating system it is possible to vary the substrate temperatures during the coating process. A pre-depositonal cleaning can be done by a HF-plasma etching process. When the deposition is done with a DC-plasma a bias-voltage can be put to the substrates. By adding reactive gas a reactive coating can be deposited.

In the first experiments the influence of biassing and heating on the substrate and the coating was investigated. The coating system was Ti-B. The coating was deposited with a DC-plasma and a titanium diboride-target without adding any reactive gas. For characterization microhardness, substrate hardness, roughness, critical load were measured and the structure was analysed with X-ray and REM.

The Ti-B coating showed the best mechanical properties by deposition without biassing and heating. The rupture structure is shown in Figure 1.

In a next step a titanium interlayer with a thickness of a few hundred nanometers was used to increase the adhesion of the Ti-B top layer. For a qualitative evaluation of the adhesion the critical load was examined in the scratch test. By using a titanium interlayer the critical load could be increased from 40–70 N.

Another examined system was silicon carbide. The coating was done with a silicon carbide target without any reactive gas, biassing or heating. Also a titanium interlayer was used to increase adhesion, but only a critical load of 11 N could be achieved. The rupture structure is shown in Figure 2.

Coatings of the systems Al-O, Zr-O and Si-N were as well deposited in a reactive and a non-reactive deposition. In all cases a reactive deposition showed the best mechanical properties. In a next step a titanium and a system own metal interlayer were investigated for increasing the adhesion. In all cases the critical load could be increased more by using the system own metal for interlayer. The critical loads achieved for the Al-O top coating with aluminium

Figure 1 *Rupture structure of a Ti-B coated bearing ring*

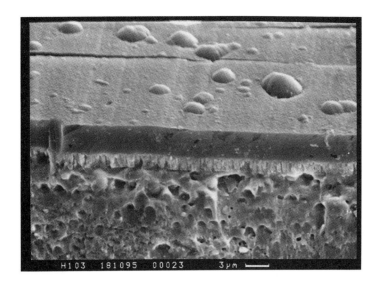

Figure 2 *Rupture structure of a silicon carbide coated ball bearing with titanium interlayer*

interlayer was 70N, for the Zr-O top coating with zirconium interlayer 50 N and for the Si-N top coating with Silicon interlayer 20 N. Raster electron microscopic examination of the rupture structure is shown in Figures 2–5.

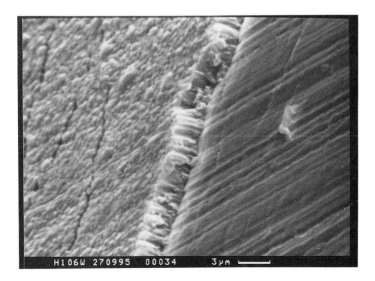

Figure 3 *Rupture structure of a Al-O coated bearing ring with aluminium interlayer*

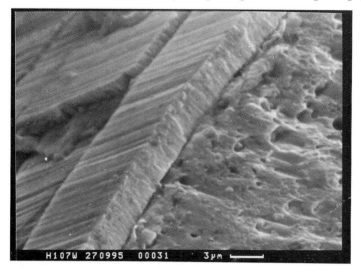

Figure 4 *Rupture structure of a Zr-O coated bearing ring with zirconium interlayer*

3 DEVELOPMENT OF A TEST BENCH FOR COATED ROLLER BEARINGS

By starting the construction of a test bench for coated roller bearings several restrictions were formulated. A selection of the most important restrictions is listed below:
- move the parts of the specimens in a defined way against each other,
- load the specimen with a static force,
- application of sensors,

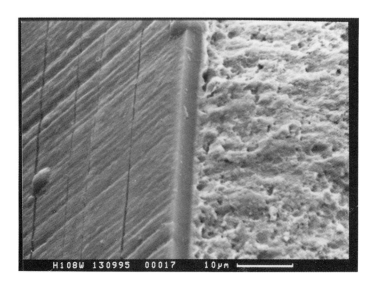

Figure 5 *Rupture structure of a Si-N coated bearing ring with silicon interlayer*

- easy and quick exchange of the specimens,
- highest safety during operation for the operator.

The restrictions led to five modules for the test bench:
- specimen fixing module,
- propulsion module,
- force loading module,
- sensoring module,
- rack module.

In the beginning of the constructing phase the specimen was defined as an axial ball bearing. These bearings are not only loaded by the rolling movement but also by a drilling movement of the balls. Figures 6 and 7 show the behaviour of a ball in a force loaded axial ball bearing.

An axial load deforms the contact area between the ball and the raceway into an ellipse. Within this elliptic area the velocity is not equal. The velocity gradient induces sliding, which causes a higher load on the coating. Therefore an axial ball bearing was chosen for test specimen.

A schematic test bench is shown in Figure 8. The maximum diameter for a test specimen is 140 mm. The propulsion is done by a belt connected electro motor with a static number of revolutions per minute. For test specimen an axial ball bearing (shortsign 51107, Co. NTN, Japan) was chosen. In a first stage of development the test course was defined as shown in Figure 9.

In a first step the load was chosen to achieve a contact Hertzian pressure of 1000 MPa. The test was carried out for a period of 30 minutes and the conditions of the balls and raceways were checked visually each 15 minutes. By showing no sign of damage in any part of the ball bearing the force was incremented to a second level load. There the Hertzian pressure per ball

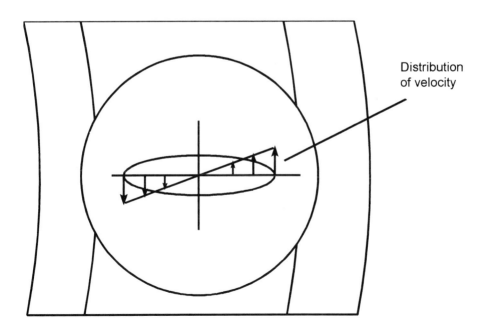

Distribution of velocity

Figure 6 *Distribution of velocity in a force loaded axial ball bearing[1]*

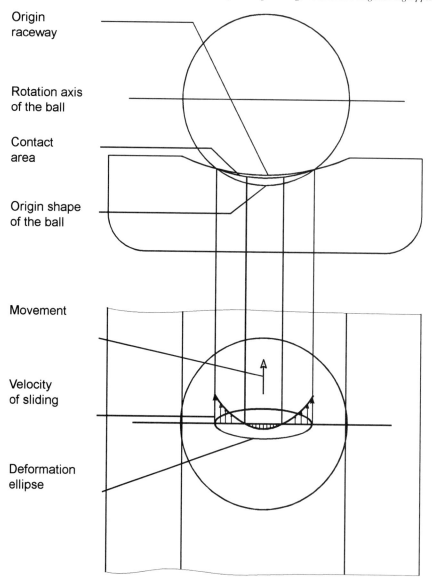

Origin
raceway

Rotation axis
of the ball

Contact
area

Origin shape
of the ball

Movement

Velocity
of sliding

Deformation
ellipse

Figure 7 *Sliding caused by a deformation ellipse[1]*

achieved 2000 MPa. At this level the test was carried on until lifetime of one of the bearing components was achieved, stopped only every 15 minutes for a visual examination. During the test the force and the bearing temperature were measured. The Ti-B coating with and without titanium interlayer and the Si-C coating, with titanium interlayer, were tested that way.

In the first tests it was found that a high load on the coated raceways was induced by the rapid starting phase. For elimination of this "starting load" a controlled starting and stopping

Figure 8 *Mounted test bench*

sequence was applied. In the second stage of development the test was modified as shown in Figure 10. The starting and stopping sequence was controlled by a programmed function. The second level load was decreased to a Hertzian pressure per ball of 1500 MPa and the controlling intervals at the first level load were decreased to ten minutes.

4 RESULTS OF TESTING COATED BALL BEARINGS

The lifetime of the coated ball bearings differed gravely depending on the coating system. An uncoated ball bearing shows damage in its raceway after a 30 minutes run at a Hertzian pressure of 1000 MPa per ball. A decrease of lifetime was recorded in the following coating systems: Si-C top coating with titanium interlayer, Si-N top coating with silicon interlayer and Al-O top coating with aluminium interlayer. Opposed to that, an increase of lifetime was recognized by the Zr-O top coating with zirconium interlayer. First cracks and delamination were detected after a 15 minutes run at 1500 MPa Hertzian pressure per ball. The Ti-B coated ball bearing showed first damage in its raceway after a 15 minute run at a Hertzian pressure of 2000MPa per ball. The highest lifetime was observed for the Ti-B top coating with titanium interlayer. An overview is given in Table 1.

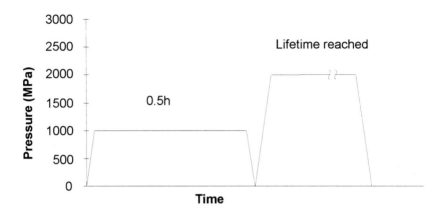

Figure 9 *First stage development of test course*

5 SUMMARY

Coating of tempered bearing steel 100 Cr 6 without reducing its hardness is possible with the MSIP process. For lubricant-free applications the bearing rings of an axial bearing (shortsign 51107, Co NTN, Japan) were coated with several hard ceramics. The examined coating systems were Ti-B, Si-C, Si-N, Al-O and Zr-O. Metallic interlayers were investigated for increasing

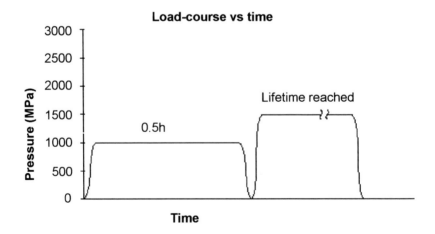

Figure 10 *Second stage development of test course*

Table 1 *Results of lifetime testing*

Coating system	First level load	Second level load
Uncoated	First damages in the raceway after a 30 min run at a Hertzian pressure of 1000 MPa per ball	
Ti–B	No damages detectable after a 30 min run at a Hertzian pressure of 1000 MPa per ball	Delamination of the coating in the raceway after a 15 min run at 2000 MPa Hertzian pressure per ball
Ti–B with titanium interlayer	No damages detectable after a 30 min run at a Hertzian pressure of 1000 MPa per ball	Delamination of the coating in the raceway after a 30 min run at 2000 MPa Hertzian pressure per ball
Si–C with titanium interlayer	Delaminations in the raceway after a 15 min run at a Hertzian pressure of 1000 MPa per ball	
Si–N with silicon interlayer	Delamination and cracks in the raceway after a 10 min run at a Hertzian pressure of 1000 MPa per ball	
Al–O with aluminium interlayer	Delamination and cracks in the raceway after a 10 min run at a Hertzian pressure of 1000 MPa per ball	
Zr–O with zirconium interlayer	No damages detectable after a 30 min run at a Hertzian pressure of 1000 MPa per ball	Isolated areas of delamination in the raceway after a 15 min run at a Hertzian pressure of 1500 MPa per ball

the adhesion between substrate and top coating. The bearing rings were coated with Ti-B with and without titanium interlayer, Si-C with titanium interlayer, Si-N with silicon interlayer, Al-O with aluminium interlayer and Zr-O with zirconium interlayer. The coated ball bearings were tested in a test rig by using silicon nitride balls. A reliable comparability was guaranteed by defining the load with the Hertzian pressure per ball. The test was divided up into two levels. In a first level the Hertzian pressure per ball achieved 1000 MPa. At this level the bearings ran 30 minutes or until first damage was detected. In the second level the Hertzian pressure was increased up to 2000 MPa and 1500 MPa per ball respectively. The bearings ran at that level until lifetime of one bearing part was achieved. The highest lifetime were obtained with a Zr-O top coating with zirconium interlayer and with the Ti-B coating with and without titanium interlayer.

Acknowledgement

This work is supported by the German Bundesministerium für Bildung und Forschung (BMBF) with support number 13N6219 in corporation with Cerobear (Herzogenrath), SHM (Aachen),

MAT (Dresden), IKTS–FhG (Dresden). Project manager is the VDI-TZ.

References

1. Die Wälzlagerpraxis, Eschmann, Hasbargen, Weigand, R. Oldenbourg Verlag GmbH München,1978.

Section 3.4 Cutting Tools and Manufacture

3.4.1
Successful Products through Surface Engineering

M. Sarwar

School of Engineering, University of Northumbria at Newcastle, Newcastle upon Tyne, UK

1 INTRODUCTION

Surface Engineering is a key technology in many sectors of manufacturing industry including automotive, aerospace, biomedical, chemical, optical and electronics. For many applications such as electronics and optics, surface engineering is an essential ingredient of component manufacture in order to achieve the desired product performance. For other applications, such as decorative products, cutting tools, surface engineering gives products a combination of added value, improvement in performance and life with increase in component quality. Table 1 gives a summary of conventional, diffusion/conversion and deposition processes associated with surface treatments. Figures 1 and 2 show the breakdown of market shares for emerging high technology processes and established low technology processes. For the established technologies the market is dominated by surface/heat treatment and electroplating processes. The emerging technology market share is only £11.5m[1] and this sector is dominated by CVD, PVD, plasma spraying and plasma nitriding processes. The line of sight high technology processes such as laser/electron beam and ion implantation, whilst preferred for some market areas such as biomedical do not command a major market share. However, 90% of the research funding in the UK[2] is directed towards the emerging technologies and only 1% of these technologies are actually used in the UK manufacturing industry. In Europe, particularly Germany, there is widespread awareness of the advanced surface engineering technologies with commercial coating centres being spread around manufacturing cities. The serious imbalance in the UK needs to be addressed to improve the UK's competitive position in the manufacturing sector.

Surface engineering technologies are very wide ranging with diverse applications. They span the conventional technologies such as electroplating, thermal spraying, weld surfacing and thermochemical treatments, such as, chemical vapour deposition (CVD), physical vapour deposition (PVD), ion-implantation, laser assisted deposition and many variants of plasma based processes. Some of these techniques are used in combination to give superior surface characteristics.

Whatever the choice of technology in order to achieve the optimum results from any surface engineering technology a "system approach" is essential[3]. This involves integrating component design, material selection, manufacturing and process specification/control, surface treatment/ engineering, handling/packaging into a total unit process. Far too often the process of surface

Table 1 *A summary of conventional, diffusion/conversion and deposition processes associated with surface treatments*

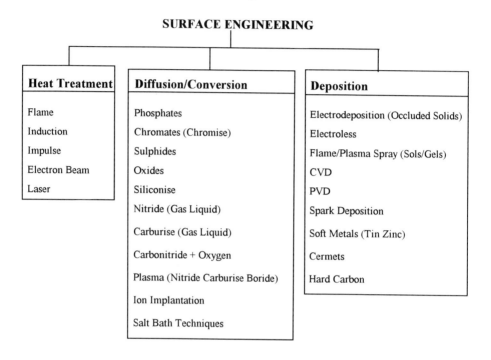

SURFACE ENGINEERING

Heat Treatment	Diffusion/Conversion	Deposition
Flame	Phosphates	Electrodeposition (Occluded Solids)
Induction	Chromates (Chromise)	Electroless
Impulse	Sulphides	Flame/Plasma Spray (Sols/Gels)
Electron Beam	Oxides	CVD
Laser	Siliconise	PVD
	Nitride (Gas Liquid)	Spark Deposition
	Carburise (Gas Liquid)	Soft Metals (Tin Zinc)
	Carbonitride + Oxygen	Cermets
	Plasma (Nitride Carburise Boride)	Hard Carbon
	Ion Implantation	
	Salt Bath Techniques	

engineering has been treated as a bolt on process. This has resulted in disappointing results where the product has not performed to its expectations.

2 APPLICATIONS IN CUTTING TOOLS

Metal Cutting is a major activity and forms a large segment of manufacturing industry[4]. Metal cutting tools form approximately 80% of the total activity. Since it involves the removal of metal and thus has been considered to be a wasteful process especially if expensive material is being machined, considerable effort has gone into shaping materials using cold forging, precision casting and powder forming. Although some success has been achieved by the above processes there is little evidence that these are replacing metal cutting activities in any significant form. Practical cutting speeds have approximately doubled every ten years during this century. This has been largely due to developments in tool materials which have been further improved by the application of advanced surface engineering technologies. These have arrested or slowed down the wear associated with cutting tools and enhanced the tool performance life.

A major success in coatings in industrial applications[5] is Titanium Nitride (TiN), with other single and hybrid coatings such as Titanium Carbide, Titanium Carbonitride, Titanium Aluminium Nitride and others gaining popularity. However such coatings are not 100% successful in applications.

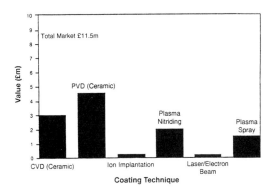

Figure 1 *Market for Surface Engineering emerging technologies*

Whilst considerable attention has been given to the characteristics of the hard film, as is evident from the numerous publications in the area, very little attention has been given to the condition of the substrate and the interface. It is now apparent and recognised by the cutting tool technologist and the materials engineers that, in order for these treatments to be fully successful, a total systems approach has to be adopted.

3 INFLUENCE OF MANUFACTURING QUALITY

Figures 3 and 4 highlight the conditions of the cutting edges prior to surface treatment. These fine burrs under normal heat treatment and cutting would not be detrimental to the overall tool performance or life; however, when surface coated these fine burrs become detached exposing the substrate material leading to reduced tool life. It is clear from the figures that the cutting edges produced on multi-point tools using current manufacturing methods are inferior in quality compared to those on single-point tools. It is not surprising that manufacturers of

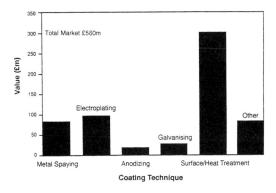

Figure 2 *Market for Surface Engineering established technologies*

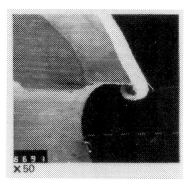

Figure 3 *Ground HSS circular saw cutting edge × 50*

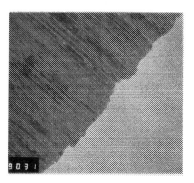

Figure 4 *Typical ground tool cutting edge × 350*

multi-point cutting tools have not had the same level of improvement in the product as single-point cutting tools. Multi-point cutting tools are far more complex and are not dependent on the cutting edge alone but the gullet geometry and surface characteristics of the rake face. Examination of cutting edges reflects the higher level of control on cutting edge geometry associated with powder metallurgy techniques as compared to solid high speed steel tools.

Figure 5 shows the effect of surface roughness on broach life. To demonstrate the importance of the surface characteristics alone to product performance no advanced surface engineering technologies were applied to the tools. A smoother surface was produced by CBN grinding and surface finish to about 0.3μm Ra compared to a standard surface of 0.8μm Ra. As can be seen, even without a coating, improvements in broach life were observed. The smoother broaches lasted nearly twice as long as the standard ones. These results clearly demonstrate

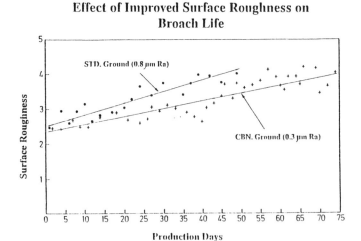

Figure 5 *Effect of improved surface roughness on broach life*

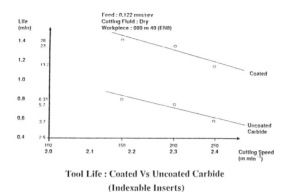

Figure 6 *Tool life: coated vs uncoated Carbide (Indexable Inserts)*

the value of the quality of the substrate prior to coating. Therefore, the reasons for mixed results sometimes achieved with advanced coating technologies are self evident. Tool manufacturers who use surface engineering as an add-on process without considering the substrate preparation could be missing out on its real benefits.

4 SINGLE POINT TOOL PERFORMANCE

Figures 6 and 7 show the results of Titanium Nitride coatings applied to cemented carbide tips and high speed steel tips. The coatings are effective in delaying certain types of wear and giving improved life but are less effective in delaying other kinds of wear. In both HSS and Carbide tool tips, the results showed that the advantage in life gained by coating diminishes

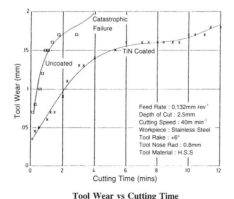

Tool Wear vs Cutting Time

Figure 7 *Tool wear vs. cutting time*

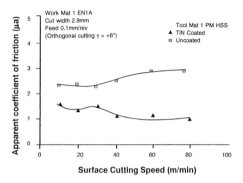

Figure 8 *Influence of surface coating on the apparent co-efficiency of friction*

under several conditions (i.e. at high feeds and speeds). The application of a coating can affect the friction properties of the surface (Figure 8) which will affect chip flow characteristics, reduction in forces and hence the temperature, increasing the tool performance and life.

5 CIRCULAR SAW PERFORMANCE

Examination of uncoated HSS circular saws revealed at the primary stages wear scars on the corners of the finishing teeth, and at a later stage, wear sears on the roughing teeth. Flank

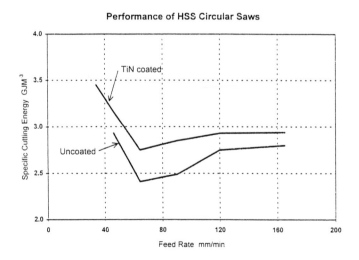

Figure 9 *Performance of HSS circular saws*

Performance of H.S.S. Drills (Ref)

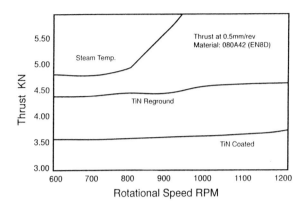

Figure 10 *Performance of HSS drills*

wear was not predominant in coated or uncoated saws. Figure 9 shows comparative performance of steam tempered and TiN coated circular saws as measured using the specific cutting energy parameter. There was a reduction in both forces and specific cutting energy with TiN coated circular saws indicating superior performance.

The performance and failure modes in circular saws are far more complex than those in single point tools. The method of manufacture does not allow the individual cutting edges to be produced with sufficient accuracy (e.g. edge sharpness, surface finish, geometrical accuracy) and the process variation can cause early failure. Furthermore, multi-point cutting tools need consistency in quality at all cutting the points as variation in a single tooth (surface roughness, Tooth height, pitch, etc.) can cause premature failure.

6 PERFORMANCE OF HSS DRILLS

The twist drill is geometrically one of the most complex of cutting tools. The manufacturing industry is constantly improving and placing high demands of consistency in drill life and performance. Therefore a combination of new drill design, materials, and coatings have been developed. The most popular and successful coating is still titanium nitride, although other coatings, TiCN, TiAlN are also proving to be successful. Figure 10 shows the performance of HSS twist drills using the thrust force as an assessment parameter. Clearly the TiN coated drill is far superior than the conventional drill and even after regrind, has a better performance.

7 CONCLUSION

The results presented show evidence of improvement in tool life and performance improving manufacturing standards. Currently, the tooling industry lacks manufacturing specifications/ standards for coatability, particularly associated with multi-point tools. Surface engineering is an expanding industry underpinning a wide range of products an applications. It is essential for the manufacture of a wide variety of products, and in other applications it gives products added value. The importance of a 'total systems approach' with tight quality control against design and manufacturing standards is vital in order to achieve the maximum benefit from any surface engineering process. This approach has been considered for particular cutting tools presented in this paper. It has been found that advanced surface coatings enhance performance, arrest or slow down wear and improve product quality.

References

1.   D. Vaughan, Future of Surface Engineering in the UK, Report by Centre for Exploitation of Science and Technology (CEST) 1990.
2.   D. Whittaker, Surface Enineering Report, (CEST) 1990.
3.   M. Sarwar and W. Ahmed, *Materials World*, March 1994.
4.   E. M. Trent, 'Metal Cutting', Butterworth, Oxford (1994).
5.   M. Sarwar and D. Gillibrand, 'The Application of Advanced Surface Engineering Technologies to Cutting Tools' Proc. IMCC, Hong Kong Polytechnic, March 1993.

3.4.2
Indirect Monitoring of Carbide Tool Wear in the Facing of Low Carbon Steels Through Measurements of the Roughness of Machined Surfaces

K. N. Strafford[1], J. Audy[1] and L. Blunt[2]

[1]LAN WARK RESEARCH INSTITUTE, THE UNIVERSITY OF SOUTH AUSTRALIA, THE LEVELS, S.A 5095, AUSTRALIA

[2]SCHOOL OF MANUFACTURING AND MECHANICAL ENGINEERING, UNIVERSITY OF BIRMINGHAM, BIRMINGHAM B 15 2TT, UK

1 INTRODUCTION

In recent times there has been a tendency to optimise technological processes by means of a control computer, using accurate data concerning the most important changes in these processes. Tool wear is a major problem in metal machining operations because of its importance to the final quality of the machined surfaces. To analyse the state of a particular cutting tool it is necessary to identify a mutual relationship between tool wear, surface roughness and dynamic phenomena developed through the process as shown in Figure 1.

Introducing this data as a mathematical model in a control computer it is possible, in principle, to appreciate and control the wear of the cutting tool. Mechanisms of tool wear occurring during machining have been widely researched and published[1,2], as well as their influence on the final quality of machined surfaces.

A number of papers has reported and reviewed such issues. For example, Fombarlet[6] cut low carbon engineering steels with P30 carbide inserts at various cutting conditions. He monitored the types of tool wear such as abrasion, adhesion and diffusion, which occurred at the cutting edges of the carbide insert, in relation to changes in cutting speed, depth of cut and feed rate. The author found that these three phenomena varied widely according to cutting conditions but, primarily, were influenced by cutting speed. Thus adhesive and abrasive wear occurred on the tool when the cutting speed was in the range of 30 to 280m/min. At faster cutting speeds tool wear took place by diffusion.

Smith[3] published micrographs of component surfaces machined at different cutting speeds. When a carbide insert was used for machining alloy steel it was found that, while the depth of cut and feed rate had only minor effects on surface finish, cutting speed was the most influential parameter. Micrographs of surfaces created at 2.6m/min exhibited features associated with discontinuous chip formation, and fracture. A substantial built- up edge on the tool was also observed. Higher cutting speeds in the range of 11m/min to 59m/min generated, for the most part, continuous chips, while the quality of surface finish improved over this range. However, some swarf debris was generated associated with the built- up edge. Surface finish appeared to be good at 112m/min, while significant tool (flank) wear occurred at 212m/min[3].

The principles underlying many techniques which have been invented over several years to evaluate the roughness of machined surfaces have been the subject of extensive discussion in the literature[4,5]. However, there has been no attempt to correlate tool wear, and the approach to with work-piece surface roughness and other in-process phenomena generated while machining.

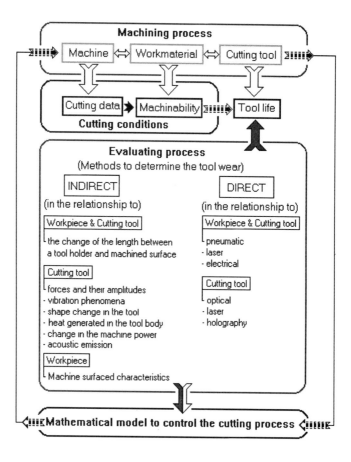

Figure 1 *Parameters, indirect and direct, which may be used to ascertain tool wear, and hence tool life in a model cutting system under adaptive control*

Information about the topography of machined surfaces is critically important to a better understanding of the relationship between tool wear and the dynamic characteristics developed in machining processes. However, "topography" is defined in many different ways within different standards. There is also a high variation in the accuracy of particular methods and devices. Other problems are associated with the different ways adopted for evaluating the roughness of machined surfaces. From the literature it is evident that the topography of machined surfaces has been characterised mainly by using optical metallography, SEM metallography and two-dimensional surface evaluating devices. It is necessary to note, however, that all surfaces interact in three dimensions, and not in two dimensions, as it has been presented in many papers[1–3,7,8].

Problems associated with metrology equipment and methods used for assessing machined

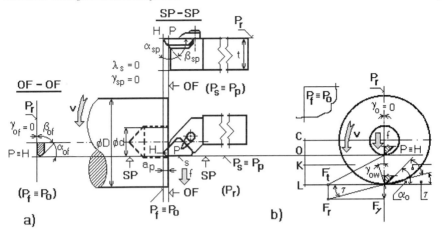

Figure 2 *(a) Basic parameters and nomenclature relating to the facing process, and (b) changes in the cutting angles as the facing process proceeds*

surfaces have been reviewed by Stout and Blunt[5]. Stout et al.[9] have also discussed the principles and development of profilometers to generate 3-D topography.

The present investigation has been with the generation and analysis of such information, to establish, inter alia, the possibility of detecting tool wear and the approach to failure by determining the relationship between tool wear, surface roughness. Others papers[13-15] have been concerned with assessment of other significant parameters in machining such as cutting force components and their amplitudes.

2 EXPERIMENTAL

2.1 Materials

Machining-facing tests using uncoated carbide inserts were conducted using a HMT lathe. The work-piece material was a low-carbon free machining steel in bar form of nominal composition 0.2%C, 1.6%Mn, 0.6%Cr. It was supplied and used in an annealed condition.

The homogeneity of the steel was assessed indirectly from variations in hardness, which were determined over three separate cross-sections cut from the bar material. The Vicker's hardness test was conducted according to Australian Standard Specification No.B 106–1954[10]. Measurements were made using a diamond pyramid involving a force equal to 50N applied for 20 secs. It was found that hardness measured across all the cross sections of the samples along a diameter varied from 142.1 to 143.3HV. When the same measurements were made along another diameter normal to the first, hardness varied in the range of 138.6HV to 145.2HV. It is evident that the observed maximum 4.6% variation in the Vicker hardness was low, and in the range recommended by ISO/R 683-3 Standard. This variation can probably be explained by the presence of small material defects, such as shrinkages, cavities and scattered porosity, associated with the manufacturing conditions.

The facing tool used was a standard, uncoated P–30 carbide insert-Type TPUN 160308 (TPU322), identified in literature[12].

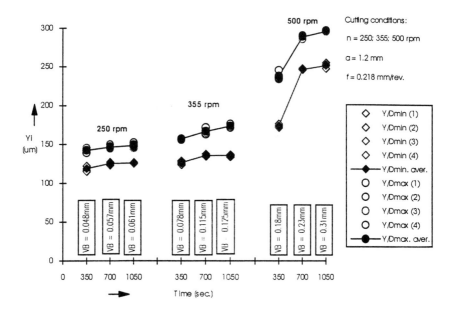

Figure 3 *Individual and average Yi values at D_{min} (100mm) and D_{max} (130mm) observed at increasing cutting speeds and times, and associated with flank wear (VB) data*

2.2 Experimental Procedure

Figures 2 (a) and (b) define, respectively, the geometry of the tool-work-piece system used, and its alteration as cutting proceeded. Machining parameters were: constant feed rate of 0.218mm/rev., constant depth of cut of 1.2mm and variable cutting speeds of between 255 to 500rpm. Tool (flank) wear at 5 minute cutting time intervals within an overall standard cutting time of 30 minutes with an accuracy 0.025μm by using an optical microscope equipped with a filar eyepiece. The topography of the machined surfaces was monitored by using two dimensional surface analysing equipment – Mitutoyo /Surftest 211/ Surface Roughness Tester No. 21 1, an optical microscope, and three dimensional surface analysing equipment with accuracies ± 0.01μm, ± 0.025μm and ± 0.0005μm, respectively.

3 RESULTS AND DISCUSSION

3.1 Cutting Conditions, Tool Wear and Work-Piece Surface Roughness

Variations in the roughness of machined surfaces were recorded until tool wear reached 0.3mm the critical value as recommended in the ISO standard 3885/77[11]. Figure 3 shows the measured spacings (four independent readings at the minimum(100mm) and maximum(150mm) bar diameters) between the individual amplitude peaks (Yi) and associated flank wear widths (VB_B) of the insert at speeds of 250, 355 and 500rpm, at increasing cutting times. It is evident

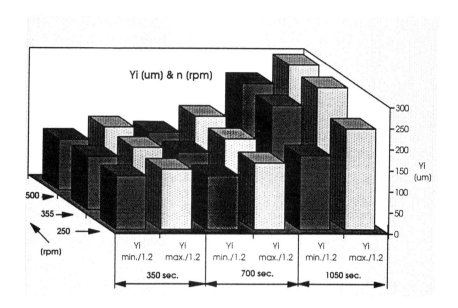

Figure 4 *Average minimum and maximum Yi values observed at cutting times of 350, 700 and 1050 seconds, and at facing speeds of 250, 355 and 500 rpm*

that average values of the spacings increase, as does the tool wear, with facing speed, and cutting time. It is believed that the small scatter was caused mainly by the presence of work-piece material defects as mentioned earlier.

Figure 4 also demonstrates the increases in spacing between amplitude peaks ($Yi_{min/max}$) with varying facing speed and cutting time. 33.4% and 75.4% increases in Yi_{min} and Yi_{max} values at the 350sec cutting time, were caused by an increase in facing speed from 250 rpm to 500 rpm. For 700 sec, the measured increases in these variables were 62.6% and 73.1 %, while for 1050 sec cutting time, the increases were only 52.9% and 59.3% for Yi_{min} and Yi_{max}, respectively. Figure 4 clearly demonstrates the increase in the minimum and maximum Yi values observed as the cutting speed was raised from 250 to 355 and 500 rpm at each of the three cutting time intervals, as well as confirming the increase in these amplitude values with increase in elapsed cutting time at each of the chosen cutting speeds. The Yi values were always larger when associated with cutting the large diameter.

Figure 5 summarises the observed corresponding individual and calculated average (Ra) roughness values, and the associated flank wear widths (VB_B) of the carbide insert. Ra values clearly increase at each facing speed as time increased. This increase is more marked as $\underline{n}$ rises from 250 to 355 rpm at given time values: the rise is less as $\underline{n}$ increases from 355 to 500 rpm. Indeed, the Ra values for t = 350 sec are similar, or even less at 500 rpm than 355 rpm. 43.5% and 69.85% increases in Ra_{min} and Ra_{max} were calculated for 350sec of cutting time associated with an increase in facing speed from 250rpm to 500 rpm. For 700 sec, these variables increased by about 35.3% and 26.6%, while for 1050 sec cutting time, it was about 38.2% and 38.2% for Ra_{min} and Ra_{max}, respectively, for this doubling of facing speed.

Figure 6 also shows the dependence of the average (Ra) values observed at the minimum

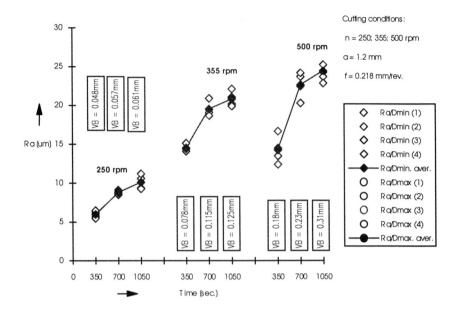

Figure 5 *Individual and average Ra values for Dmin (100mm) and Dmax (130mm) at increasing cutting speeds, times and associated flank wear (VB_B) data*

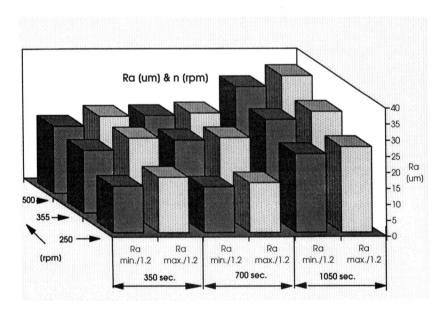

Figure 6 *Average and maximum Ra values observed at cutting times of 355, 700 and 1050 seconds and facing speeds of 250, 355 and 500 rpm*

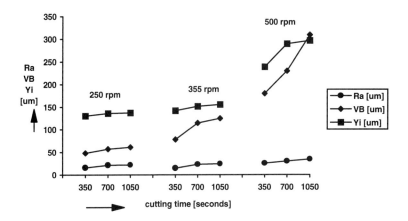

Figure 7 *Roughness (Ra), flank wear (VB), amplitude peaks (Yi) versus cutting times and conditions*

and maximum bar diameters, as the facing speed was increased at increasing elapsed cutting times, confirming the trends noted in Figure 5. The relatively steep rise in Yi and Ra values in the early stages of cutting may be assumed to be characteristic of the "running in" period for the tool, when the flank wear is also high. In the later stages of cutting both the surface roughness parameters are associated with much less increase in flank wear.

Figure 7 conveniently summarises average surface topography (Ra and Yi) values, as well as the corresponding tool flank wear (*VB*) levels. It is clear that work-piece surface roughness increases as both facing speed and cutting time are increased. However, whereas tool Ra values are seen to increase relatively slowly, but steadily, as time and revolution are raised, Yi values and especially tool wear, are observed to increase more rapidly as time and particularly facing speed increase (in the range of 355 to 500rpm).

3.2 Cutting Conditions, Tool Wear and Chip Formation

Figures 8 (a) and (b) illustrate the pattern of chips formed under the chosen cutting conditions at the facing speeds of 250, 355 and 500 rpm at increasing cutting times. It was noted earlier that defects in the surface topography – manifest in the increased Ra values – were mainly associated with the high speed in the facing process. In harmony with this it was found that chips produced at 250 rpm were relatively shorter than those produced at the 500 rpm facing speed. The length of these chips depended mainly on the level of tool wear. Small, separate chips were also obtained when the tool wear was slight, in the range of 0.048 to 0.06lmm. Increase in the tool wear (via increased speed) caused the creation of short helical chips, a state characteristic of the facing speed of 355rpm. A speed of 500 rpm produced relatively long helical chips but only when tool wear was less than approximately 0.28mm. Snarled chips began to be produced when tool wear (VB) was close to the critical value 0.3mm. Analysis of surfaces, produced at speeds of 250 rpm and 500 rpm, suggested that discontinous chip formation and fracture was preceded by a build- up of material in the gap between the engaged tool and the work-piece. Chips began to be formed in a snarled pattern and were pressed into the space between tool and work-piece as illustrated schematically in Figure 9. It may be concluded that the rougher surface topography developed as a consequence of shearing taking place in the contact area between the chips and the tool. A facing speed equal to 500 rpm generated continuous long helical chips. The machined surfaces which developed under these conditions were of a very poor quality.

4 CONCLUSIONS

The main conclusions to be drawn from this study may be summarised as:

1. Linkages have been demonstrated in the facing of a mild steel with an uncoated carbide tool between tool flank wear and work-piece surface roughness under a range of cutting conditions.

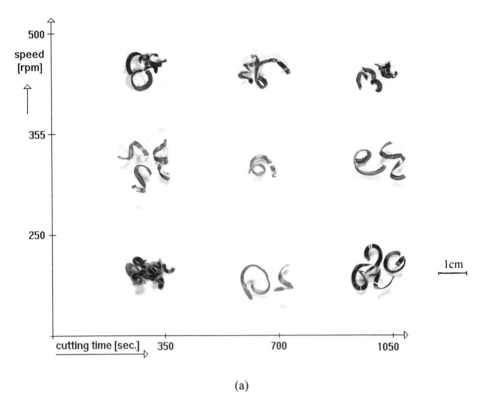

(a)

Figure 8 *The influence of facing conditions on chip morphology obtained from (a) 100mm diameter and (b) 130mm diameter of the work-material to be cut. Short pieces of material – small discrete chips – were associated with early steps of cutting (t = 355 rpm or 700 seconds) and low facing speeds (~ 250 rpm). As it increased to 1050 seconds, and then cutting at higher speeds → 500 rpm, there was a tendency for long continuous spiral chips to form, with an associated increase in work-piece surface roughness*

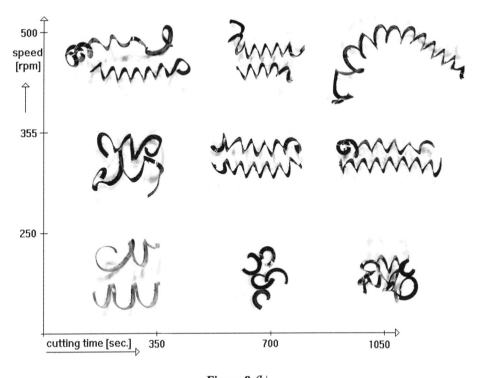

Figure 8 *(b)*

2. Surface roughness, measured both in levels of Ra values and amplitude (Yi) levels, increased with increasing cutting (facing) speeds, in the range 250 to 500 rpm, for a constant feed rate (0.218 mm/ ev) and depth of cut (1.2 mm).
3. The sensitivity of surface roughness to cutting speed became more evident over the higher speed interval 355 → 500 rpm.
4. At constant cutting speed these surface roughness parameters are also increased with increased cutting time although this machining variable is much less important than speed of cutting.
5. Swarf (chip) morphology has also been shown to be strictly dependant on cutting speed. Shorter, separate chips are favoured by higher cutting speeds (→ 355 rpm) essentially being replaced by a tendency to more continuous swarf in the form of long helical chips at the highest speed (500 rpm).
6. Such changes in work-piece surface quality (and in the nature of the swarf) have been shown to be systematically associated with tool flank wear. Small changes in tool wear are associated with significant deterioration in work-piece quality. Tool wear has been shown more critical to surface quality the higher the facing speed. Deterioration also becomes increasingly rapid as cutting time increases.
7. When measured and understood in a particular cutting work-piece/ tool system, such parameters, in conjunction with knowledge of these relationships to cutting forces, could be used to monitor and control the cutting process to optimise manufacturing efficiency.

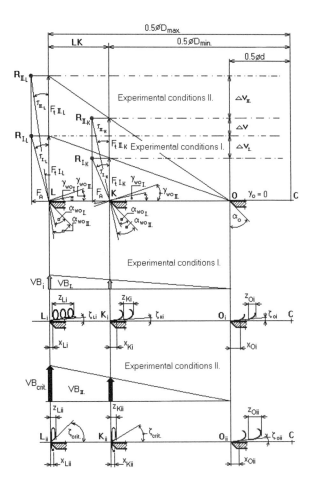

Figure 9 *Change of clearance angle (α) and cutting rake angle (γ) in relation to cutting conditions, and their influence on the pattern of chips. As the tool wears, α and γ increase. Swarf tends to jam in the gap, the work-piece surface finish roughens and the tendency to failure increases*

Also, in fully investigated and characterised machining systems, such information could assist in the systematic design of advanced e.g. coated cutting tools and the development of QA/QC standards.

References

1.   E. M. Trent, 'Metal Cutting', Butler & Tanner Ltd., Great Britain, Second Edition, 1989, p. 190.

2. S. Kalpakjian, 'Manufacturing Processes for Engineering Materials', Addison-Wesley, Second Edition, USA, 1991, p. 515.
3. G. T. Smith, 'Advanced Machining', EFS Springer-Verlag, Exeter, U.K. 1989, p. 216.
4. I. Kazimir and J. Audy, 'Pocitacove Vyhodnocovanie Technologickych Javov', Trend-Vuma, Alpha Technicka a Ekonomicka Literatura, Nove Mesto nad Vahom, 1/1990.
5. K. J. Stout and L. Blunt, Nanometers to Microns Three- Dimensional Surface Measurement in BioEngineering, 2nd Australian International Conference on Surface Engineering, Volume 1, Adelaide, 1994, p. 1.
6. J. Fombarlet, Improvement in the Machinability of Engineering Steels Through Modification of Oxide Inclusions, American- Society for Metals, Metals Park, Ohio, 44073, 1983, p. 369.
7. T. Araki, M. Osawa and S. Yamamoto, 'Review on the studies for improving machinability of steel in Japan', American Society for Metals, Metals Park, Ohio, 44073, 1983, p. 7.
8. J. N. Duta, G. S. Patki, 'Influence of Banded structure on machinablity with particular reference to case hardening steels', American Society for Metals, Metals Park, Ohio, 44073, 1983, p. 427.
9. K. J. Stout et. al., 'The Development of Methods for the Characterisation of Roughness in Three Dimensions', University of Birmingham, Edgbaston, Great Britain 1993, p. 11, 31, 55 and 257.
10. Vicker's Hardness Testing, Australian Specification No.B 106–54.
11. ISO Standard-3685, Tool Life Testing, 1970.
12. Turning Tools, Metal- working Products, 931 94, Sandvik Coromant, 1993.
13. M. A. McPhee, C. Subramanian, K. N. Strafford, T. P. Wilks and L. R. Ward, "Evaluation of Coated Twist- Drills Using Frequency Domain Analysis", 2nd Australian International Conference on Surface Engineering, Adelaide, March 1994.
14. K. N. Strafford and J. Audy, 'Indirect Monitoring of Carbide Tool Wear in the Machining of LowCarbon Steels Through Measurement of Cutting Process Parameters', 2nd Australian International Conference on Surface Engineering, Adelaide, Australia, March 1994.
15. K. N. Strafford and J. Audy, 'Indirect Monitoring of Machinability in Carbon Steels by Measurement of Cutting Forces', International Conference on Mechanics of Solids and Materials Engineering, Singapore, June 1995.

Appendix

Symbols used in Figures 2 (a)–(b) and 9.

v	Facing speed (m/min)
v_f or f	Feed rate (mm/rev.)
a	Depth of cut (mm)
γ	Cutting rake angle
λ	Back rake angle
α	Clearance angle
χ	Approach angle

ε	Nose angle
r	Nose radius
ΦD	Maximum diameter of the work-material to be faced
Φd	Minimum diameter of the work-material to be faced
Pf	Assumed work plane

3.4.3

Potential of Physical Vapour Deposited Coatings of a Cermet For Interrupted Cutting

Giampaolo E. D'Errico, Emanuele Guglielmi

ISTITUTO LAVORAZIONE METALLI CONSIGLIO NATIONALE RICERCHE. VIA FREJUS 127 10043 ORBASSANO-TO. ITALY

1 INTRODUCTION

Cermet cutting tools can conveniently be used for machining work materials such as carbon steels, alloy steels, austenitic steels and grey cast iron[1-9] and their resistance to wear mechanisms is expected to be improved by use of appropriate hard coatings. But a scarce literature is currently available such that the potential of coated cermet tools is not still assessed. In fact, positive results are obtained with application to limited cutting conditions, but negative results can also be found when applications to interrupted cutting processes, like milling and interrupted turning are investigated[10,11,17,19].

The present paper deals with the influence of some Physical Vapour Deposition (PVD) coatings on the performance of a cermet tool when milling blocks of normalised carbon steel AISI-SAE 1045.

Some TiN, and Ti_2N monolayers, and TiCN+TiCN, TiCN+Ti_2N, and TiN+TiCN multilayers were deposited on a commercial cermet insert by a cathode arc deposition technique[12-16].

Dry face milling tests are performed on a vertical CNC machine tool. The cutting performance of the uncoated and coated inserts are presented and compared in terms of tool life obtained until reaching a threshold on mean flank wear, and also in terms of workpiece surface roughness. In order to try to assess the potential of these coatings, relevant results are discussed from the point of view of physical-mechanical and adhesion characteristics of the deposited layers. The paper is organised as follows. The next Section 2 describes the cermet substrate, the coatings and the cutting parameters used in the experimental work. Section 3 is devoted to the presentation and discussion of the results, mainly focusing on a comparative analysis of the cutting performance obtained in terms of tool life, resistance to wear mechanisms, effects on workpiece surface. The influence of coating adhesion characteristics is also dealt with in Section 3. The last Section 4 contains a brief summary of the experimental work performed with a list of detailed conclusions and some guidelines for future work.

2 EXPERIMENTAL CONDITIONS

2.1 Substrate

The current generation of cermet cutting tools is based on titanium carbonitride (TiCN) with

Table 1 *Main Characteristics of Ti-Based Coatings (Nominal Values)*

Characteristics	TiN	TiCN	Ti$_2$N
Optimal thickness (μm)	1/20	1/8	1/5
Hardness (HV 0.01)	2200/2400	3000/4000	2400/2700
Critical load on high speed steel (N)	60/80	50/70	50/70
Friction coefficient against AISI-SAE 52100	0.67	0.57	-
Oxidation resistance, 1 hour in air ($^\circ$C)	450/500	450/500	450/500

addition of vanadium carbide (VC), cobalt (Co), nickel (Ni), tantalum carbide (TaC), niobium carbide (NbC), tungsten carbide (WC) and molybdenum carbides (Mo$_x$C). TiCN controls resistance to diffusion and adhesive wear; VC controls fatigue strength; Co and Ni contribute to the resistance to plastic deformation; TaC and NbC increase resistance to thermal shock; WC is a binder element between TaC/NbC and TiC; Mo$_x$C increases toughness.

The cermet used for substrate in the present work was a commercial square insert with a chamfered sharp cutting edge preparation (SPKN 1203 ED-TR) for milling applications (ISO grade P25–40. M40)[18]. This cermet has the following percentage volume composition: 52.04 TiCN, 9.23 Co, 5.11 Ni, 9.41 TaC, 18.40 WC, and 5.80 Mo$_2$C, and a hardness of 91 HRA[2,3].

2.2 Coatings

The coatings used in this study are obtained by a cathodic arc PVD process with large area sources (150 mm $\times$ 800 mm), and arc sources with a special magnetic-field control system (based on magnetic arc confinement).

An important feature of this PVD process is the low internal stress level of the coatings that allows without risk of peeling off or cracking the deposition of thick films of TiN or TiCN. Some main physical-mechanical characteristics of such Ti-based coatings are provided in Table 1, according to industrial data relevant to this technology.

Using the above coating technology, two monolayers (thickness ~3.5 mm): TiN, and Ti$_2$N and three binary coatings (thickness 6/7 mm): TiCN+TiCN, TiCN+Ti$_2$N, and TiN+TiCN are deposited on a cermet insert to be tested in milling experiments. The coatings were characterised in terms of adhesion to the substrate by means of scratch tests (Figure 1).

2.3 Cutting Parameters

Dry face milling operations were performed using a vertical CNC milling machine (28 kW) in the following conditions: cutting speed v_c = 180 m/min, feed f_z =0.20 mm/tooth, axial depth of cut a_a =2 mm and radial depth of cut a_r =100 mm. The geometry of the milling cutter (for six inserts) is: cutter diameter ϕ =130 mm; corner angle κ_r = 75°, orthogonal rake angle γ_o = 2°, axial rake angle γ_p =7°, radial rake angle γ_f =0°. Blocks of normalised carbon steel AISI-SAE 1045 (HB 190$\pm$5) were used for workpieces (100 mm $\times$ 250 mm $\times$ 400 mm). In such conditions, the length of a pass was L = 400 mm, and the time per pass T = 45.6 s.

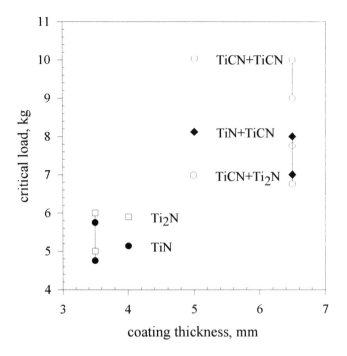

Figure 1 *Results of scratch tests*

These conditions are the same, but the cutting speed, used in a previous related work[17] where it was found that a relatively high cutting speed (v_c was 250 m/min) might be a cause of scarce performance obtained by PVD coatings. Consequently in the present work a cutting speed value v_c =180 m/min was used, even if this value is lower than the range of cutting speed that is usually recommended for this cermet insert.

3 RESULTS AND DISCUSSION

3.1 Tool Life Comparison

Maximum flank wear VB_B measured every 9.12 minutes (i.e. every 12 cuts) during the milling operations is plotted in Figure 2.

Means of six VB_B values (with standard deviations) measured after 63.84 minutes (i.e. 84 cuts) on each insert type are reported in Table 2 along with tool lives calculated from intersections of the straight line $VB_B = 0.20$ mm with each curve plotted in Figure 2. Percentage variations of tool life obtained by use of coatings are also shown in Table 2, with reference to the uncoated insert.

The experimental results point out that the influence of coatings was generally negative in terms of flank wear VB_B, with the exception of the Ti_2N coating which gives a tool Ue increment of 9.4 percent with respect to the substrate.

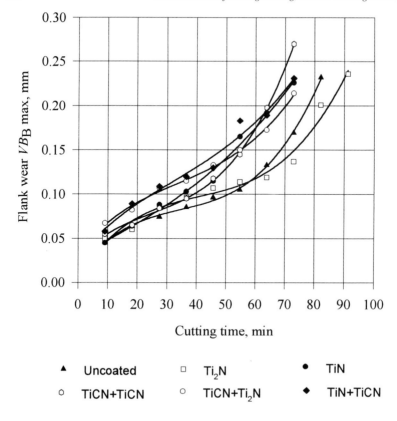

Figure 2 *Wear vs. time*

3.2 Resistance to Wear Mechanisms

Figure 3 contains two photographs (low magnification) taken after 72.96 minutes (i.e. 96 cuts) on a TiCN+TiCN coated insert with a flank wear VB_Bmax = 0.35 mm (Figure 3a), and

Table 2 *A synopsis of experimental results*

Coating	VB_B max: Mean	(Std.Dev) (mm)	Tool life (min)	% variation
Uncoated	0.133	(0.019)	77.7	base line (100)
TiN	0.119	(0.018)	66.7	85.8
Ti$_2$N	0.192	(0.021)	85.0	109.4
TiCN+TiCN	0.197	(0.029)	64.2	82.6
TiCN+Ti$_2$N	0.172	(0.015)	70.0	90.1
TiN+TiCN	0.189	(0.027)	65.2	83.9

Note: VB_Bmax measured at time t = 63.84 minutes; tool life calculated at VB_B=0.20 min.

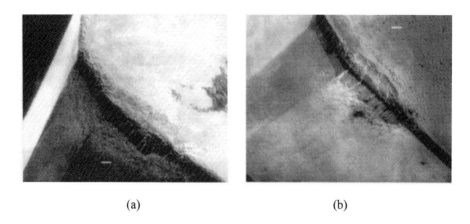

(a)   (b)

Figure 3 *Wear morphology of a TiCN+TiCN coated insert (a) and a substrate (b)*

on a substrate which shows a flank wear VB_Bmax = 0.23 mm and   also a crater wear (Figure 3b).

It is worth noting from Figure 3a that a coated insert may be free from crater wear that instead affects a cermet insert during a milling operation, as shown in Figure 3b.

The surfaces of the same inserts shown in Figure 3 are illustrated by profiles in Figure 4, where craters are longitudinally inspected by means of a Form Talysurf surface analyser (Rank Taylor Hobson). Maximum depth of craters are respectively K_T = 0.015 mm (TiCN+TiCN: Figure 4a) and K_T = 0.036 mm (substrate: Figure 4b). Close to the crater border, Figure 4b shows evidence also of the existence of a built-up-edge (a temperature related wear phenomenon) produced by pressure-welding of workpiece material, due to shearing of chip. The built-up-edge is comparatively very small in Figure 4a (coated insert).

This means that these PVD coatings are useful to reduce crater and built-up-edge both due to shearing of chip on the insert's rake face. On the other side, the damage on the insert flank (flank wear) caused by the abrasive action of the material being machined was raised by the use of such coatings. Since the damage due to flank wear was predominant in the experimental conditions, the coating efficacy was not well exploited, such that the uncoated insert generally performed better than coated inserts. The Ti_2N coating was an exception that perhaps may be related to its resistance to diffusion wear, which is a very important wear mechanism acting at high cutting temperatures generated during the cutting process (results of static diffusion tests reported in an independent work[10] show that nickel and cobalt diffuse from cermet into steel). It should also be stressed that milling is an interrupted cutting process and therefore the insert is subject to intermittent thermal shock even in dry operations.

3.3 Effects on Workpiece Surface

Form the point of view of the roughness, the surface finish obtained by coated inserts was always better than the finish obtained by the substrate. Roughness is measured in terms of R_a

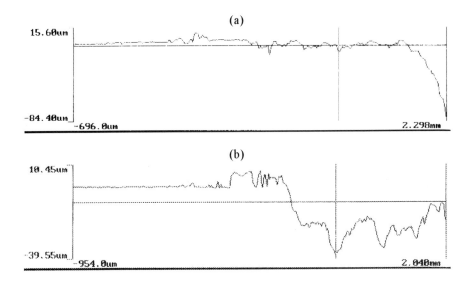

Figure 4 *Profiles of a TiCN+TiCN coated insert (a) and a substrate (b)*

on the workpiece surface by means of a portable surface roughness instrument: it should be noted that such measurements are affected by thermal state and height of the block being machined. This is shown in Figure 5, where roughness R_a is plotted vs. cutting time for each type of insert.

The better results obtained in term of roughness by use of coated inserts is in accordance with the observation that coatings reduce formation of built-up-edge, which negatively affects the texture of workpiece surface.

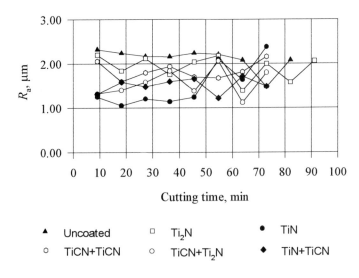

Figure 5 *Roughness vs. cutting time*

3.4 Influence of Coating Adhesion Characteristics

From the point of view of the adhesion characteristics, no direct correlation can be found between the cutting performance and the results of the laboratory scratch tests, since the Ti_2N coating, which is the best performing in terms of tool life, exhibits the second lower critical load in Figure 1.

4 FINAL REMARKS

4.1 Overview of Experimental Work

The influence of some Physical Vapour Deposited coatings (TiN, Ti_2N, TiCN+TiCN, TiCN+Ti_2N, and TiN+TiCN single- or multiple-layered) on the performance of a commercial cermet tool was investigated in milling carbon steel AISI-SAE 1045. The coatings were characterised in terms of adhesion to the substrate by means of scratch tests. Dry face milling tests were performed on a vertical CNC machine. The cutting performance of the uncoated and coated inserts was presented and discussed in terms of tool life and workpiece surface's roughness. Results of experiments performed point out that the influence of coatings may be somehow negative on the wear behaviour.

4.2 Conclusions

In the light of the experimental results presented and discussed the following conclusions can be drawn.

1. Generally, PVD coatings enhance the substrate resistance to crater wear, and improve the workpiece surface finish.
2. Among other PVD coatings, Ti_2N provides a better barrier to diffusive wear.
3. The potential of PVD coatings in terms of resistance to abrasive wear is not fully exploited in interrupted cutting: this applies both to single and multiple layered structures.

4.3 Guidelines For Future Work

While the above points apply to the experimented milling conditions, it is worth noting that they are also in accordance with independent, previous, and related works focused on interrupted cutting processes. Further investigation focused on substrate/coating interface and interlayer behaviour is expected to give a deeper insight on the reasons why some PVD coatings result in a decrement of the substrate cutting performance.

References

1. I. Amato, N. Cantoro, R. Chiara and A. Ferrari, in P. Vincenzini (ed.), *Proceedings of 8th CIMTEC-World Ceramics Congress and Forum on New Materials*, Firenze (Italy), 28 June–4 July, 1994, **3D**, p. 2319.

2.   S. Buglios, R. Chiara, R. Calzavarini, G. E. D'Errico and E. Guglielmi, in L. A.
      Dobrzanski (ed.), *Proceedings 14th International Conference on Advanced Materi-
      als and Technologies AMT'95*, Zakopane (Poland), May 17–21, 1995, p. 67.
3.   G. E. D'Errico, S. Bugliosi and E. Guglielmi in M. S. J. Hashmi (ed.), *Proceedings
      14 th International Conference on Advances in Metarials and Processing
      Technolgies AMPT'95*, Dublin (Ireland), 8–12 August, 1995, **3**, p. 1278.
4.   J. Destefag *Tooling and production*, 1994, **59/10**, 59.
5.   H. Dot in E. A. Almond, C. A. Brookes and R- Warren (eds.), *Proceedings 1st
      International Conference on the Science of Hard Materials*, Rhodes, 23–28 Septem-
      ber, 1984, p. 489.
6.   R. Porat and A. Ber, *CIRP Annals*, 1994, **3911**, 71.
7.   H. Thoors, H. Chadrasekaran and P. Olund, *Wear*, 1993, **162-164**, 1.
8.   H. K. TbnshofiF, H. G. Wobker and C. Cassel, *CIRP Annals*, 1994, **43/1**, 89.
9.   C. Wick, Manufacturing Engineering, December, 1987, p. 35.
10.   W. Konig and R. Fritsch, *Surface and Coatings Technology*, 1994, **68–69**, 747.
11.   S. Novak, M. S. Sokovic, B. Navinsek, M. Koniac and B. Pracek, in M. S. J. Hashmi
      (ed.), *Proceedings International Conference on Advances in Materials and Process-
      ing Technologies AMPT'95*. Dublin (Ireland), 8–12 August, 1995, **3**, 1414.
12.   H. Randawa , P. Johnson, *Surface and Coatings Technology*, **31**, 1987, 303.
13.   P Helhnan, D Franchi, and H Curtins, *Proceedings 9th International Congress on
      Heat Treatment and Surface Engineering*, Nice (1994).
14.   H. Curtins and W. Blbsch, *Proceedings International Conference on Metallurgical
      Coatings and Thin Films ICMCTF '95*, San Diego-CA, USA, 1995.
15.   R. F. Bunshah et al., *Deposition technologies for films and coatings*, Noyes Publica-
      tions, 1982.
16.   B. Rother and J. Vetter, 'Plasma coatings methods and hard coatings', Deutscher
      Verlag für Grundstoffindustrie, Leipzig, 1992.
17.   G. E. D'Errico, R. Chiara, E. Guglielmi and F. Rabezzana, to be presented to *Inter-
      national Conference on Metallurgical Coatings and Thin Films ICMCTF '96*, San
      Diego-CA, USA, April 22-26, 1996.
18.   ISO standard No. 513-1975 (E), "Application of carbides for machining by chip
      removal – Designation of the main groups of chip removal and groups of applica-
      tion".
19.   G. E. D'Errico, R Calzavarini and B Vicenzi, to be presented to *4th International
      Conference on Advances in Surface Engineering*, Newcastle upon Tyne, UK, May
      14–17, 1996.

3.4.4
Sliding Wear Behaviour of TiN-Based PVD Coatings on Tool Steels

C. Martini, E. Lanzoni and G. Palombarini

INSTITUTE OF METALLURGY – UNIVERSITY OF BOLOGNA, VIALE RISORGIMENTO, BOLOGNA, ITALY

1 INTRODUCTION

PVD TiN coatings are now widely used to increase the wear resistance of cutting and forming tools. The demand for improved performances of tools and engineering components, as well as the availability of new deposition techniques, has led to the development of second generation TiN-based ternary PVD coatings: Ti(C,N), (Ti,Al)N, (Ti,Nb)N and many others. Moreover, a considerable interest is addressed to new advanced materials, such as metastable Ti_2N, gradient or multilayer coatings[1] and triboactive coatings designed to promote the in-service formation of protective surface layers. Process improvements provided advantages like a better control of the microstructure, increase in the coating/substrate adhesion and decrease in the deposition temperature. Cathodic arc deposition technology is particularly attractive because of its great versatility and flexibility. The development of the arc deposition processes has been hindered so far by; (i) a lack of reproducibility mainly due to unsatisfactory control of arc source and film properties and (ii) high costs. Now, adherent multicomponent or multilayer coatings are successfully produced by arc evaporation processes[2].

The need for an improved knowledge on the in-service wear and failure mechanisms of hard coatings is widely recognized. The aim of the present work was to study the tribological behaviour of TiN-based ternary coatings deposited on tool steel with two different PVD techniques: Electron Beam Hollow Cathode Discharge (EB-HCD) and Arc Evaporation (AE).

The study was carried out by unidirectional dry sliding tests using a slider-on-cylinder contact geometry which, if compared to the more widespread pin-on-disc configuration, has some advantages such as a wider range of materials selection, ease and low cost of both test conduction and specimen preparation.

Most sliding tests on TiN-based coatings are carried out selecting steel as counterfacing material[3,4]. This choice is generally justified on the basis of a presumably closer simulation of the more common working conditions of a coated tool in machining. In this work, a hard ceramic material was chosen as counterface in order to; (i) concentrate wear on the material under examination, and (ii) avoid the presence of iron oxides in the 'third body' forming between the first bodies in relative motion.

2 EXPERIMENTAL

Bars (5x5 mm cross-section) of AISI M35 tool steel (nominal composition, wt.%, C 0.80, Cr 4.00, V 2.00, W 6.00, Mo 5.00, Co 5.00) were coated with TiN-based films using EB–HCD and AE processes carried out by industrial facilities.

In the EB–HCD process Ti was evaporated by an electron beam coming from a hollow cathode gun. The specimens were heated at the deposition temperature (~500°C), ion-etched for 15 min in Ar⁺ at 1 Pa, coated with Ti for 10 s and then coated with TiN at 200 V accelerating voltage and 670A electron beam discharge current intensity (standard conditions) by introducing a controlled flow of N_2 into the deposition chamber. Ti(C,N) deposition was carried out introducing a flow of C_2H_2 and N_2 after 150s of TiN deposition. Thus, layered coatings containing Ti-TiN and Ti-TiN-Ti(C,N) films respectively, were produced. Some TiN and Ti(C,N) coatings were also produced by higher discharge current (710A).

In the AE process Ti was evaporated by a spark-ignited arc that was guided on the Ti source by magnetic fields. The highly ionized metal vapour thus obtained was used to ion-etch steel specimens for 5 minutes at 600V (standard conditions) or 800V bias ('energized' conditions). Then the bias was gradually lowered (in ~1 minute) to 80V (standard) or 150V in order to allow deposition of a thin Ti film, and N_2 was introduced in the chamber to start TiN deposition. Some steel samples were coated with TiN at 300°C such as to evaluate the possibility of lowering to a significant extent the deposition temperature. Again, Ti(C,N) was produced by introducing C_2H_2 and N_2 after the deposition of a TiN film, the total Ti(C,N) thickness being about twice the TiN sublayer one. In the final phase of the Ti_2N coating deposition, the N_2 flow was gradually increased to the TiN deposition value, thus producing a thin TiN outer layer. The main differences between EB-HCD and AE deposition processes are summarized in Table 1.

The as-deposited coatings were characterized by means of scanning electron microscopy (SEM), X-ray diffraction analysis (XRD), electron probe microanalysis (EPMA), X-ray photoelectron spectroscopy (XPS), microhardness (MHD) and surface roughness measurements. Some results are reported in Table 2 and Table 3 for AE and EB–HCD deposited coatings, respectively. Thickness values were measured on samples fractured by V-notching and bending the steel substrate. XRD analyses were carried out by a computer controlled goniometer and using the Cu-K_α radiation. Vickers MHD was measured on the external surfaces with 1N load and 15s dwell time. Roughness R_a (centre line average) and surface profiles were measured using a stylus profilometer with 0.5μm tip radius. Roughness and MHD values are the average of at least three measurements.

The tribological behaviour of coated tool steel samples was studied under unidirectional dry sliding conditions, using a slider-on-cylinder tribometer schematized in Figure 1. The flat slider was the coated sample, while the counterfacing cylinder was an AISI 1040 steel bar 40

Table 1 *Deposition process parameters*

Process	Deposition temperature	Deposition pressure (mbar x 10³)	Deposition rate (μm/min)	Ionization degree (%)	Etching
AE	450	5	0.05	80	By Ti^{n+}
EB–HCD	500	8	0.1	5–10	By Ar⁺

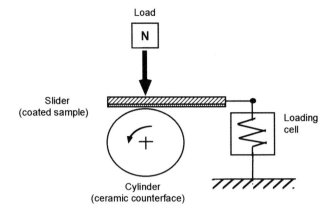

Figure 1 *Schematic of the slider-on-cylinder tribometer*

mm in diameter, coated by plasma spray with an 87 vol% Al_2O_3–13 vol% TiO_2 ceramic coating (Rockwell D hardness 60 HRD, roughness $0.5\mu m$ R_a). The tests were carried out in air with 33N applied load and 0.3m/s relative speed, for sliding distances up to 25km. A load of 33N was chosen after preliminary tests, in order to induce friction transitions in every tested coating. Both friction resistance and total wear were continuously recorded. At least two repetitions of each test were carried out.

Profiles and maximum depth of wear tracks were measured by surface profilometry on both sliders and cylinder. The morphology of wear tracks and the features of both third-body formed between counterfacing materials and wear debris were studied by means of SEM, EPMA and XRD.

3 RESULTS AND DISCUSSION

3.1 Characterization of As-Deposited Films

As shown by EPMA and XPS analyses, the composition of TiN films is nearly stoichiometric irrespective of the deposition process. In Ti(C,N) films, about 40% of N atoms are substituted by C, while in (Ti,Al)N films about 9% of Ti atoms is substituted by Al. The ceramic part of all coatings consists of single NaCl-type cubic phases, except for Ti_2N which is tetragonal. TiN, Ti(C,N) and (Ti,Al)N films display (111) preferred crystallographic orientations irrespective of deposition process (Figure 2).

As shown by SEM observations on fracture surfaces almost normal to the external surface, all EB-HCD and AE deposited films display a dense fibrous microstructure, with a morphology typical of the zone-T region in the Messier model[5] (Figure 3). This result is the consequence of processing conditions such as pressure and ion bombardment energy, which are relatively high, and temperature which, instead, is relatively low if compared with the melting temperature of the ceramic material being deposited. A microstructural refining of EB–HCD deposited films was obtained by increasing the discharge current intensity, along with a reduction in film thickness caused by the corresponding increase in the sputtering rate of the previously deposited material[6].

Table 2 *Features of the AE deposited coatings*

	Coating	Deposition temperature (°C)	Bias (V)	Thickness (μm)	Microhardness (HV$_{0.1}$)	Surface roughness R$_a$ (μm)
A	TiN	300	600–80	2.0	1723	0.27
B	TiN	450	800–150	3.0	1745	0.23
C	(Ti,Al)N	450	600–80	3.5	1765	0.31
D	TiN	450	600–80	1.5	1682	0.24
E	Ti(C,N)	450	600–80	3.0	2050	0.26
F	Ti$_2$N	450	600–80	1.5	1465	0.22

Ti(C,N) films display hardness values which are considerably higher than those displayed by TiN films, irrespective of deposition process. It is to be noted that micro-hardness values reported in Table 2 and Table 3 are composite values, being influenced by substrate compliance. The real values, which are considerably higher than the experimental ones, might well be calculated by different models[7]. In the present work, however, the differences among the experimental values measured for substrates coated with films of different composition were large enough to be effectively used in the evaluation of tribological performances. Moreover, taking into account film thickness, it may be deduced that, in the case of films deposited under standard conditions, microhardness is higher for AE coatings (both Ti-TiN and Ti-TiN-Ti(C,N)) than for EB–HCD coatings of the same composition.

3.2 Tribological Behaviour

Friction transitions leading the dynamic friction coefficient to increase from 0.10–0.20 to 0.75–0.85 were generally observed after sliding distances depending on film composition and properties. The transition starts when the wear of the TiN-based coating causes the underlying steel to come in contact with the counterfacing ceramic material; phenomena of steel transfer from slider to cylinder, as well as of iron oxidation, occur to an increasing extent and, in a relatively short time, the friction coefficient raises to values typical of steel on steel contacts. As shown by the line profiles reported in Figure 4, the ceramic coating on the counterfacing cylinder undergoes no damage before the friction transition but, subsequently, it undergoes damage in the form of grooves produced at first by hard particles of carbides emerging from the steel surface, and then also by hard fragments detached from the counterfacing ceramic. These results are supported by EPMA and XRD analyses carried out on wear debris.

Table 3 *Features of the EB-HCD deposited coatings*

	Coating	Deposition temperature (°C)	Discharge current intensity (A)	Thickness (μm)	Microhardness (HV$_{0.1}$)	Surface roughness R$_a$ (μm)
G	Ti(C,N)	500	670	6.5	2054	0.30
H	Ti(C,N)	500	710	4.5	3100	0.37
I	TiN	500	670	5.5	1714	0.31
L	TiN	500	710	4.0	1854	0.49

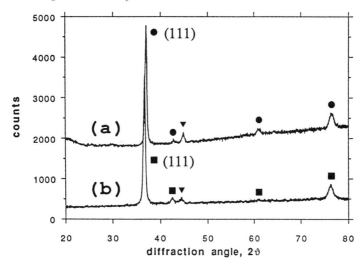

Figure 2 *XRD patterns for PVD coatings: (a) EB-HCD TiN, and (b) AE (Ti,Al)N. (●) TiN, (■) (Ti,Al)N, (◆) substrate Cu-K$_\alpha$ radiation*

3.2.1 AE deposited coatings. Figure 5 shows typical friction curves recorded for TiN-based coatings deposited by AE process. Ti$_2$N films were the only ones to pass over 10 km of sliding distance (although by a few hundreds of meters) without undergoing transition. On the other hand their curves, as compared to those of other films, were characterized by a remarkable increase in the friction coefficient before the transition. This behaviour can be explained considering that the friction resistance is determined at first by the outer part of the film, constituted by TiN, and then by the inner layer of Ti$_2$N up to transition.

With regard to the TiN-based films, it can be seen that the wear resistance of TiN coatings deposited in standard conditions is higher than those of the other TiN coatings: about 4 km of

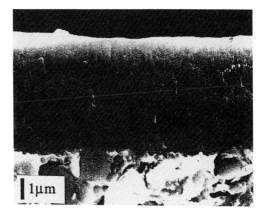

Figure 3 *SEM fractograph of an EB-HCD TiN coating ~ 5.5μm thick*

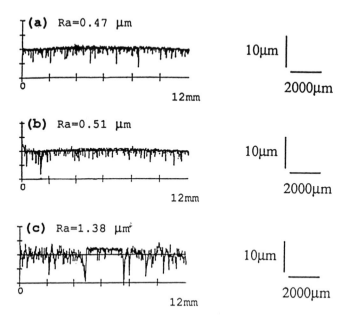

Figure 4 *Line profiles recorded on the counterfacing cylinder, (a) before testing, and (b) after tests interrupted (b) before friction transition and (c) after transition*

sliding distance before transition (curve D) vs. 1.5 km (curve A) and vs. 3 km (curve B). The worse behaviour of coating A can be explained by a loss of resistance caused by an excessive lowering of the deposition temperature (from 450°C standard to 300°C, Table 2). For the coating B, unexpectedly no improvement in the tribological behaviour was produced by the use of a higher deposition accelerating voltage and the consequent microstructural changes. The increase in current intensity might have caused overheating of the small-size specimens and softening of the substrates.

The longest distance before transition among the TiN-based AE coatings was displayed by Ti(C,N) (curve E in Figure 5), mainly as a consequence of its considerably higher hardness.

The introduction of Al in AE films did not lead to any significant improvement in wear resistance (curve C in Figure 5). The sliding speed adopted in these tests (0.3 m/s) was probably too low to allow the protective effects of Al against film oxidation to occur[8]. Further experiments are in progress at values of sliding speed as high as 1.2 m/s.

3.2.2 EB–HCD Coatings. Figure 6 shows friction curves recorded for TiN and Ti(C,N) coatings deposited by EB-HCD process. The longest distances before transition (18–20 km) were displayed by TiN coatings (curves I and L vs. curves G and H), in spite of their considerably lower hardness as compared with Ti(C,N) coatings (Table 3).

Moreover, as shown in Figure 6, the wear resistance of EB-HCD coatings was considerably

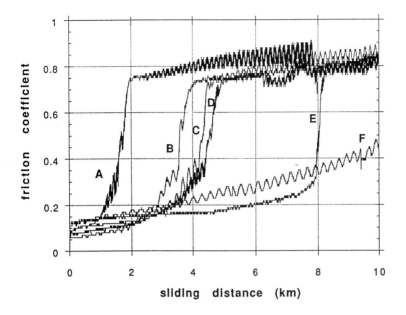

Figure 5 *Friction curves recorded for AE coatings at 33N and 0.3 m/s (see Table 2 for key letters)*

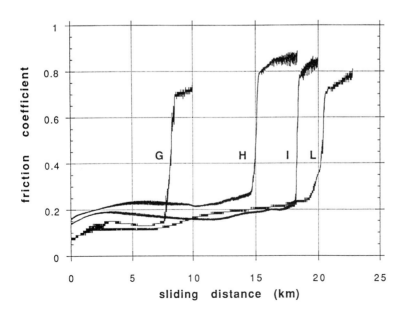

Figure 6 *Friction curves recorded for EB-HCD coatings at 33N and 0.3 m/s (see Table 3 for key letters)*

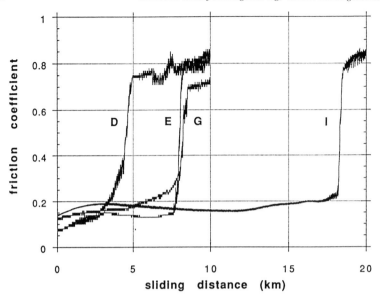

Figure 7 *Comparison among friction curves recorded at 33N and 0.3 m/s for TiN and Ti(C,N) coatings deposited by AE and EB-HCD under standard conditions (see Table 2 and Table 3 for key letters)*

higher than that displayed by all AE films. This can be explained taking into account the results of analyses carried out on the film-substrate interfacial regions and reported elsewhere[9,10]. In fact, it has been shown that considerably higher amounts of Fe-Ti interaction products formed at the EB-HCD TiN-substrate interfaces. In particular, the formation of Fe-Ti solid solutions is to be considered determining for attaining a satisfactory adhesion at the interface[9]. Much lower amounts of these solid solutions were found at the interface of AE deposited films of TiN and Ti(C,N)[10]. In conclusion, among the tested coatings the best combination between microstructural properties and interface adhesion was obtained for EB–HCD TiN.

The tribological behaviour of Ti(C,N) coatings (which, as reported above, actually are Ti-TiN-Ti(C,N) coatings) needs further discussion. As shown in Figure 7, in fact, the resistance to sliding wear of EB-HCD Ti(C,N) is considerably lower than that of EB-HCD TiN (curve G vs. curve I), notwithstanding (i) the higher hardness (Table 3), and (ii) the similarity between the interfacial regions with the substrate, produced by the same process of deposition. In regard to this point, it is to be considered that Ti(C,N) layers tend to undergo fracture at the interfacial regions with the underlying TiN. In this way, detachment of hard fragments of Ti(C,N) occur during sliding, giving rise to severe abrasive wear and early transition in the friction behaviour. The increase in wear resistance observed for EB–HCD Ti(C,N) deposited at higher discharge current intensity (curve H vs. curve G in Figure 6) can be ascribed to the improved properties of the coating[6], and in particular to a considerably higher hardness (Table 3).

On the other hand, the behaviour of AE Ti (C,N) is better than that of AE TiN (curve E vs. curve D in Figure 7) mainly because of a considerably higher hardness. In this case both a lack of adhesion at the interface and formation of hard fragments of Ti(C,N) during sliding lead to an early transition in the friction behaviour.

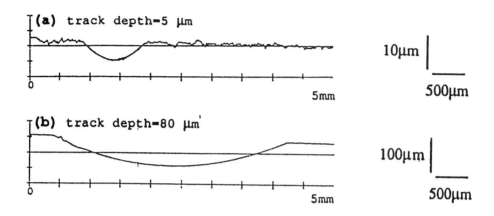

Figure 8 *Line profiles measured for the wear tracks produced after 10 km sliding distance on sliders coated with (a) EB-HCD TiN (coating I in Table 3), and (b) AE TiN (coating D in Table 2)*

In any case, the sliding distance at the friction transition characterizes the wear resistance of the coating. As shown in Figure 8, in fact, the after-transition wear rate of the unprotected slider strongly increases up to values which are typical of steel-ceramic contacts.

4 CONCLUSIONS

The microstructural analyses and tribological tests carried out on TiN-based coatings deposited on tool steel by two different PVD techniques (arc evaporation and electron-beam hollow cathode discharge) allow to draw the following conclusions:
- The slider-on-cylinder geometry is very useful to evaluate the wear resistance of coatings as the sliding distance before the friction transition occurring when the coating itself is worn and a direct contact establishes between substrate and counterfacing material.
- The highest resistance to dry sliding against a hard ceramic material, observed for the EB–HCD TiN coatings, is due to a satisfactory combination among microstructural characteristics and an interface adhesion adequately promoted by deposition of an intermediate film of titanium.
- The tribological behaviour of the AE-deposited films was determined by a lack in adhesion, attributable to inadequate interactions between the intermediate layers of Ti and the substrate. This clearly indicates that further efforts are required to optimize the deposition conditions.
- The introduction of carbon in theTiN lattice, carried out in order to produce Ti(C,N) films on TiN sublayers, leads to harder coatings. The increase in wear resistance, however, was limited by the tendency of the coating to undergo first fracture at the Ti(C,N)-TiN interfacial regions, then severe abrasive wear.
- No significant improvement in the tribological behaviour was obtained by the introduc-

tion of aluminium in the TiN lattice. Probably the adopted sliding speed was too low for the protective action of Al against oxidation to develop.

- The coatings constituted by substoichiometric Ti_2N displayed an interesting behaviour under dry sliding condition. However, further investigation is required to find processing conditions suitable to limit the increase in the friction coefficient with the sliding distance and to delay the friction transition.

Acknowledgements

The authors wish to thank Dr. R.Mandelli, Samputensili S.p.A., Ortona, Italy, for making available the industrial PVD facilities (ULVAC and PLATIT); Dr. P.Lavalle, Samputensili, for carrying out depositions and for help and advice. This work was carried out with financial support from MURST, Rome, Italy.

References

1. A. Matthews and A.Leyland, *Surf. Coat. Technol.*, 1995, **71**, 88.

2. H. Curtins, Proc. Conf. 22nd Int. Conf. Metallurgical Coatings Thin Films 1995, San Diego, paper G1.04.

3. K. Holmberg and A. Matthews, 'Coatings Tribology', D. Dowson Ed., Elsevier, Amsterdam, 1994 Tribology Series No. 28.

4. E.Vancoille, J. P. Celis and J. R .Roos, *Wear*, 1993, **165**, 41.

5. R. Messier, A. P. Giri and R. A. Roy, *J. Vac. Sci. Technol.*, 1984, **A2**, 500.

6. R. Bertoncello, A. Casagrande, M .Casarin, A. Glisenti, E. Lanzoni, L. Mirenghi and E. Tondello, *Surf. and Interf. Anal.*, 1992, **18**, 525.

7. A.Thomas, *Surf. Eng.*, 1987, **3**, 117.

8. M. Bromark, M. Larsson, P. Hedenquist, M. Olsson, S. Hogmark and E. Bergmann, *Surf. Eng.*, 1994, **10**, 205.

9. M. Carbucicchio and G. Palombarini, *Appl. Surf. Sci.*, 1993, **65/66**, 331.

10. M. Carbucicchio, C. Martini, G. Palombarini and M. Rateo, submitted for publication to *Phil. Mag.*

3.4.5

Performance of Physical Vapour Deposited Coatings on a Cermet Insert in Turning Operations

Giampaolo E. D'Errico[1], Roberto Calzavarini[1], Bruno Vicenzi[2]

[1]Istituto Lavorazione Metalli, Consiglio Nazionale Ricerche, via Frejus 127, I-10043 Orbassano–TO, Italy

[2]Centro Sviluppo Materiali S. P. A., Corso Perrone 24/a, I–16152 Genova–GE, Italy

1 INTRODUCTION

The current generations of cermets for cutting tools are materials composed of both ceramic and metallic phases based on TiC, TiN and Ni.[1] Such tools can conveniently machine a variety of work materials such as carbon steels, alloy steels, austenitic steels and grey cast iron.[2-10]

Cermets, like hard metals, are composed of a fairly high amount of hard phases, namely Ti(C,N) bonded by a metallic binder that in most cases contains at least one out of Co and Ni. The carbonitride phase, usually alloyed with other carbides including WC, Mo_2C, TaC, NbC and VC, is responsible for the hardness and the abrasive wear resistance of the materials. On the other hand, the metal binder represents a tough, ductile, thermally conducting phase which helps in mitigating the inherent brittleness of the ceramic fraction and supplies the liquid phase required for the sintering process

A question of recent interest is to assess if the resistance of cermet cutting tools to wear mechanisms may be improved by the use of appropriate hard coatings. Controversial conclusions are available in the relevant technical literature, since positive results are obtained in some cutting conditions but negative results may also be found, especially with application to interrupted cutting processes.[11-15]

The intended goal of the present paper is to contribute to a deeper insight on the influence of some Physical Vapour Deposition (PVD) coatings on the performance of a cermet insert for turning applications. The cutting insert was produced using a cermet grade purposely composed and prepared aiming at a high toughness. Monolayers of TiN, and Ti(C,N) thin films were deposited by means of an industrial coating process based on a novel cathodic arc technique. Machining trials were performed using steel AISI-SAE 1045 for work material.

Results of continuous and interrupted dry turning operations are presented and discussed in order to propose a comparative evaluation of the wear performance of coated and uncoated inserts. Since from a machining point of view the insert resistance to mechanical impacts (which relates to toughness) is affected by the microgeometry of the cutting edge preparation, both sharp and chamfered inserts were prepared and tested.

Table 1 *Weight Percentage Composition of the Cermet Substrate*

Component	Co	Ni	W	TiC	TiN	WC
Wt%	8	11	12	23	34	12

2 EXPERIMENTAL WORK

2.1 Cermet Substrate

The cermet grade prepared for this work was designed with a binder rich composition in order to obtain a high toughness.[16] The hard phase is based on titanium carbonitride and tungsten carbide, whilst the binder phase contains cobalt, nickel and tungsten (Table 1). A high content of titanium nitride was used (34% wt) in the composition because it is known to enhance toughness, and also WC was added for the same reason.[17]

This cermet grade was obtained by usual powder metallurgy routes based on cold isostatic pressing and gas pressure sintering.[18-20]

A typical microstructure of a sintered sample is shown in the SEM image of Figure 1.

The phases have been identified by X-ray diffraction and the results have been cross-checked with the results of EDAX during SEM sessions. Letters in Figure 1 indicate respectively, *A*: Ti(C,N) phase, rich in Ti and N; *B*: binder phase, containing Co, Ni and some W; *C*: reacted (Ti,W)(C,N) phase, rich in C; *D*: segregated (Co,W)C.

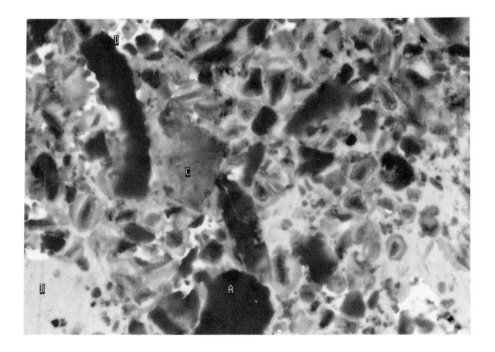

Figure 1 *Cermet microstructure × 5000*

The substrate hardness is 1350 HV_{20} and its indentation toughness is 14 MPa.[15]

The sintered product was finished by diamond tool grinding. Square inserts, ISO code SNUN 120308 F and T, for turning applications were produced in two types: without a chamfer (type designated by F in the ISO code F) and with a chamfer (designated by T) on the cutting edge. Surface roughness of the finished inserts was about 0.04 μm R_a.

2.2 Coatings

Some substrates were PVD coated using a recently developed commercial process, the PLATIT® arc-PVD process.[21-25] The coatings were made at a substrate temperature of 480°C and were either a TiN or a Ti(C,N). The soaking time at deposition temperature was 2 hours, and deposition pressures were 4×10^{-1} Pa and 5×10^{-1} Pa, respectively. After coating, the roughness R_a was 0.04 μm for TiN coatings, and 0.06 μm for Ti(C,N) coatings.

The surfaces' appearance can be seen in Figure 2a (a TiN coated sample) and in Figure 2b (a Ti(C,N) coated sample).

The orthogonal sections taken across the coatings of the surfaces relevant to the samples in Figure 2a–b are shown in Figure 3a–b respectively. The TiN coating is thinner so that it does not cover the surface lines caused by grinding. In both coatings many pores are visible. A figure for the thickness may be about 2 μm for TiN, and 3 μm for Ti(C,N).

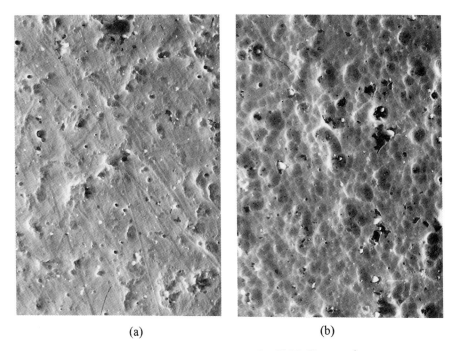

(a)   (b)

Figure 2 *Surface × 3000 of TiN (a) and Ti(C,N) (b) coated inserts*

(a)

(b)

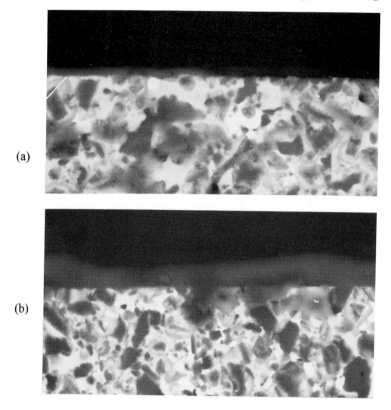

Figure 3 *Orthogonal sections × 5000 of TiN (a) and Ti(C,N) (b) coated inserts*

2.3 Machining Conditions

Longitudinal turning trials were carried out on a CNC lathe (nominal power 22 kW) for machining steel AISI-SAE 1045 using SNUN 120308 F and T inserts mounted on a tool holder Sandvik Coromant$^{\delta}$ CSRNL 3225 P12-1C. Dry continuous and interrupted operations were performed in the following conditions.

2.3.1 Continuous turning. The workpieces were cylinders of length l=450 mm, and an initial diameter ϕ =100 mm. Cutting parameters:

Table 2 *Continuous Turning, VB_B max=0.40 mm, Various v_c (m/min): Sharp Inserts.*

			Tool life (s)		
Coating	v_c=200	v_c=250	v_c=300	v_c=400	v_c=500
Substrate	840	270	60	20	4
Ti(C,N)	-	1200	300	45	9
TiN	-	-	135	-	-

Table 3 *Continuous Turning, $VB_B max=0.40$ mm, $v_c=300$ m/min: Sharp Inserts*

	Tool life (s) (4 iterations)				Mean
Substrate	60	90	55	80	71.25
Ti(C,N)	300	230	120	126	194.00
TiN	135	180	160	210	171.25

(a)

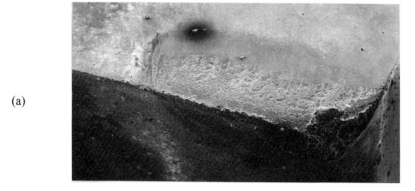

(b)

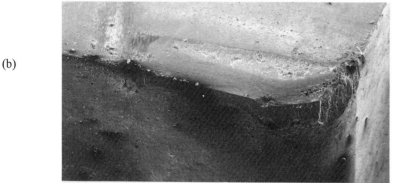

(c)

Figure 4 *Wear morphologies × 40: substrate (a), Ti(C,N) (b), and TiN (c) coated inserts*

Cutting speed: v_c=200; 250; 300; 400; 500 m/min;
Feed rate: f=0.25 mm/rev;
Depth of cut: a_p=1.5 mm.

2.3.2 Interrupted turning.[26] The workpieces were composed of 12 rectangular parallelepipedes clamped every 30° on a base, with a cylindrical symmetry (maximum external diameter ϕ = 224 mm). Cutting parameters:

Cutting speed: v_c= 100; 200 m/min;
Feed rate: f = 0.25 mm/rev;
Depth of cut a_p = 2 mm.

3 RESULTS AND DISCUSSION

3.1 Continuous Turning Tests

Sharp cutting inserts were used with application to continuous turning. Results of relevant tests are summarised in Tables 2–3, in terms of tool life observed until reaching a flank wear threshold VB_Bmax=0.40 mm.

   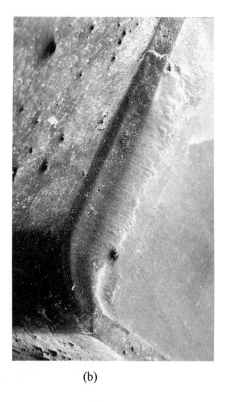

(a)   (b)

Figure 5 *Condition of a sharp (a: × 25) and a chamfered (b: × 40) insert*

Table 4 *Interrupted Turning (cutting time 120 s), v_c=250 m/min: Chamfered Inserts*

Coating	Substrate	Ti(C,N)	TiN
VB_B max, mm	0.52	0.60	0.59

Table 2 reports individual data obtained varying the cutting speed in the range v_c=200÷500 m/min.

Figure 4 shows the wear morphology of an uncoated insert (a), after 40 s of cutting time, a Ti(C,N) coated insert (b), after 270 s, and a TiN coated insert (c), after 120 s of cutting in the conditions of Table 2 (v_c=300 m/min).

Table 3 reports data relevant to 4 iterations with their mean values for v_c=300 m/min.

(a)

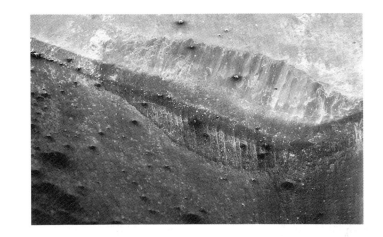

(b)

Figure 6 *Conditions (× 40) of an uncoated (a) and Ti(C,N) coated (b) insert*

3.2 Interrupted Turning Tests

In interrupted turning operations it was found that the sharp insert type was prone to fracture on the cutting edge, as shown in Figure 5a which illustrates the condition of a sharp uncoated insert after 5 seconds of cutting time (v_c=100 m/min). As a counter example, Figure 5b shows the condition of a chamfered uncoated insert after 1200 seconds at the same cutting speed (flank wear VB_B is negligible). This is the reason why chamfered inserts were tried in interrupted turning.

Results of these tests are summarised in Table 4, in terms of flank wear measured after a cutting time t=120 seconds at v_c=250 m/min.

A comparison of the inserts condition is illustrated in Figure 6a (uncoated insert) and Figure 6b (Ti(C,N) coated insert).

3.3 Use of Chamfered Insets in Continuous Turning

In order to better understand the effect of the chamfer on the behaviour of the coated inserts, some experiments were performed using also chamfered inserts in continuous turning

(a)

(b)

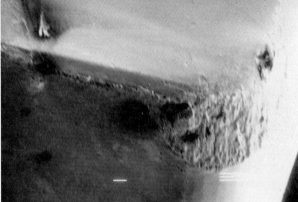

Figure 7 *Chamfered insert condition (× 40) after 30 s (a), and 60 s (b)*

tests. Results of these tests are summarised in Table 5 which reports data relevant to 4 iterations along with their mean values, for v_c =300 m/min.

Figure 7 shows the progress of the flank wear on a chamfered, Ti(C,N) coated insert, after 30 seconds (Figure 7a) and after 60 seconds (Figure 7b).

For comparison purpose, Figure 8 shows instead the progress of the flank wear on a sharp, Ti(C,N) coated insert, after 60 s (Figure 8a) and after 270 s (Figure 8b).

3.4 Cutting Performance

An interpretation of the efficacy of the coatings on the insert's wear behaviour is not straightforward.

The results of continuous turning tests show that the Ti(C,N) as well as the TiN coatings provide sensible improvements in cutting performance. Data reported in Table 2 show that in terms of tool life coated inserts perform better than uncoated inserts: this applies to all the experimented cutting conditions. A more detailed analysis of the results of four trial iterations points out that a certain scatter of data may be found (Table 3): but, in terms of individual values, coated inserts always exhibit a longer tool life with respect to the uncoated inserts,

(a)

(b)

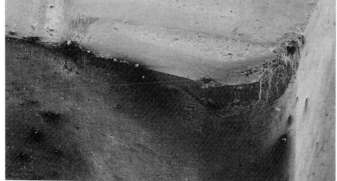

Figure 8 *Sharp insert condition (× 40) after 60 s (a), and 270 s (b)*

Table 5 *Continuous Turning, $VB_B max = 0.40$ mm, $v_c = 300$ m/min: Chamfered Inserts*

	Tool life (s–4 iterations)				Mean
Substrate	90	75	80	60	76.25
Ti(C,N)	60	90	60	50	65.00
TiN	90	75	70	55	72.50

and in terms of mean values the best performing is the Ti(C,N) coating. The above considerations apply to the use of sharp inserts. From a comparison of data in Table 5 (relevant to chamfered inserts), with data in Table 3 (relevant to sharp inserts), it can be seen that the chamfered edge preparation is at detriment of the coated insert's performance, while the substrate is almost insensitive to this variation of the cutting edge microgeometry.

On the other hand, the results of interrupted cutting tests show that the insert performance is driven by the substrate properties only. Data in Table 4 and wear morphology in Figure 6 point out that no improvement is obtained by use of such coatings in interrupted turning. These results are relevant to the chamfered insert type. Actually, as far as the influence of cutting edge preparation is concerned, from Figure 5 it can be observed that in interrupted turning, the chamfered microgeometry of the insert has proved to be crucial to the performance obtained in the experimental conditions. Even if no evidence is available, it is plausible that the effect of the coating is not fully exploited since the film may be prematurely removed due to the frequent mechanical impacts caused by intermittent cutting processes. The binder rich substrate has anyway proved beneficial for the interrupted cutting performance.

4 CONCLUSIONS

In the light of the experimental results presented and discussed, the following conclusions can be drawn.

1. Positive results were obtained in continuous turning by use of PVD coatings on sharp inserts.
2. The chamfered edge preparation is at detriment of the coated insert's performance.
3. Negative results were obtained in interrupted turning by use of such coatings on chamfered inserts.

Negative results may be due to the problems introduced by the chamfer required in the experimental conditions, as well as to the possibility that such coatings are not an efficient barrier to wear mechanisms originated during the cutting process, that become more severe in interrupted cutting. Up to now, this is an unresolved question to be investigated in future work.

Acknowledgements

The authors wish to thank Ing. P. Lavalle of SU Coatings S.p.A. (Ortona, Italy) for the PVD coating and G. Andalò of G. Andalò s.r.l. (Imola-BO, Italy) for the grinding of the inserts. Part of this work has been carried out within the programme "Piano Nazionale di Riferimento sui Materiali Innovativi Avanzati (PNR-MIA)" committed to Centro Sviluppo Materiali S.p.A. by the Italian Ministero della Università e Ricerca Scientifica e Tecnologica (MURST).

References

1. P. Ettmayer, W. Lengauer, *Powder Metallurgy International*, 1989, **21**, 37.
2. I. Amato, N. Cantoro, R. Chiara and A. Ferrari, in P. Vincenzini (ed.), Proceedings of 8th CIMTEC-World Ceramics Congress and Forum on New Materials, Firenze (Italy), 28 June- 4 July, 1994, Vol. 3, Part D, p. 2319.
3. S. Bugliosi, R. Chiara, R. Calzavarini, G.E. D'Errico and E. Guglielmi, in L.A. Dobrzanski (ed.), Proceedings 14th International Conference on Advanced Materials and Technologies AMT'95, Zakopane (Poland), May 17-21, 1995, p. 67.
4. G.E. D'Errico, S. Bugliosi and E. Guglielmi, in M.S.J. Hashmi (ed.), Proceedings 14th International Conference on Advances in Metarials and Processing Technologies AMPT'95, Dublin (Ireland), 8-12 August, 1995, Vol. III, p. 1278.
5. J. Destefani, *Tooling and Production*, 1994, **59/10**, 59.
6. H. Doi, in E.A. Almond, C. A. Brookes and R. Warren (eds.), Proceedings 1st International Conference on the Science of Hard Materials, Rhodes, 23-28 September, 1984, p. 489.
7. R. Porat and A. Ber, *CIRP Annals*, 1994, **39/1**, 71.
8. H. Thoors, H. Chadrasekaran and P. Olund, *Wear*, 1993, **162-164**, 1.
9. H.K. Tönshoff, H.-G. Wobker and C. Cassel, *CIRP Annals*, 1994, **43/1**, 89.
10. C. Wick, *Manufacturing Engineering*, December 1987, 35.
11. W. König and R. Fritsch, *Surface and Coatings Technology*, 1994, **68-69**, 747.
12. S. Novak, M.S. Sokovic, B. Navinsek, M. Komac and B. Pracek, in M.S.J. Hashmi (ed.), Proceedings International Conference on Advances in Materials and Processing Technologies AMPT'95, Dublin (Ireland), 8-12 August, 1995, Vol. III, p. 1414.
13. G.E. D'Errico, R. Chiara, E. Guglielmi and F. Rabezzana, to be presented to International Conference on Metallurgical Coatings and Thin Films ICMCTF '96, San Diego-CA, USA, April 22-26, 1996.
14. G.E. D'Errico, E. Guglielmi, to be presented to 4th International Conference on Advances in Surface Engineering, Newcastle upon Tyne, UK, May 14–17, 1996.
15. B. Vicenzi, M. Capurro, P. Cirillo, R. Calzavarini, to be presented to European Conference on Advances in Hard Materials Production Euro PM96, Stockolm, Sweden, May 27–29, 1996.
16. L. Risso, B. Vicenzi, F. Monteverde, G. De Portu, C. Melandri, in A. Bellosi (ed.), Proceedings of Fourth Euro Ceramics, Faenza Editrice, Italy, 1995, Vol. 4, p. 319.
17. P. Ettmayer, H. Kolaska, W. Lengauer, K. Dreyer, in H. Bildstein and R. Eck (eds.) Proceedings of the 13th International Plansee Seminar, Metallwerk Plansee, Reutte, 1993, Vol. 4, p. 191.
18. P. J. James (ed.), 'Isostatic Pressing Technology', Applied Science Publishers, Barking, Essex (GB), 1983.
19. P. Ettmayer, H. Kolaska, K. Dreyer, *Powder Metallurgy International*, 1991, **23**, 224.

20. W. D. Kingery, J. M. Woulbroun, F. R. Charvat, in 'Sintering Key Papers', (S. Somiya and Y. Moriyoshi eds.), Elsevier Appl. Sci., Essex & New York, 1990, p. 405.
21. H. Randawa , P. Johnson, *Surface and Coatings Technology*, 1987, **31**, 303.
22. P. Hellman, D. Franchi, and H. Curtins, Proceedings 9th International Congress on Heat Treatment and Surface Engineering, Nice (1994).
23. H. Curtins and W. Bleosch, Proceedings International Conference on Metallurgical Coatings and Thin Films ICMCTF '95, San Diego-CA, USA, 1995.
24. R.F. Bunshah *et al.*, Deposition technologies for films and coatings, Noyes Publications, 1982.
25. B. Rother and J. Vetter, 'Plasma coatings methods and hard coatings', Deutscher Verlag für Grundstoffindustrie, Leipzig, 1992.
26. Italian Standard UNI 9961, Part 1, 1992.

3.4.6

Coating of Silicon Carbide Matrix Composites for Industrial Use in Corrosive Environments

S. J. Bull[1] and A. R. McCabe[2]

[1]MATERIALS DIVISION, HERSCHEL BUILDING, THE UNIVERSITY, NEWCASTLE UPON TYNE, UK

[2]AEA TECHNOLOGY, B552, HARWELL, OXON, UK

1 INTRODUCTION

For very high temperature applications, where the strength of alloys and refractories is reduced to low levels and sintered ceramics are too brittle, the only materials which can be used are Ceramic Matrix Composites (CMCs). Such materials are already used in a range of applications such as heat exchangers, aero-engines and casting nozzles, but these applications do not generally involve high mechanical and thermal stresses for which many composites are unsuitable. Composites based on C and SiC fibres with a SiC matrix have been developed which do have sufficient strength for such applications[1,2] and can be made in a sufficient range of shapes to be useful. However, in order to achieve their benefits it is essential to protect such CMCs from oxidation and corrosion[3]. There are clearly a large number of potential operating environments with different protection requirements so in this study we have focused on two potential applications:

1. Cast iron pipe manufacture – in this process open channels are used to carry molten metal to rotating moulds. These are often made from cast iron or steel and coated with a carbon varnish to provide a thermal barrier and prevent wetting by the molten iron. Such coatings need to be replaced or repaired frequently so there is potential for replacement with a thermal shock resistant CMC with an appropriate coating to prevent wetting or corrosion by molten iron.
2. Radiant tube furnaces – gas fired furnaces are used in the ceramics industry which generally contain a radiant tube made from reaction bonded silicon carbide that is limited to use at temperatures $\leq 1250°C$. For use in this application CMCs would have to operate above this temperature and maintain their structural integrity through thermal cycling when the gas burner is switched on and off or the furnace door is opened. They should also be stable against oxidation and corrosion in the products of gas combustion.

This paper discusses some of the factors affecting the selection of potential coating materials for these applications and the problems associated with coating a composite material. The results of laboratory testing of candidate coatings and limited testing in the industrial environment are also highlighted.

2 SCHEMES FOR PROTECTION OF SILICON CARBIDE MATRIX COMPOSITES

Silicon carbide is resistant to oxidation under isothermal conditions at temperatures up to 1650°C in air but requires substantial protection for cyclic operations and use at higher temperatures[4]. At temperatures >1200°C the formation of silica from the silicon carbide matrix with a volume expansion can act to close microcracks and seal the composite from further attack. However, at lower temperatures the rate of silica formation is too low and an alternative means of protection is necessary. C/SiC and SiC/SiC composites are particularly vulnerable in the range 800°C to 1100°C where oxidation can burn out the carbon fibres or carbon interfacial layers[3]. Because of the very high strain to failure inherent in CMCs (by virtue of their design) any scheme for protection must be able to accommodate high strains, and bridge or close matrix microcracks which form above a critical applied stress but which do not lead to component failure. There are thus two basic approaches to protection.

1. Modifying the SiC matrix to improve oxidation resistance – this can be difficult without compromising mechanical properties.
2. Coating the entire composite with an appropriate protective coating. This is more attractive as the coating can be chosen to match the requirements of different applications.

To accommodate the micro cracking which occurs during service and the damage induced by thermal cycling a number of different coating approaches are possible.

1. Self healing coatings – coatings which form reaction products at temperatures below the service temperature which can plug microcracks. An example might be a coating which reacts with oxygen to produce a glassy deposit at lower temperatures than the normal protective silica formation. Composites can be protected at intermediate temperatures (up to 1100°C) by the formation of a boric oxide glass with low viscosity by oxidation of a TiB_2 coating[5]. This is molten in the critical temperature range and can fill cracks and reduce the ingress of oxidants. At higher temperatures the boron oxide is volatile and protection is limited.
2. Coatings with high strain to failure – coatings of ceramic materials produced by plasma spraying have a structure which consists of lenticular splats of material stacked on each other in a random fashion. The bonding between splats is not perfect and there can be considerable porosity between them. These coatings often deform by intersplat slippage and have a very high strain to failure under conditions of tensile loading. There will be a path for the environment through the coating but if the coating is thick enough (>100 mm) this is tortuous and can be blocked by reaction products. These coatings have been investigated in some detail in this study.
3. Deformable coatings – another way that the coating can accommodate the strain at the tip of the microcrack is by plastic deformation. A soft metallic coating is ideal in this case but these do not have the required oxidation resistance. High temperature oxidation-resistant alloys, such as MCrAlYs(M is a metal such as Co, Fe or Ni) do offer some protection but are limited in service temperature to less than 1200°C. An added disadvantage is the large thermal expansion mismatch between the CMCs and most metallic coatings which leads to considerable thermal stresses and coating detachment during thermal cycling. This is therefore not the preferred protection route.
4. Protective multilayers – in most applications no one coating has the ideal combination of

properties and thus a multilayer coating approach is necessary to derive the maximum protection. In addition, by introducing interfaces parallel to the composite surface it is possible to deflect microcracks and reduce the chance of surface cracking leading to environmental ingress. One concept for a protective multilayer has been outlined by Strife and Sheehan[6] which consists of three layers:

(a)   An outer oxide layer for erosion and volatilisation protection.
(b)   An inner material molten at and below the operating temperature to act as a crack filler and a barrier to oxygen.
(c)   A layer to protect the composite from the melt.

Clearly this is a complex coating which would need considerable development of each layer so in this study we have concentrated on development of simpler protection schemes.

3 SELECTION OF COATING MATERIALS

The main criteria for assessing the suitability of potential protective materials are:

1.   Stability at the proposed operating temperature (taken to be 1400°C in this study)
2.   Environmental stability – for the two applications studied here this means resistance to oxidation, water vapour attack and corrosion by the products of combustion for the radiant tube furnace and resistance to oxidation and attack by molten iron for the casting application.
3.   Compatibility with the substrate. The coating should not react excessively with the SiC substrate at the operating temperature. Furthermore there should not be an excessive thermal expansion mismatch between coating and substrate to prevent coating cracking and detachment.

Hillig[7] has ranked a wide range of oxides, nitrides, carbides, borides and silicates for their suitability in high temperature composite materials. Of the oxide materials zirconia, hafnia, alumina, chromia and magnesia have the necessary oxidation resistance and stability to water vapour. MgO has a very high thermal expansion coefficient compared to the other oxides and can be excluded on this basis. Water vapour can reduce the protective properties of chromia[8] and under some conditions volatile oxides can be produced at temperature (e.g. CrO_3) which limit the coatings to service below 1200°C. Zirconia has a high permeability to oxygen at elevated temperature[9]; it is generally stabilised with oxides such as calcia, yttria or ceria which are susceptible to attack by water vapour although this effect is not major in the case of ceria[10]. Non oxide coatings rely on the formation of protective oxide scales to afford protection but can be used as oxygen diffusion barrier layers to prevent reactions with the substrate.

Thus for the radiant burner application coatings of alumina or zirconia are candidate materials based on literature data and cost and availability. However the thermal expansion mismatch with C/SiC or SiC/SiC composites is still quite large (Table 1) and this will lead to problems with coating cracking and detachment. Mullite and zirconium silicate coatings have much closer expansion coefficient and the required environmental stability and substrate compatibility. These layers would not be suitable for the casting application since iron oxide and silica show a low temperature eutectic at below the service temperature[11]. This problem can be solved by applying an alumina or zirconia capping layer, effectively grading the coating composition from the substrate to the coating surface.

Thus four different coating materials have been identified which could protect C/SiC and

Table 1 *Thermal expansion coefficients of CMCs and candidate coatings*[6,7,12] *(mean value in the range 500–1500°C)*[6,7,12]

Material	Thermal expansion coefficient, α (°C^{-1})
C/SiC	3.8×10^{-6}
SiC/SiC	4.5×10^{-6}
Alumina	8.9×10^{-6}
Zirconia	1.4×10^{-5}
Chromia	7.8×10^{-6}
Magnesia	1.4×10^{-5}
Hafnia	1.4×10^{-5}
Yttria	8.5×10^{-6}
Silica	0.5×10^{-6}
Mullite	5.1×10^{-6}
Zirconium Silicate	5.4×10^{-6}
AlN	6.5×10^{-6}
TiC	8×10^{-6}
ZrC	8×10^{-6}
HfC	8×10^{-6}
HfB$_2$	8×10^{-6}
ZrB$_2$	8.5×10^{-6}
TiB$_2$	8.5×10^{-6}
SiC	4.5×10^{-6}

SiC/SiC CMCs. The following sections describe the experimental work carried out to determine the effectiveness of protection of all materials.

4 EXPERIMENTAL

4.1 CMC Substrates

Both SiC/SiC and C/SiC substrates were manufactured by SEP by chemical vapour infiltration. The SiC/SiC composite behaviour is limited by the SiC fibre stability (loss of properties at 1250°C) and the fact that it has a 2D weave structure with low interlaminar shear strength. The C/SiC composite has a 3D weave to avoid this problem but the high strength, thermally stable carbon fibres are easily oxidised unless protected. After composite manufacture test pieces were machined from larger plates. This leaves open fibre ends which can easily be oxidised. For the C/SiC composites a thick sealing layer of CVD SiC was applied to the test pieces after machining to reduce this problem.

4.2 Coating Deposition

Coatings were deposited by atmospheric plasma spraying using commercially available powders as the source material. Prior to coating the samples were grit blasted with alumina to promote adhesion – grit blasting conditions were chosen so as to minimise gross damage to the composite. The substrates were pre-heated to 100°C prior to coating and agitated in a specially designed hopper to ensure that all surfaces were coated with at least 100μm of coating material.

4.3 Oxidation Testing

Coated samples of both C/SiC and SiC/SiC were isothermally oxidised in air for 1 hour at 1000°C in alumina crucibles to determine the effectiveness of protection under conditions where silica is known not to be protective. Selected samples were isothermally oxidised at 1000°C in a sensitive microbalance to determine the oxidation kinetics. Coated samples were also oxidised at 1400°C to determine the performance of the coatings under conditions where protective silica scales are known to form. Oxidation tests were also carried out using a simulated combustion products mixture (72% N_2, 17% H_2O, 9% CO_2, 2% O_2) in the temperature range 1100°C to 1400°C. Thermal cycling tests were carried out between 1400°C and 900°C for 500 cycles with a dwell time of five minutes at each temperature.

5 RESULTS

Table 2 shows the effect of short term isothermal exposure of CMCs coated with APS layers of 100μm thickness. In most cases the weight losses of coated CMCs are less than their uncoated counterparts showing that the coatings do afford some protection. For the C/SiC composite tested at 1000°C the best protection is offered by a zirconia coating whereas at 1400°C the mullite shows the best performance. Thermal expansion mismatch has induced tensile cracking in the alumina and zirconia coated composites on cooling (Figure 1) which is much worse for the higher exposure temperature. For the SiC/SiC composite there is little benefit to oxidation resistance at 1000°C from any coating and the benefits at 1400°C are marginal. At both temperatures mullite and zircon coatings perform better than alumina and zirconia. The poor performance of the coatings on SiC/SiC composites is due to difficulties encountered in coating composites with a weave structure (Figure 2). Complete encapsulation

Table 2 *Percentage weight loss after exposure for 1 hour in air of coated and uncoated CMCs*

	C/SiC	SiC/SiC	C/SiC	SiC/SiC
Uncoated	3.6 ± 0.4	0.40 ± 0.05	1.4 ± 0.1	0.04 ± 0.01
Mullite	3.3 ± 0.4	0.31 ± 0.05	0.2 ± 0.05	0.02 ± 0.01
Zircon	3.3 ± 0.4	0.33 ± 0.05	0.6 ± 0.08	0.02 ± 0.01
Al_2O_3	3.1 ± 0.4	0.64 ± 0.18	1.1 ± 0.08	0.1 ± 0.04
Zirconia	2.6 ± 0.3	0.42 ± 0.06	1.1 ± 0.1	0.3 ± 0.05

Figure 1 *Through thickness cracking due to thermal expansion mismatch in zirconia coated C/SiC*

Figure 2 *Coating detachment after high temperature exposure associated with the weave structure of SiC/SiC*

of the composite is almost impossible and thermal expansion mismatch cracking associated with the weave structure is visible even at modest temperatures. Hence coatings with the lowest thermal expansion mismatch give the best performance.

Isothermal exposure of mullite and zircon coated C/SiC at 1000°C in a sensitive microbalance shows an approximate linear weight loss with time (Figure 3) until the weight loss reaches 24 to 27% when the carbon fibres have been completely burnt out. The linear weight loss indicates that performance is not diffusion controlled but depends on the permeation of gas through cracks and coating porosity via a convoluted path. Mullite coatings perform better than zircon under these conditions.

In the as-deposited form air plasma-sprayed coatings have a porosity of around 10%. The porosity, and hence gas permeability of the coating can be reduced by vacuum plasma spraying but usually at the expense of some strain tolerance. An alternative method of improving performance is to use a refractory carbide barrier layer which has a low oxygen permeability and when it is oxidised can fill some of the porosity in the sprayed layer. A thin (5μm) sputtered TiC layer was used here to dramatically improve performance (Figure 3). Clearly further work is needed to optimise this barrier layer to get the best performance out of sprayed coatings.

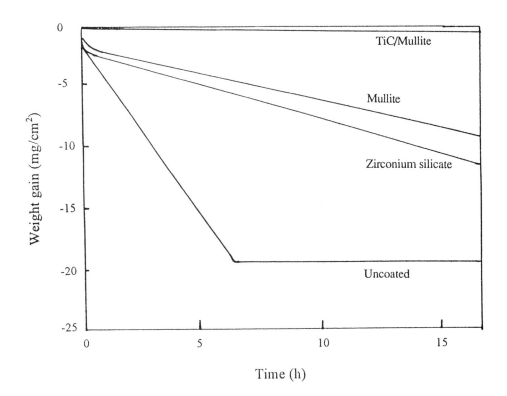

Figure 3 *Microbalance weight loss traces for C/SiC samples coated with mullite and zirconium silicate compared to uncoated materials*

Both mullite and zircon coatings on C/SiC survived the thermal cycling test without spallation with a final weight loss of about 4%. Considerable spalling of alumina and zirconia was observed under the same conditions. On the SiC/SiC composite little or no weight change was observed for the mullite and zircon coatings (~ 0.2%) but again the alumina and zirconia spalled.

In the tests in simulated combustion products the mullite coatings were found to outperform the zircon coatings confirming the isothermal air oxidation results.

6 DISCUSSION

Previous workers have demonstrated that coatings of mullite, alumina and yttria can be effective in reducing the corrosion of silicon carbide heat exchanger components which survived 2000h plus ten thermal cycles up to 1200°C[13]. The results in this study are consistent with these observations. Indeed a C/SiC ring was coated with APS mullite in this study and was exposed in a radiant tube furnace for 1200h at 1200°C with little or no weight loss[14]. The low thermal expansion mismatch between mullite and the C/SiC composite was critical to the success of this component.

The better performance of alumina coatings in isothermal tests at 1000°C is probably due to oxygen permeability. Silica (and hence mullite) has the lowest permeability to oxygen above 1300°C but alumina has a lower permeability below this temperature[9]. The detailed coating microstructure, phase distribution and porosity will also play a role and there is scope for optimising the coating structure for better oxidation protection.

The structure of the composite clearly has an effect on the protectiveness of the coating; the weave structure of SiC/SiC is almost impossible to encapsulate completely in a thick plasma sprayed coating and contains large open channels which probably need to be sealed before coating. The stresses induced by thermal expansion mismatch can also generate interlamellar failure within the composite if they are not controlled adequately.

7 CONCLUSIONS

There is a number of solutions for the protection of high temperature composites from environmental attack of which coating is potentially the most attractive. However, it is almost impossible to generate a crack free coating and microcracks will be generated in service which will limit performance so some sort of crack management approach is necessary. Minimising thermal expansion mismatch between coating and substrate is also necessary to reduce cracking.

The heterogeneous nature of the composite surface (e.g. exposed fibre ends) is an additional problem for coating which can disrupt the coating/substrate interface and promote coating detachment. However, it is possible to deposit thick strain tolerant ceramic coatings by air plasma spraying which have sufficient adhesion to withstand thermal cycling and low operating stresses. More work is needed to improve interfacial adhesion if the coatings are to be used in high stress applications.

Although some improvement in component life can be achieved by plasma sprayed coatings alone, maximum benefit will probably be derived from a multilayer coating involving an oxygen diffusion burner and glassy crack sealant. A considerable amount of work will be needed to develop such a coating.

Acknowledgements

This work was performed at AEA Technology as part of the COSIC BRITE/EURAM programme (BREU 0346) for which the CEC provided 50% funding. The authors would like to thank the other partners (SEP, Pont à Mousson, CERECO, British Gas and Naples University) for their contributions to the work and useful discussions.

References

1. J. M. Jouin and F. Christin, in 'Designing Ceramic Interfaces II', ed S. D. Peteves, CEC, 1993, p. 191.
2. E. Pestourie, P. Taveau and .F Christin, 'Proc. 4th Int. Symp. Ceram. Mat. & Components for Engines', Goteborg, Sweden, June 10-12, 1991.
3. J. Thebault, *J. de Physique IV Colloque C9*, 1993, **3**, 831. G. G. Gnesin in 'Refractory Carbides', ed G V Samsonov, Consultants Bureau, 1974, p. 97.
4. C. Courtois, .J Desmaison and H. Taivil, *J. de Physique IV Colloque C9*, 1993, **3**, 843.
5. J. R. Strife and J. E. Sheehan, *Ceramic Bull.*, 1988, **67**, 369.
6. W. G. Hillig, *Mat. Sci. Res.*, 1986, **20**, ed R E Tressler et al, Plenum Press, New York, 1986.
7. P. Kofstad, *Materials Science Forum*, 1994, **154**, 99.
8. E L. Courtright, *Surf. Coat. Technol.*, 1994, **68/69**, 116.
9. C. Leach and N. Khan, *J. Mat. Sci.*, 1991, **26**, 2026.
10. E. M. Levin and H. E. McMurdie, *Phase Diagrams for Ceramists*, American Ceramic Society, Columbus, Ohio, 1975
11. MCIC Report, *"Oxides"* (1979).
12. J. R. Price and M. van Roode, *Ceram Trans.*, 1990, **10**, 469.
13. Final Project Synthesis Report, BRITE-EURAM contract No BREU 0436, COSIC, 1994.

3.4.7

Implementation of TiAlN and CrN Coatings and Ion Implantation in the Modern Plastics Moulding Industry

E. J. Bienk and N. J. Mikkelsen

DTI TRIBOLOGY CENTRE, DANISH TECHNOLOGICAL INSTITUTE, TEKNOLOGIPARKEN 8000
AARHUS C, DENMARK

1 INTRODUCTION

Plastics materials are gaining new application areas every day and their use is growing rapidly all around the world. Modern plastic industries operate with new plastics materials, large series and still shorter cycle times while at the same time endeavouring to meet the equally vital requirements of good and constant product quality and low production costs. Industries are increasingly aware of productivity and product quality and since surface treatment has proved able to improve these competition parameters, its use is rapidly gaining ground in the plastics as well as in the die casting industry[1].

2 BACKGROUND

Until very recently, coating of plastic moulding tools was – except perhaps by galvanic processes – practically non-existent due to the relatively high temperatures at which the treatments are performed, normally above 400°C. The majority of tool steels used for moulds are cold working steels annealed at relatively low temperatures (180–300°C), which makes most surface treatments non-applicable[2]. Tools for moulding of plastics are mostly very expensive, consisting of many parts with very complicated shapes and narrow tolerances, which precludes the hot surface treatments due to the risk of dimensional changes. The hard protective coatings are seldom taken into consideration because the bulk material of the mould parts, which are seldom hardened to more than 54 HRC, if hardened at all, or which may be made of very soft materials like copper or aluminium alloys, gives insufficient support to the coatings.

Attempts to coat these tools anyway frequently result in quality and adhesion problems. These problems are very often caused by the presence of residual plastics material on the tool surface, which is difficult to remove even by very careful cleaning. Further difficulties arise from the fact that most mould parts are spark eroded. The quenched defect zone, the so-called *white layer*, left on the surface by this process will normally cause problems with adhesion of surface coatings produced by any surface treatment process, therefore a special surface preparation before the treatment is often needed.

Taking all these aspects into account, surface treatment of tools for moulding of plastics is not an easy task. Howewer, due to R & D work in the area and based on the steadily increasing amount of professional experience, this type of job is on the way to becoming routine work.

A very important step towards implementation of advanced surface treatment is the recent development of the sputtering technique, allowing growth of good quality films at relatively low temperatures[3]. Also, application of very effective RF sputter cleaning is of great importance to improved adhesion. Another surface treatment process which is well suited for treatment of tools for moulding of plastics is ion implantation. Although state-of-the-art commercial ion implantation is available only in very few places in the world and despite the fact that general knowledge of the process is relatively poor[4], this process has quite a large potential application area in the moulding of plastics due to its low process temperature.

In the following, the mechanisms of surface improvement by coatings and ion implantation are described briefly, the tribological problems arising in the moulds during operation are listed and examples of solutions to these problems by optimal surface treatments are given from the job treatment experience of the Tribology Centre at the Danish Technological Institute.

3 SURFACE IMPROVEMENT BY ION IMPLANTATION AND PVD COATING

Ion implantation is a technique by which single ions are accelerated and hit the target surface at speeds of thousands of kilometers per second and with energies in the order of hundreds of keV. Depending on ion type and energy, the ions penetrate the target to a depth of up to $0.5\mu m$ and cause re-alloying of the surface[4,5]. Technically, there are almost no limits to what materials can be implanted, in practice however, nitrogen and chromium are the most frequently used ions for treatment of plastics moulding tools. Nitrogen ions, which can be implanted to about 33 atomic per cent i.e. much higher concentrations than for an ordinary nitriding process, partially form interstitial solution and partially react with the target material to form nitrides. The resulting nitrides of vanadium, chromium and other alloying elements are harder than iron nitrides, therefore, when a high-alloyed tool steel is used as a target, the effect of the ion implantation will be more pronounced. Thus, the mechanism of the surface improvement obtained by nitrogen ion implantation is a hardening and passivating of the outermost surface layer, which becomes more wear resistant. By ion implantation with chromium it is possible to increase the amount of chromium in the surface layer up to 30–40 atomic per cent, making the steel highly corrosion resistant.

The mechanism of surface improvement by PVD coatings is different. A thin, typically 1–5 μm coating is deposited on top of the surface, and since the deposition temperature is in the 180–500°C range, alloying or diffusion zones will normally not occur. To ensure good adhesion of these coatings, absolute cleanness of the surface to be coated is crucial. The typical hardnesses of the PVD coatings range from 1800 to 3300 HRC i.e. several times higher than any material used for producing mould parts. Therefore, choosing the coating material most appropriate to the purpose is of the highest importance, and aspects such as the degree and type of load, the brittleness of the coating and the ability of the bulk material to support the coating should also be considered.

Normally, TiAlN, which is very hard (3300 HV) and characterized by having a stable low coefficient of friction against steel, will be chosen to protect hard bulk materials subjected to evenly distributed load. For softer tool materials, which give less support to the coating, the softer but also more ductile CrN coating will be chosen, especially when the load is localized to specific areas.

4 PRACTICAL SURFACE TREATMENT OF MOULDS

The commonest and most pronounced problem in tools for moulding of plastics is wear of the moving parts, e.g. cores, ejectors, die sets or steering elements[6]. The cyclic movement in the direct metal-to-metal contact causes seizing and severe adhesive wear of the parts[7], and in many cases the use of lubricants is forbidden, e.g. with moulds for medical applications. In this case, a thin hard PVD coating is sufficient to isolate the two metal parts from each other, reducing or virtually stopping the wear[8]. Unless other factors, such as the need to improve the slip of the plastics material, point towards another best choice, TiAlN would be selected for coating steel parts and CrN for coating copper alloy parts. Normally, the low hardness of the parts causes no difficulties since it is able to withstand the low contact pressure in this type of sliding contact.

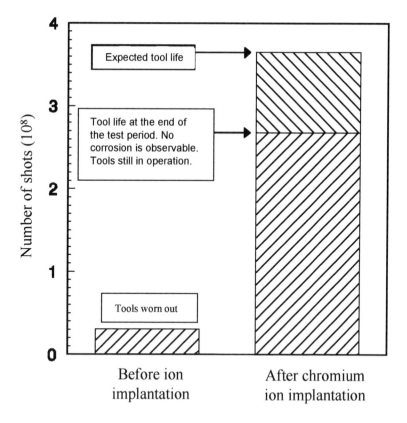

Figure 1 *Injection mould inserts from LEGO System Ltd are subjected to corrosive wear at the air outlets. The life of the inserts is increased by about 12–13 times by Cr⁺ implantation*

Many plastics materials are filled with hard particles such as glass fibres or minerals which cause more or less severe abrasive wear of all parts with which they come into contact[6]. A properly chosen wear resistant PVD coating can reduce or eliminate this type of wear[8,9]. In the case of mild abrasive wear, ion implantation is an efficient treatment[10]. Also, tool surfaces protected by a wear resistant layer will keep the original surface finish longer, whether mirror finish or rough according to the purpose. Coating type and thickness depends on the hardness and size of the abrasive particles. Normally, TiAlN would be chosen, except in the case of very large abrasive particles, where thick CrN or a multilayer coating would be preferred. For very small particles, ion implantation may be considered. Nitrogen ion implantation can also be a very feasible solution when the gates of the moulds are subjected to mild abrasive wear. By proper alignment of the ion beam the critical surfaces, which are deep down at the bottom of the nozzles at the gates, can be surface treated. Due to geometrical factors such surfaces cannot be treated successfully by PVD.

The correctly chosen wear resistant PVD coating is also suitable for protection of parts made in very soft materials like unhardened steel or copper alloys. For this purpose, the softer and more ductile CrN is usually the best solution[8].

For some types of plastics, e.g. ABS, PP and PVC, corrosion of the tool parts can be a major problem. The corrosive wear may be reduced by choosing a different tool material. If this is impossible or insufficient and if the corrosion attack area is relatively large, the problem may to some extent be helped by PVD coating. In the case of local corrosion attack, e.g. at air outlets on large tools, the best known treatment is selected area implantation of chromium ions, see Figure 1. By this, local realloying of the tool steel to high-chromium corrosion resistant steel only at the critical areas will eliminate corrosive wear and may be a very cheap and very effective solution for the expensive moulding tools[4].

Of the problems connected with moulding of plastics, the ones most difficult to handle are the slip properties and the friction between the plastics and the tool material. The tendency of the plastics material to stick to the tool surface depends on the plastics type, the tool material, the moulding process parameters, e.g. temperature, and it can even be influenced by the colour of the plastics within one and the same plastics type. Therefore, there is no single answer to the best choice of treatment for a given tool with regard to reduction of the slip problem. At DTI Tribology Centre intensive research, including both laboratory tests and field tests, has been undertaken in order to find the optimal solutions for the most widely used plastics materials. At present, the treatments most often used by DTI Tribology Centre to improve the slip properties of the mould parts are ion implantation, TiAlN coating, CrN coating and ion implanted TiN and CrN coatings. Research is still going on in order to find the optimal combinations of materials and treatments. Table 1 summarizes some of the results obtained at DTI Tribology Centre.

Under present development, surface treatment of aluminium alloys is worth mentioning because of its potential for extending tool life and product quality of aluminium moulds for prototypes or short series. Nitrogen and oxygen ion implantation can be used to form hard AlN or Al_2O_3 respectively in the outermost layer of the tool surface. Coatings, plasma nitriding and duplex coatings are also considered as solutions. Extensive research is undertaken at DTI Tribology Centre in connection with appropriate surface treatments of aluminium alloys, i.e. materials for moulds for prototypes and short series.

Excessive tool size and weight are limiting factors to the use of surface treatment since only a few plants allow treatment of e.g. extrusion screws of several meters or moulds weighing several hundred kilos, and obtaining good uniform coating on such large tools may also be a problem. This remains to be solved.

Table 1 *Effects of surface treatment of mould parts for moulding of plastics. Examples from job treatment at the Tribology Centre, Danish Technological Institute*

Tool	Material	Problem	Treatment	Effect
Standard ejectors and cores	Steel	Adhesive wear, seizing	TiAlN	No adhesive wear observed in long series
Cores for moulding small details of PE	Hot working steel, annealed 390°C, 54 HRC	Sticking of PE to the tool	N^+ implanted CrN, N^+ implanted TiN	Sticking eliminated
Cores and moulds for details in rubber types plastics	Cold and hot working steels, 50-58 HRC	Sticking of rubber to the tool	TiAlN	Sticking eliminated
Large mould	Cold working steel, annealed 300°C, 56 HRC	Corrosive wear around air outlets	Cr^+ implantation of critical areas	Corrosion eliminated, tool life enhanced by more than 10 times
Gates for injection moulding tool	Cold working steel, annealed 250°C, 58 HRC	Mild abrasive wear	N^+ implantation	Wear reduced, tool life enhanced by 3.5 times
Mould and core for injection moulding of tubes	Hot working steel, annealed above 400°C, 52 HRC	Products (PP and polyester) difficult to remove using both air blasting and ejecting	TiAlN	Slip of the plastics improved, air blasting enough to remove products
Core for injection moulding of packing material	Cold working steel, annealed below 250°C, 52 HRC	Abrasive filler particles eroding several mm deep crater in core around gate	CrN	*Wear reduced, improvement by a factor of 8, tool still running
Nozzles and calibration gates for extrusion moulding of profiles	Corrosion resistant steel, 300 HB	Large (up to 20 μm) abrasive filler particles eroding deep tracks along the flow, re-polishing necessary	Thick CrN	*Improvement by a factor of three (no sign of wear over a period where tool used to be polished 3 times
Core for injection moulding of buckets	Ampco (copper alloy)	Abrasive wear from the white TiO_2 filling, re-polishing every week	CrN	*No sign of wear after running for 4 months without re-polishing

*Preliminary reports

References

1. O. Knotek, F. Löffler and B. Bosserhof, *Surf. Coat. Technol.*, 1993, **62**, 630.

2. Menges, Mohren, 'How to Make Injection Moulds', Hanser Publishers, Munich Vienna New York (1986).

3. A. A. Voevodin, C. Rebholz, J. M. Schneider, P. Stevenson, A. Matthews, *Surf. Coat. Technol.*, 1995, **73**, 185.

4. C. A. Straede, N. J. Mikkelsen, to be published in *Surf. Coat. Technol.*, 1996.

5. 'Proc. Eighth Int. Conf. on Ion beam Modification of Materials', eds. B.D. Sartwell and A. Matthews, *Surf. Coat. Technol.*, Vols. 65–66 Elsevier Science S.A., Holland, 1994.

6. G. Menning and G. Paller, *Mat.-wiss. u. Werkstofftech.*, 1993, **24**, 152.

7. D. Severin, D. Petersohn, H. Sander and J. Schulz, *Mat.-wiss. u. Werkstofftech.* 1993, **24**, 160.

8. E. J. Bienk, H. Reitz and N.J. Mikkelsen, *Surf. Coat. Technol.*, 1995, **76–77**, 475.

9. H. Freller, H.P. Lorenz, Beschichten mit Hartstoffen, VDI Verlag, ISBN 3-18-400986-0, 19.

10. J. K. Hirvonen, 'Surface Alloying by Ion, Electron and Laser Beams', 1985 ASM Materials Science Seminar, Eds. L.E. Rehn, S.T. Picraux, H. Wiedersich, 1985, 373.

3.4.8
Thin Film Characterization Methods

F. Löffler

PHYSIKALISCH-TECHNISCHE BUNDESANSTALT, BUNDESALLEE 100, 38116 BRAUNSCHWEIG, GERMANY

1 INTRODUCTION

The designation "thin film" is usually derived from the manufacturing process. For example, coatings deposited from the vapour phase are referred to as "thin films". In the past years, many processes and modified processes were further developed for vapour phase deposition. As the film properties depend considerably on the manufacturing process, brief reference is in the following section made to the manufacturing processes applied.

Thin films are above all used for decorating, in the fields of optics and microelectronics, for wear and corrosion protection, and to achieve specific surface properties. Thin films are to an increasing extent also used in commercial production and in industrial series processes. In addition to the increased use of thin films by industry, there are also demands for the establishment of a quality system (QS). Many firms have already set up QSs in compliance with the series of standards DIN EN ISO 9000 foll. These standards require the traceability of measuring and testing devices to national standards to guarantee a uniform quality standard and identical measurement results in inspections carried out on receipt and on delivery. This leads to a great need for informative measurements and calibrations which can be traced back to national standards. To meet the increasing demands in this field within as short a time as possible, cooperation with the metrology institutes is useful on the international level.

2 COATING METHODS AND PRODUCIBLE FILM SYSTEMS

The processes of thin film production are classified according to the principle applied to vaporize solid material. Accordingly, a distinction is made in particular between

- physical vapour deposition (PVD)
  - thermal evaporation
  - magnetron sputter ion plating
  - arc ion plating
  - thermionic arc
  - electron beam
  - anodic arc
  - laser PVD
  - ...

- chemical vapour deposition (CVD)
  - thermal CVD
  - plasma CVD
  - microwave CVD
  - ...
- others
  - ion implantation
  - plasma surface treatment
  - plasma polymerisation
  - ...

An exact separation of the different production principles is not always possible, as the processes listed above have various sub-principles and as there are – to an increasing extent – combinations of processes. Whereas in the PVD processes the solid materials are vaporized according to physical principles (e.g. sputtering), a chemical reaction (usually at high temperatures of up to 1600 °C) for vapour generation takes place in the case of the CVD principles. The particles can be applied to the substrate either as cluster (particle compound) or as atoms. An example of atomic growth is the sputter principle shown in Figure 1. An argon particle is ionized in the plasma space and hits a negatively polarized target (cathode). Through its impact on the target surface, the argon ion removes a solid matter particle, e.g. a metal atom, which makes use of the transferred energy and flies to the substrate. On its path, the atom can collide with other gas atoms or solid matter atoms, and this leads to an energy exchange or to phase formation. The film proper is formed when several layers of atoms have reached the substrate surface. Diffusion processes inside the film may take place both during and after coating.

Figure 2 compares the most essential PVD process principles. Different energy states of the particles or atoms result from the different principles. These energy states are responsible for the differing structures of the films and the film properties which often differ substantially even when the film composition is the same. In the case of all PVD methods, reactive gases can be added, and in this case the film is formed from the components of the reactive gas and from the components of the solid material. When titanium material vaporizes in methane-nitrogen atmosphere, a $Ti_xC_yN_z$ film is obtained, the x-, y- and z-components depending on

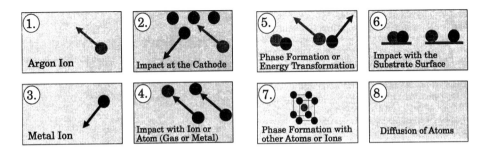

Figure 1 *Phase formation using PVD technology*

the parameters set such as gas flow rate and cathode voltage. The PVD methods also have in common that it is possible to apply a bias to the substrates and to heat the substrates. All processes take place in the high vacuum, at a pressure of about 10^{-5} mbar.

Thermal vaporization allows metal films to be sputtered on bands and foils up to 5 m in width. These films are especially suitable for optical and decoration purposes and may be subjected only to low mechanical stress.

Almost all film systems can be deposited by the magnetron sputter ion plating (MSIP) method. Figure 3 shows the film systems so far developed and investigated by the author, and it thus documents the versatility of the method. Any combination of reactive gas atoms (non-metals) and solid matter atoms (usually metals) can basically be deposited. Although the degree of ionization of the solid matter particles is only up to 5%, very good adhesion of films and good mechanical properties of these can be achieved when bias voltages in the range from 30 to 200 V are applied. To ensure good all-round coating, the components and, if necessary, the cathodes, too, must be moved during the coating process. Among the variants of the sputter principle are unbalanced sputtering, rf-sputtering and dc-sputtering with gas separation.

The main advantage of the arc ion plating (AIP) method is the high degree of ionization of the solid matter particles of up to 95%. This allows the components to be uniformly coated on all sides. The adhesive strength and the mechanical stability of the films are high. In view of the high energy density at the cathode, high demands must be placed on the homogeneity and

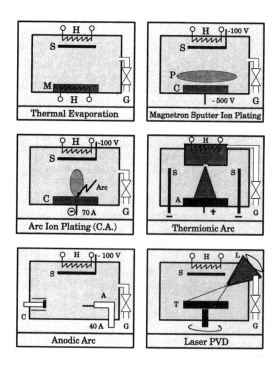

Figure 2 *PVD processes[1]*

Gas / Target	Ar	Ar,N$_2$	Ar,CH$_4$	Ar,O$_2$	Ar,N$_2$,CH$_4$	Ar,N$_2$,O$_2$
Ti	Ti	Ti$_2$N, TiN	TiC	TiO, TiO$_2$	TiN$_x$C$_y$	
TiC	TiC	TiC$_x$N$_y$				
TiN	TiN					
TiAl$_x$		TiAl$_x$N	TiAl$_x$C	TiAl$_x$O$_y$	TiAl$_x$C$_y$N$_z$	TiAlON
TiZr$_x$		TiZr$_x$N	TiZr$_x$C		TiZr$_x$C$_y$N$_z$	
TiHf$_x$		TiHf$_x$N				
Al	Al	AlN		Al$_2$O$_3$		
Al$_2$O$_3$	Al$_2$O$_3$	AlO$_x$N$_y$	AlO$_x$C$_y$	Al$_2$O$_3$	AlO$_x$N$_y$C$_z$	
Ti-Al$_2$O$_3$	TiAl$_x$O$_y$	TiAl$_x$O$_y$N$_z$				
TiAl$_x$V$_y$		TiAl$_x$V$_y$N	TiAl$_x$V$_y$C	TiAl$_x$V$_y$O$_z$	TiAl$_x$V$_y$N$_z$C$_v$	
Cr	Cr	Cr$_2$N, CrN	Cr$_x$C$_y$			
CrAl$_x$		CrAl$_x$N	CrAl$_x$C			
W		W$_2$N, WN	W$_2$C, WC			
WCr$_x$		W-Cr-N	WCr$_x$C			
Ta		TaN	TaC		TaN$_x$C$_y$	
Si		Si-N	Si-C	Si-O		
SiC	Si-C	Si-C-N	Si-C			
Si$_3$N$_4$	Si-N	Si-N	Si-C-N			
Al$_x$Si$_y$		Al-Si-N				Si-Al-O-N
TiB$_2$	TiB$_2$	Ti-B-N	Ti-B-C		Ti-B-N-C	
carbon	a-CH diamond		a-CH diamond			

Figure 3 *Coating systems by target and gas combination*[2]

density of the cathode. Inhomogeneous cathodes may break during coating. Owing to the high energy density of the arc on the cathode, particles of mm size, so-called "droplets", may get loose, which deposit on the substrate surface and lead to an inhomogeneous structure of the film. Variants of the AIP method are the steered arc, the random arc (which is most frequently used) and rf-arc operation.

The thermionic arc evaporation (TEA) process is closely related to the electron beam (EB) method. In both cases, an electron beam is focused on the anodized material to be vaporized which is usually available in the form of powder. The vaporous material is then deposited on the components. The electron beam is often generated via a separate vacuum chamber equipped with a resistance-heated filament. Depending on the solid materials used, ionization rates of up to 95% can be achieved by this method. Variants of these basic principles are various combinations of these, such as the combination of sputtering with TEA or the use of two EB sources, where one source effects evaporation and the other is used for substrate heating.

In the anodic vacuum arc (AVA) process, too, the high degree of ionization ensures good coating on all sides of the components. As a result of indirect deposition through a cathode-anode system, very homogeneous layer structures of high surface quality are obtained by this method. This qualifies the method for optical and electronic applications, metal layers being chiefly applied.

The laser PVD method allows non-conducting ceramic materials to be deposited at high coating rates of up to 30 mm/h. Via a lens system, a laser beam is focused on to the rotating material to be vaporized. To date, the method has been applied for Al_2O_3 and ZrO_2 coatings, a CO_2 TEA laser being used for target vaporization.

The chemical reactions in the CVD processes require high activation energies which are made available in the form of thermal energy (up to 1600 °C) or as a combination of thermal energy and plasma-excited energy. In the case of a superposed plasma excitation (e.g. microwave plasma), the coating temperature can be reduced down to 500 °C, which keeps distortion low and makes only slight after-treatment of the coated materials necessary. Each coating material has a reaction formula of its own and different reagents which are usually added in the gaseous state into a heated boiler. Typical kinds of reaction are:

Chemical synthesis:

$$TiCl_4 \text{ (g)} + CH_4 \text{ (g)} + H_2 \text{ (g)} \xrightarrow[10-150mbar]{800-1000^{\circ}C} TiC \text{ (s)} + 4 HCl \text{ (g)} + H_2 \text{ (g)}$$

Pyrolysis:

$$SiH_4 \text{ (g)} \xrightarrow{>650^{\circ}C} Si \text{ (s)} + 2 H_2 \text{ (g)}$$

Disproportionation:

$$2 \, GeI_2 \text{ (g)} \longleftrightarrow Ge \text{ (s)} + GeI_4 \text{ (g)}$$

Photopolymerisation: Treatment by ultraviolet radiation.

Whereas almost any coating material can be deposited in the PVD processes, the deposition capability in CVD processes is limited to the reactions which are possible from the thermodynamics point of view. Aluminium matter can be included into a TiN layer only in a very complex process. In addition, toxic chemicals are produced in the CVD processes, part of which must be disposed of with much effort and outlay.

3 CLASSIFICATION OF THE MEASURING TECHNIQUES CHARACTERIZING THIN FILMS

The methods for the characterization of thin films can be classified according to:

- the measurand to be determined,
- the methods' informational content as regards film and compound properties,
- their relation to practical application, and
- the traceability to national standards.

Figure 4 describes the first three of the above criteria by some examples. Information on the film plasticity is obtained by application of a nano-indenter to the surface. As the depth of the indentations produced can be only 5% of the film thickness, influences from the basic

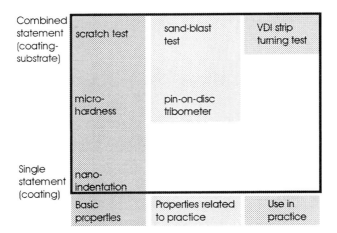

Figure 4 *Classification of measuring techniques*

material (substrate) can be excluded; this has been proved by several investigations. Such an investigation provides information on the pure film properties and is referred to as *single statement*. In a scratch test, a diamond is moved across the film surface to cause the film to chip or peel-off the substrate. In this case, the measurement result is obtained as a *combined statement* (coating-substrate). When a microhardness measuring device is used, the indentation can penetrate into the basic material or can at least be influenced by it. It cannot, therefore, be ensured that the pure film property is determined. The measuring method can then be classified between a pure single statement and a clear combined statement. With test methods closely related to practice and methods directly used in practice, a single statement cannot usually be made any longer since compound properties are concerned here.

Measuring methods which allow a single statement to be made for film characterization are basically suited to be traced back to national standards. Film thickness standards can be used as standards for the calibration of film thickness measuring instruments, or standardized hardness test blocks with a defined nano-hardness can, if necessary, be used to calibrate and test material testing machines. Only very few standards are so far available for this measurand, and the need in the measurement laboratories and coating firms will certainly increase considerably in the years to come. The calibration hierarchy represented in Figure 5 illustrates how a rapid expansion of calibration work can be achieved on the international level.

The above classifications allow an assessment of the measuring instruments to be made according to their suitability for:

- the development of films,
- the development of procedures,
- acceptance criterion,
- tracing-back,
- release criterion after primary development, and
- testing in practical use.

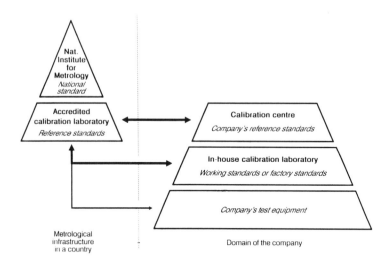

Figure 5 *Calibration hierarchy from the national standard to the test equipment*[3]

The user will have to determine on the basis of this assessment scheme which of the measuring techniques he requires in practice taking the input of time and money into account.

The methods of film characterization will be presented in the following section on the basis of Figure 4, according to the fundamental properties of thin films, realistic test methods and practical use. In view of the large number of potential examination principles, only a selection of the methods will be given below.

4 FUNDAMENTAL PROPERTIES OF THIN FILMS

In the following, some methods are described which provide fundamental information on the film properties. The possibility of tracing the methods back to standards is discussed on the basis of some examples, and the state of normalization and standardization is described.[4]

4.1 Film Thickness

The film thickness is determined at all stages of the development, production and application of films. The film thickness is determining for both the rest time of the coated component, the time required for coating and the costs involved. The film thickness must, therefore, be optimized for the intended use and be continuously monitored during the production phase. The methods applied to determine the film thickness can be divided into destructive and non-destructive methods:

- Destructive methods
  - Mechanical methods
    - profilometry

- • difference thickness (dial gauges)
- • difference weight (weighing)
- ◆ Chemical methods
  - • analytical methods
    - • titration
    - • atomic absorption spectroscopy (AAS)
  - • gravimetry
  - • electrochemical methods (coulometry)
- ◆ Metallographic methods
  - • cross grinding
  - • spherical cup grinding
- • Non-destructive methods
  - ◆ Radiometric methods
    - • X-ray fluorescence
    - • beta-ray backscattering
  - ◆ Electromagnetic methods
    - • adhesive force
    - • magnetoinduction
    - • eddy current

The great variety of methods available for the determination of film thickness documents the importance of this measurand in the application and development of films. Some of these methods are briefly described in the following:

Spherical cup grinding is a method very frequently applied; it is represented in Figure 6. A sphere with defined diameter is placed on a sample. The sphere is rubbed with diamond paste (approx. 6 μm) and rolled on a defined section of the sample surface. The reflected light

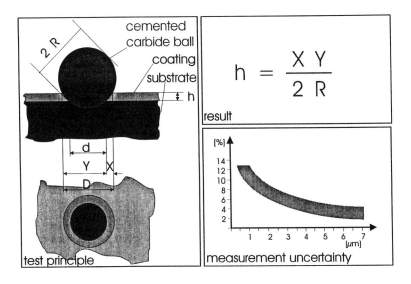

Figure 6 *Determination of film thickness by means of spherical cup grinding*

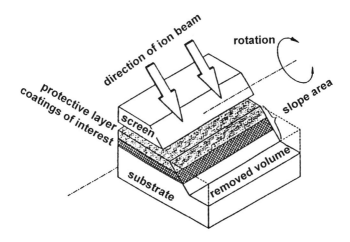

Figure 7 *SEM-based investigation of film thickness[5]*

microscope shows in the resulting ground spherical section the area of the basic material and of the film in ground condition. On the basis of the geometric conditions, the film thickness can then be determined using an approximation formula. While the uncertainties of measurement are high with very small film thicknesses (<1 μm), reliable measurement results can be obtained for film thicknesses of 3 μm and more. It is the great advantage of this method that it can be universally applied for almost all combinations of films and basic materials. The spherical cup grinding method is described in the European preliminary standard DIN V ENV 10071 part 2.

Profilometry is also described in a preliminary standard (DIN V ENV 1071 part 1). First a step must be produced between film surface and surface of the basic material. This can be done by covering part of the surface during coating or by subsequently removing part of the film by chemical or mechanical means. A contact stylus instrument then passes across this step and measures the step height and thus the film thickness. Depending on the coating material used and the film thickness to be expected, the tip radii (between 2 and 10 mm) and the load applied via the tracing diamonds (usually 10 mN) must be varied.

A method for determining the film thickness with very low uncertainty of measurement is *cross grinding*, provided ion beam paring is used to produce the ground section. Via a system of diaphragms, the ion beam is focused on to the section to be removed and causes the film to be removed down to the basic material as shown in Figure 7. This ground section is then measured by means of a microscope, the highest resolution of the scanning electron microscope being achievable[5]. When a coating method is applied, which allows the application of a uniform film to be adjusted (e.g. sputter methods or anodic arc), the above method can be made use of to manufacture film thickness standards. This will make it possible to meet industry's great demand for material measures for film thickness. This will also pave the way for the establishment of calibration laboratories for the measurand "film thickness".

In the case of the *X-ray fluorescence method* (ISO 3497), the sample surface is exposed to X-rays; this produces fluorescent radiation with a line spectrum. With the *beta-ray backscatter method*, beta particles hit the film surface, proportion of the particles being backscattered. In

	AES	XPS	SIMS	SNMS	ESMA	GDOS	RBS
information depth in nm	0,5 - 10	0,5 - 10	0,3 - 100 a)	0,3 - 100 a)	2000	1 - 10 e)	3 - 10
rel. uncertainty in ppm	< 0,1	0,1	1 - 0,001	0,1	---	---	1
lateral disintegration in μm			1 - 0,0001 a)	1 a)	100	1	
element indication	Z > 4	all	all (+ isotopes)	all (+ isotopes)	Z > 4	all	all (+ isotopes)
indication of compounds	in special cases	yes	in special cases	in special cases	no	no	no
picture of element composition	yes: SAM	yes b)	YES: IMMA	in parts	yes	no	no
analysis of insulator	in special cases	yes	in parts	in special cases	no	yes d)	yes
quantitative sketch	+	+		+	++	++	+++
measurement time	high	high	high	high	low	very low	high
costs	high	high	high	high			high
industrial use	+	++	+	+	+++	+++	+
depth profile	yes	yes b)	yes a)	yes	no	yes	yes f)

a) in dynamic operation   b) small spot ESCA-device   c) in following measurements
d) with rf operation   e) physical not exact defined   f) not destructive

Figure 8 *Characteristic data of some surface-analysing methods*[6]

both cases, a statement on the film thickness can be made on the basis of the backscatter intensities. For this purpose, the density of film and basic material and – in certain cases – a number of other parameters must be very precisely known. This method is, therefore, above all suited to assure the quality of known coatings applied in series production.

4.2 Chemical Composition

The determination of the chemical composition is – on the one hand – very important, for the development of films in particular, and – on the other hand – very cost-intensive considering the high input of instruments and staff required. A routine check of the composition of the films, for example within the framework of controls made upon receipt and delivery, is therefore usually not made.

Apart from the pure analysis intended to identify the elements contained in a film, for the development of films information is important about the compounds found, about the bond character of the atoms and about the distribution of elements over a larger surface area. Especially in the case of the hard material layers frequently used it is of utmost importance to know whether the metals and non-metals are found side-by-side or whether they exist in the form of a hard material. This information will allow important details of the behaviour of films to be obtained. Figure 8 gives a survey of some surface-analysing methods frequently used and describes these methods on the basis of some essential characteristic data.

Some draft standards concerning surface-analysing methods exist (e.g. ENV draft 38 "Determination of chemical composition"); moreover, many material measures required for the calibration of the instruments are available.

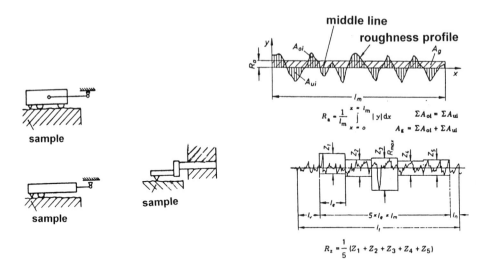

Figure 9 *Mechanical roughness measuring method*

4.3 Roughness, Porosity, Morphology and Topography

As the determination of the roughness of a film surface does not differ from the determination of the roughness of a compact material, the same methods can be applied here. The only difference is the often rather great hardness of the coated surfaces, which may lead to a rapid abrasion of the styluses and thus to incorrect measurements. The surface roughness is determined by:

- mechanical tracing methods,
- optical non-contact methods.

Details of the profile method and the roughness quantities which can be determined are represented in Figure 9. It is important for a description of the roughness of film surfaces to state several roughness parameters (e.g. R_a and R_z), as misinterpretations are otherwise possible. The mechanical methods of roughness determination are described in several standards (e.g. DIN 4772, DIN 4768, ISO 4287 or ISO 4288).

The optical roughness measuring methods, i.e. non-contact scanning of the surfaces at high measuring rates, are specially suitable for automatic quality control. Depending on the radiation source used and the direction of measurement, a distinction is made between:

- light section method,
- interference microscopy,
- electron beam interferences,
- speckle contrast method,
- stray light method, and
- scanning tunneling microscopy.

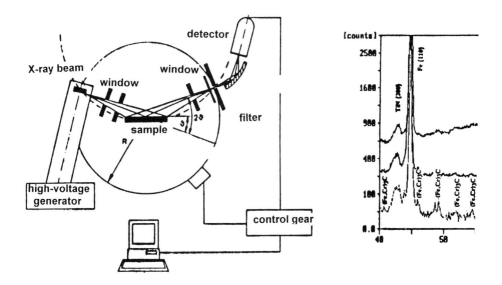

Figure 10 *Radiographic investigation of structure and texture*

While the resolution is up to 1 μm with the first-mentioned method, the atomic structure on the film surface can be made visible by means of scanning tunnelling microscopy.

Normally, the above methods also allow a statement to be made about the porosity, morphology and topography of the films. For this, the optical methods are of greater significance. Voids and fissures become visible under the scanning electron microscope, possibly after grinding of a surface section. Other methods that can be used are the electrogaphical indentation method and heat wave microscopy.

4.4 Structure, Texture and Internal Stress of the Films

The arrangement of the atoms in the solid body is chiefly described by the structure and texture. This information is often in direct correlation with the mechanical film properties so that the behaviour in practical use can well be explained on this basis. This applies to a similar extent to the films' internal stresses which may, on the one hand, increase the stability and may lead to lower adhesiveness on the other.

By means of the radiographic fine structure analysis, information can be obtained about the structure and texture of thin films, the effort and outlay required being comparatively low. As shown in Figure 10, the sample is exposed to X-radiation which is reflected, depending on the crystal structure and lattice spacings of the atoms, and evaluated in a detector. The reflection diagram shown in Figure 10 provides information on the orientation of the films and their phase composition. This method allows even the slightest differences between two surface films to be detected and thus makes an important contribution to the development of films.

The internal stresses of films are usually also determined by the method of X-radiation. For this purpose, the coated samples must be tilted in several planes and the detected back reflections evaluated through computer programs. Mechanical methods are used as well, for example

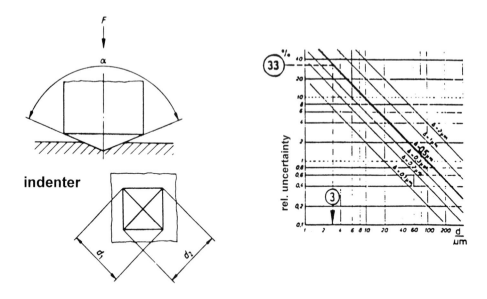

Figure 11 *Microhardness determination according to Vickers and reading error in the optical microscope*

bending and subsequent measurement of sheets after coating, or the so-called *"bore-hole"* method, where a hole of only mm size is drilled into the sample surface and the geometrical change is a measure of the internal stress.

4.5 Hardness Measurement

In addition to the film thickness, the hardness is among the essential mechanical characteristics of a film. The hardness is often used as an acceptance criterion and determined at least per batch, as it provides information on whether the film in question has the required mechanical hardness characteristics.

For the determination of the hardness of thin films, the same measuring principles can basically be applied as they are used to determine the hardness of compact materials. The main difference lies in the load applied which is much lower than in hardness measurements on compact materials. As a general rule, the penetration depth of the diamond indenter into the coating must not exceed 20% of the coating thickness in order to keep the influence of the substrate and thus the uncertainty of measurement low.

Figure 11 describes the determination of microhardness according to Vickers. A diamond pyramid penetrates into the surface, and the hardness can then be determined on the basis of the diagonals of the permanent indentation in the surface. According to Bückle's rule for the determination of the maximum permissible reading errors in the optical microscope during microhardness tests, the relationship between indentation diagonals and reading accuracy is established, and this allows the relative uncertainty of measurement to be determined. For example, for a typical reading accuracy of 0.5 mm and an indentation diagonal of 3 mm, a

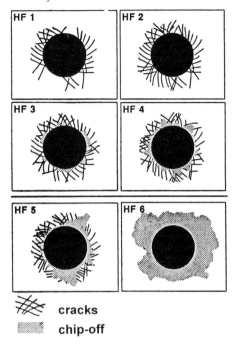

cracks

chip-off

Figure 12 *Classification of the adhesive strength of films according to Rockwell C test*

relative uncertainty of measurement of 33% results. When the diagonals were measured with the scanning electron microscope, this reading error would be considerably smaller. As an alternative to the equilateral Vickers pyramid it is also possible to use the oblong Knoop diamond pyramid.

The nano-indenters not only allow the indentation depths of the diamonds and thus the uncertainties of measurement to be considerably reduced but also the scope of information on plastic and elastic deformability to be improved. In the case of the nano-indenter, a diamond tip penetrates into the film surface down to only nm depth, and during this process the path of the indenter and the load applied are continuously determined. After a defined penetration depth has been reached, the load relief process begins, with the residual forces being picked up as a function of the penetration depth. In addition to the hardness values, other mechanical characteristics, such as the E-modulus, can be calculated.

4.6 Adhesive Strength

It depends on the adhesive strength of the films to what extent they stick to the substrates under defined loads. Although the adhesion of the films produced by modern coating methods is usually very good, considerable damage may result when the films chip off. The methods applied to determine the adhesive strength are, therefore, of decisive importance, also for the checks made on receipt and delivery. It should be pointed out in this context that the opinions concerning the methods of film characterization, which are to describe the adhesive strength, differ substantially in many fields and on the international level. Whereas one side considers the methods to be non-reproducible (e.g. various measurement conditions are used) and rejects them, the other side is in favour of them and even has drawn up draft standards covering these

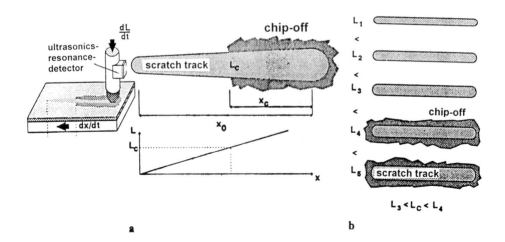

Figure 13 *Scratch test for the determination of the adhesive strength*

measuring techniques. This underlines the importance of a strict observance of the instructions for the measurements carried out for a characterization of the adhesive strength and makes it quite clear that material measures are urgently required to allow these measurements to be checked.

A qualitative method for the determination of the adhesive strength is the *Rockwell indentation test* which is frequently carried out in series production for quality assurance purposes. A conventional indentation according to Rockwell C (DIN 50 103 part 1) is produced and the apparent chip-off at the edge of the indentation assessed on the basis of a six-grade scale. For a hardness of the basic material of 54 HRC and a maximum film thickness of 4 mm, a permissible chip-off at steps HF1 to HF4 results. This grading may differ when the requirements for the hardness of basic material and substrate and those for the film thickness are different (See Figure 12).

Another method used to determine the adhesive strength is the *scratch test*. As shown in Figure 13, a Rockwell C diamond scratches over the surface of a coated sample under a defined load. The load can be increased continuously (Figure 13a) or in steps (variant b). When a certain load is reached or exceeded, chip-off takes place, a distinction being made between cohesive or adhesive forms. The load under which chip-off phenomena occur for the first time is referred to as the critical load. Since the method and thus the measurement result depend decisively on the composition of film and basic material, several conditions must be exactly met during the measurement. The scratch test has already been described in a European preliminary standard, DIN V ENV 1071 part 3.

4.7 Other Fundamental Properties

Other fundamental properties of the films are related to optics, decorating effect and reflection and can be made use of for an evaluation of film properties using the methods specified for compact materials.

Other special properties can be determined in addition, for example biological, photoelectrical, magnetic and electro-optical properties.

5 TEST METHODS CLOSELY RELATED TO PRACTICE

The measuring methods closely related to practice use so-called "model" test facilities which simulate the stress to be expected in practical use. Complex stressing mechanisms are to describe the loading capacity of the films by simple means. These measuring techniques close the gap between the test methods applied to determine the fundamental properties and the actual use in practice, which can normally be described only with considerable effort and outlay. In view of the large number of potential applications, there are numerous realistic methods for the characterization of thin films. To allow the stress during practical use to be simulated as well as possible by these methods, the stress situation arising later must be described as precisely as possible, for example by means of the components of a tribo system. On the basis of this description, one or, if necessary, several model test methods are selected and the stress parameters chosen which must be simulated by them. This way of proceeding allows practical application to be very well simulated. The test methods closely related to practical application can be classified according to the stress mechanisms acting on the surface, and it should be borne in mind that several of these can be effective simultaneously.

- Corrosion
  - ageing test
  - current-density-potential measurement

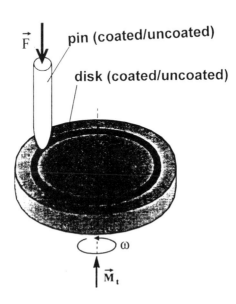

Figure 14 *Diagrammatic sketch of the pin-on-disk tribometer*

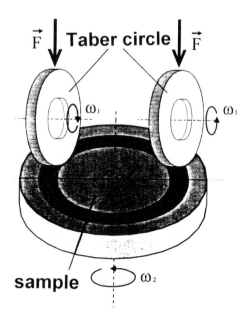

Figure 15 *Taber abraser principle*

- • pin-on-disk test with corrosive intermediate layer
- • Taber abraser with corrosive intermediate layer
- • Adhesion
  - • Scotch test
  - • pin-on-disk test
- • Thermal reactions
  - • annealing in defined atmosphere
  - • thermal shock test
  - • heat wave measuring method
- • Abrasion
  - • Taber abraser
  - • pin on disk test with abrasive intermediate layer
- • Erosion
  - • sandblasting
- • Cavitation
  - • sonotrode
- • Surface fatigue cracking (fatigue strength)
  - • impact test

This list shows typical test methods, some of which will be described in the following:

The *pin on disk tribometer* is frequently used to describe not only the endurance of thin films but also the durability of compact materials. It excels above all by its good reproducibility and its versatility as regards the parameters and variants which can be set (e.g. with corrosive or abrasive intermediate layers). The tribometer components shown in Figure 14 comprise a

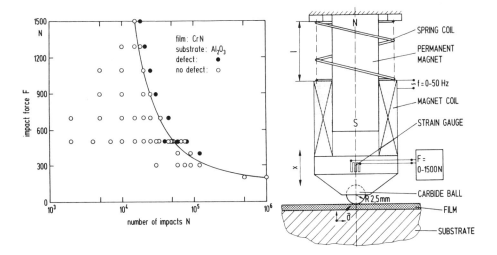

Figure 16 *Impact test with typical load diagram*

plate, a pin and several motion elements and load-application devices. They allow the disk to be rotated, with the pin rotating simultaneously to produce an elliptical path. By adding corrosive media or abrasive matter (sandpaper), or by simultaneously applying thermal energy, the test method can be closely adapted to future practical use. In the last instance, the edge of the disk can even contact the pin, or the pin can be designed as a "sphere". After the changes in length or weight have been determined (it is also possible to measure the damaged areas when layers are concerned), a diagram of the load as a function of the length subject to wear can be plotted.

A similar multi-purpose design can be realized for the *Taber abraser*. As shown in Figure 15, two disks roll on a larger disk, the contacting line being displaced at an angle in relation to the circular orbit. Any material can be used as contact material. It is also possible to add corrosive media contained in an extra pan. The amount of wear is plotted against the length subject to stress.

While the two methods referred to above simulate uniform, continuous stressing, the *impact test* means dynamic loading of the film surface. A ball 2.5 mm in diameter, made of cemented carbide, ceramic material or diamond, strikes on the surface at frequencies between 1 and 50 Hz and under loads of up to 1500 N, causing damage which depends on the number of impacts. This test method allows an endurance strength curve (similar to the Wöhler curve for compact materials) to be plotted for thin films. As films are frequently subjected to dynamic stress, this method is of particular importance for the description of the endurance of thin films. Figure 16 shows how the load is applied via a combined spring-electromagnet system. The measurement of the forces and the control of the frequencies and loads must be realized via a rapid SPC-aided computer system to meet the requirements for high-accuracy measurements[7].

Another novel measuring method is the *cavitation measurement* where the surface is subjected to stress by means of a sonotrode arrangement. Miniature microjets hit the surface and, after a certain time, separate particles from the surface. Similar to the impact test, recordings of the damage caused are made discontinuously and then used to assess the durability

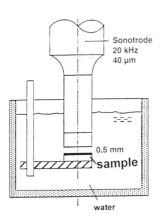
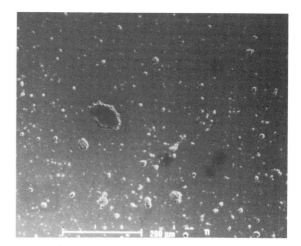

Figure 17 *Diagrammatic sketch of the cavitation test and typical recordings of sections under stress (NUTECH, Neumünster)*[9]

of the film. Typical recordings of the damage and the principle of the measuring setup are shown in Figure 17. First investigations also document a good relation between the potential adhesion of the film and the cavitation resistance.

The selection of an appropriate test method allows the stresses to be expected in practice to be exactly described; it will often be necessary to combine several test methods.

6 PRACTICAL USE

Although a combination of the methods for the determination of the fundamental characteristics with the test methods which are closely related to practice allow reliable statements to be made on the durability to be expected, a final statement can be made only in practical use. As different parameters are often found in practice, such parameters are selected for the tests, which describe a typical situation comparable to that found with other users. There is, of course, any number of applications, two of which will be briefly described in the following.

The interrupted cut means an especially high degree of stress during the cutting of materials; it is applied in milling work as well as in shaping and various turning operations. Depending on the material and cooling lubricant used, both a mechanical alternating stress and a thermal alternating stress can be superimposed. To obtain a test method typical of this stress, the *VDI strip turning test* has been defined, where laths made of the material to be tested are arranged in four grooves of a steel shaft and this "shaft of laths" is cut with a cutting tool. Here, too, different cutting speeds are set to obtain ample information about the behaviour of tool and material. A stress curve is shown in Figure 18; the films have been made of different materials, and different coating methods have been applied. There are clear differences between the films investigated. The tendency towards a behaviour of this kind in the strip turning test had

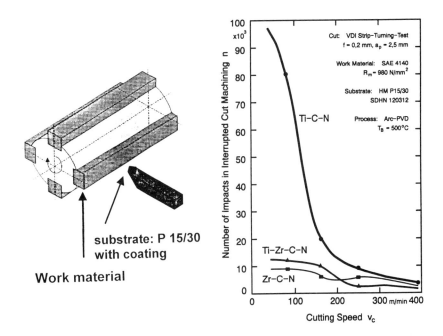

Figure 18 *Endurance behaviour of various films during the VDI strip turning test[9]*

already been predicted on the basis of a combined preliminary test for resistance to thermal changes and impacts.

In the *ball bearing test*, parameters are defined which furnish results typical of different conditions of use and which are a selection from the indefinite number of cases in which ball bearings are used. Here the attempt is made, on the basis of higher loads, to obtain information also about the endurance behaviour under lower loads and at higher numbers of revolution (which would then be possible) and/or about a longer useful life. In the case of this method which is important above all for the development of new materials and intermediate substances, the parameters set and the measurement data must be exactly checked and controlled, if necessary.

7 CONCLUSION

Many methods are available for the characterization of thin films. Their usefulness depends above all on an optimum selection of the measuring method as such and of the parameters applied. Both research and industry require for this purpose:

1. reliable strategies according to which the optimum measuring methods and measurement parameters can be devised for a concrete application,

2. reproducible measuring methods and a description of these,
3. methods for the calibration of these measuring techniques and
4. traceability of the measuring methods and the material measures required for them to national standards.

References

1. F. Löffler, *Surface and Coating Technology*, 1994, **68/69**, 729
2. F. Löffler, *Vacuum*, 1992, **43**, 397.
3. E. Fay, Calibration by Accredited Laboratories: An Element of Quality Management, In: Anais (proceedings) Seminário Internacional de Metrologia Elétrica, Curitiba, Paraná (Brasil), COPEL/PRE/LAC (1994), p. 189–201.
4. H. Jehn, G. Reiners, N. Siegel, Charakterisierung dünner Schichten, DIN–Fachbericht 39, Deutsches Institut für Normung e.V., Berlin 1993.
5. T. Ahbe, K. Hasche and K.-P. Hoffmann, Untersuchungen zur Schichtdickenmeßtechnik in der PTB, PTB-Mitteilungen, 1994, 104, p. 43, Braunschweig.
6. O. W. Madelung, Oberflächenanalyse – Verfahren, Anwendung, Anbieteradressen, VDI-Verlag, 1989, Düsseldorf.
7. E. Lugscheider, F. Löffler, Optimierung der Mikrostruktur von PVD–Hartstoff-schichten für Wälzbeanspruchung, Forschungsbericht der Deutschen Forschungs-gemeinschaft, Lo 515/5–2, DFG (1995), Bonn.
8. G. Waller, A. Schmidt-zum Berge, Vergleichende Untersuchungen von dünnen Hartstoffschichten mit Hilfe der Kavitationserosion, Werkstoffwoche, 5./6.12.1995, Bad Nauheim.
9. O. Knotek, F. Löffler, G. Krämer, *Int. J. of Refractory Metals and Hard Materials*, 1996, **14**, 195.

Section 3.5 Power Generation

3.5.1

Studies of Surface Films on High Ni-Cr-Mo Alloys in Simulated SO$_2$ Scrubber Environment

N. Rajendran, G. Latha and S. Rajeswari

DEPARTMENT OF ANALYTICAL CHEMISTRY, UNIVERSITY OF MADRAS, GUINDY CAMPUS, MADRAS, INDIA

1 INTRODUCTION

Sulphur dioxide, by far the most dangerous atmospheric pollutant, remains a subject of major concern due to its effect on air quality. Man made SO$_2$ emission from many industrial activities accounts for major atmospheric aggressivity. Indeed, SO$_2$ emissions from coal-fired power plants were looked upon as unavoidable, because coal is the only economically available fuel.

Today, Flue Gas Desulphurization (FGD) is the most prevalent technology for controlling the SO$_2$ emission from coal-fired power plants. Amongst the many systems currently available, wet lime/limestone scrubbing is the one which is most advanced and most commonly used[1-6]. Fontana[7] have reported that the capital cost incurred in installing a FGD plant, amounts to almost 25% of the total cost of installation of the whole thermal power plant. Hence, adequate measures have to be undertaken in maintaining the plant from degradation of the materials due to corrosion.

However, the materials of construction presently employed (type 316L SS) for FGD system often fail, due to the localized corrosion attack by the aggressiveness of environment encountered during SO$_2$ scrubbing[8]. The corrosion performance of the material can be improved by suitably alloying with certain elements[9] and/or modifying the corrosive environment by the application of inhibitors[10-12].

The aim of the present work is to evaluate the localized corrosion behaviour and to study the nature of the passive film of the alloys C-276 and alloy 59 in the simulated SO$_2$ environment.

2 EXPERIMENTAL PROCEDURE

2.1 Electrode Preparation

The chemical composition of the materials under study, namely 316L SS, alloy C-276 and alloy 59 is given in Table 1.

Specimens of dimension 1cm × 1cm × 0.3cm were cut from the respective alloy and each specimen was soldered to a copper rod to provide electrical contact. Then they were moulded with epoxy resin in such a way that only one side with 1cm^2 surface area was exposed. Before the experiment, the working electrode was wet ground with SiC papers down to 800 grit followed by 5 micron diamond paste to get scratch free surface.

Table 1 *Chemical Composition of the alloys (wt. %)*

Alloys	Ni	Cr	Mo	C	S	Mn	W	Si
316L SS	12.7	17.2	2.4	0.03	0.003	1.95	-	0.03
Alloy C-276	57.0	16.0	16.0	0.01	0.003	0.40	3.5	0.04
Alloy 59	59.0	23.0	16.0	0.01	0.003	0.40	-	0.04

2.2 Electrochemical Cell Assembly

A three-electrode system comprising a saturated calomel electrode as the reference, platinum foil as the counter electrode and the materials under study, as the working electrode was used, The experiment was carried out in simulated SO_2 scrubber medium, whose composition and operating conditions are given in Table 2.

2.3 Pitting Corrosion Studies

The working electrode was allowed to stabilise for 30 minutes to attain E_{corr}. Then the potential was increased in the noble direction from E_{corr} at a scan rate of 1mV/sec until the breakdown potential (E_b) was attained, where the alloy entered the pitting region. The sweep direction was then reversed after reaching an anodic current density of 3 mA/cm^2 until the potential where the reverse scan current density meets the passive region. The potential at which the reverse anodic scan meets the passive region is termed as the pit protection or the repassivation potential (E_p).

Thus the experimental parameters of interest recorded from the cyclic polarization curve were E_{corr} (corrosion potential), E_b (pitting potential) and E_p (pit-protection potential).

2.4 Crevice Corrosion Studies

The crevice corrosion studies were carried out with the help of a crevice assembly, designed by Dayal et al[13]. The crevice was created on the specimen by bringing the tip of the glass rod of the crevice assembly with the centre of the electrode surface. The critical crevice potential (E_{cc}), was determined by increasing the potential from E_{corr} in the noble direction at a scan rate of 1mV/sec until the current increased rapidly.

2.5 X-ray Photoelectron Spectroscopic Study

The nature of the passive film on 316L SS, alloy C–276 and alloy 59 was

Table 2 *Chemical composition and operating conditions of the baseline solution*

Chloride	10000 ppm
Fluoride	1000 ppm
Sulfite	2000 ppm
pH	5.0
Temperature	52 ± 2°C

The pH of the medium was adjusted with sulfuric acid.

analysed by X-ray photoelectron spectroscopy. The specimens of dimension lcm x lcm x 0.3cm were anodically polarized at 200mV in the SO_2 scrubber medium for one hour, for the growth of the passive film. Then the specimens were removed from the electrochemical cell, rinsed in distilled water, dried in a stream of flowing argon and kept inside a desiccator until they were transferred to the evacuated sample chamber of the XPS unit. The samples were irradiated with MgKα radiation with a mean energy of 1253.6 eV. All the experiments were carried out at a vacuum of 10^{-8} to 10^{-9} torrs. The analyser energy was set at a range of 50 eV for the entire analyses.

The survey spectrum of the specimen was first carried out and ion-etching using argon ions was adopted to remove the air formed contaminated layers. For depth profiling of the passive film, an argon ion source attached with the equipment was used at a current of 10–200μA for various durations of 1, 3, 5, 10 and 20 minutes continuously. The high resolution photoelectron spectra at each interval were taken for $2p_{3/2}$ levels for nickel, chromium and $3d_{5/2}$ level of molybdenum. The output of the photoelectron spectroscopic analyses was obtained as the binding energy versus intensity counts plot through the data acquisition system interfaced to the XPS unit.

The binding energy of the elements nickel, chromium and molybdenum was measured from the spectra obtained for the respective elements and then the value was corrected with respect to C 1s binding energy.

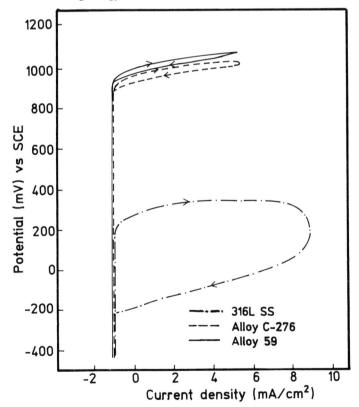

Figure 1 *Potentiodynamic cyclic anodic polarization curves for the type 316L SS, alloy C-276 and alloy 59*

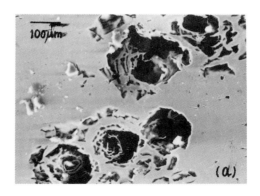

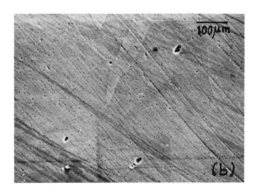

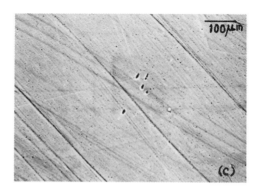

Figure 2 *Pit morphology of (a) 316L SS, (b) alloy C-276 and (c) alloy 59*

3 RESULTS AND DISCUSSION

3.1 Role of Alloying Elements on Pitting Potential

The critical pitting potential values of type 316L SS, alloy C-276 and alloy 59 were determined from the polarization curves and are presented in Figure 1. The mean value of critical pitting potential for 316L SS was 240 mV. The presence of 16% Mo, 57% Ni 16% Cr and 3.5% W in the alloy C–276 increased the E_b value to 930 mV whereas the presence of 16% Mo, 59% Ni and 23% Cr in the alloy 59 increased the E_b value to 950 mV.

Thus, it is evident that the alloys with high Ni, Mo and Cr content increased the pitting potential value and thereby improved the pitting corrosion resistance under simulated FGD environment.

Previous workers have reported that molybdenum addition to Cr and Ni bearing alloys improve pitting resistance and alter the active-passive transition to favour easier passivity[14,15]. Streicher[16] has suggested that the combined addition of chromium and molybdenum to alloys greatly improves their corrosion resistance. Yang et al[17] reported that the presence of Mo inhibits the corrosion process through the formation of a molybdenum salt film, which is apparently difficult to break down.

Morphology of the pitting attack observed from Scanning Electron Microscopy on the type 316L SS, alloy C–276 and alloy 59 are shown in Figure 2 (a–c). The 316L SS indicates larger pits over the entire surface of the specimen. The observed fact suggested that the active site for the initiation and propagation of the pit was scattered throughout the specimen and the material is highly susceptible to localized attack. The alloy C–276 and alloy 59, showed immunity towards pitting attack. This can be attributed due to the hydrolysis of nickel, which yields essentially a neutral pH, so that the dissolved nickel ions are not likely to contribute to the acidification of the solution within the pit. As a result of the reduced aggressiveness of the environment within the pit, the initiation of the pit is less favoured.

The repassivation potential values were determined for the above alloys from the polarization curves and are presented in Figure 1. The addition of Cr, Ni and Mo to 316L SS shifts the repassivation potential from –210 mV to 912 mV in the case of alloy C-276 and from –210 mV to 935 in the case of alloy 59. The significance of these observations is that new pits cannot be initiated above this potential and hence, it can be inferred that the increased Mo, Cr and Ni content hinders the development of new pits and also slows down the kinetics of the growing pits.

From the above points, it is indicated that with regard to the repassivation potential, the results obtained confirm that these alloys are superior to 316L SS.

3.2 Role of Alloying Elements on Critical Crevice Potential

The anodic polarization behaviour of the alloys in the presence of crevice is shown in Figure 3. The critical crevice potential for 316L SS was 142 mV, whereas the alloy C–276 and alloy 59 exhibited the E_{cc} value of 910 mV and 938 mV respectively. Thus, the present study has shown the beneficial effect of molybdenum, chromium and nickel in improving the crevice corrosion resistance of the alloy C-276 and alloy 59.

3.3 XPS Study of the passive film on alloy C-276 and alloy 59

The phenomenon of metallic passivity is the loss of chemical reactivity experienced by certain alloys under particular environmental conditions. The passive film on alloys consists mainly of oxides. The main elements that contribute much for the formation of such barriers are Cr, Ni and Mo. To understand the factors controlling the chemical composition and stability of these metallic oxides on the passive film, it is essential to discuss the role played by these elements.

3.3.1 Ni $2p_{3/2}$ Spectra. The specimens were scanned between 850.0 eV to 865.0 eV to determine the presence of nickel ions and their oxidation states in the passive film.

The Ni $2p_{3/2}$ peak studied for 316L SS showed qualitatively the presence of a single peak at 852.4 eV throughout the depth profile. However, constant intensification of the peaks was observed at increasing depth profile as shown in Figure 4a.

The spectra observed for the alloy C–276 and alloy 59 were resolved into three peaks at the binding energy values of 852.4 eV, 854.5 eV and 855.3 eV in the outer most region and at 1 minute of the depth profiling. Further sputtering enlarged the shoulder of the main peak at

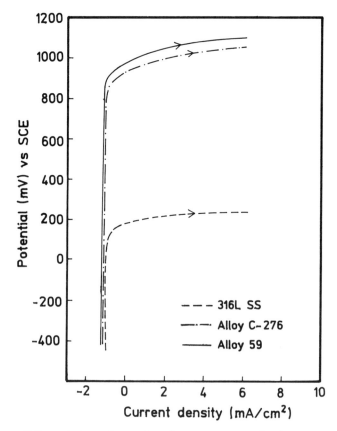

Figure 3 *Potentiodynamic anodic polarization curves for the type 316L SS, alloy C-276 and alloy 59 in the presence of crevice*

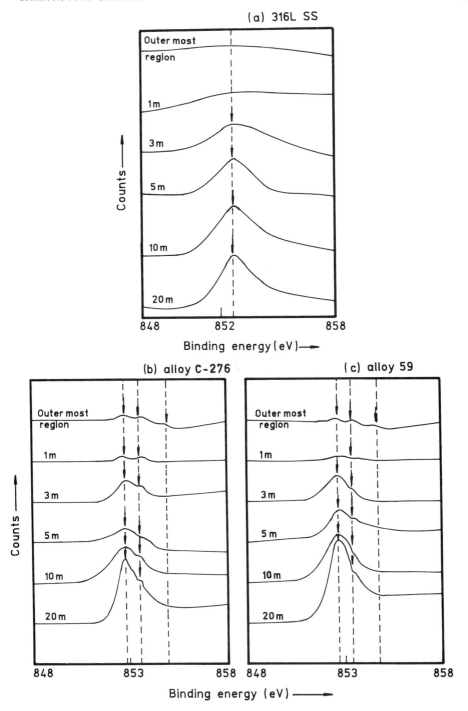

Figure 4 *XPS high resolution spectra of nickel present in the passive films*

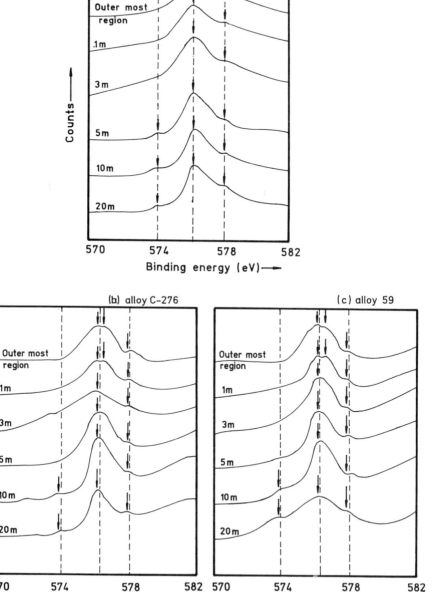

Figure 5 *XPS high reolution spectra of chromium present in the passive films*

852.4 eV with the simultaneous vanishing of the peak at 855.3 eV as shown in Figure 4b & 4c. However, the spectra showed a narrow intense peak at 852.4 eV and a slight shoulder peak at 854.5 eV.

It has been reported that the binding energy values of 852.4 eV, 854.5 eV and 855.3 eV correspond to nickel in the form of metallic state, NiO and Ni(OH)$_2$ respectively[18-21]. Alloy C-276 and alloy 59 showed the existence of NiO and Ni(OH)$_2$ while 316L SS exhibited the peak due to metallic nickel only.

3.3.2 Cr 2p$_{3/2}$ Spectra. Chromium in the form of Cr^{3+} and Cr^{6+} was observed with the corresponding binding energy values of 576.3 eV and 578.1 eV respectively in the outerlayer and at the depth profiling of 1 and 3 minutes. Metallic chromium was detected after the depth profiling of 5, 10 and 20 minutes of etching along with Cr^{3+} and Cr^{6+} as shown in Figure 5a.

In the case of alloy C-276 and alloy 59, Cr was detected at the binding energy values 576.3 eV, 576.7 eV and 578.1 eV in the outermost region and at 1 minute of depth profile. As the depth profiling continued to 3 and 5 minutes, the peak observed at 576.7 eV vanished, whereas the peaks at 576.3 eV and 578.1 eV remained, which is shown in Figures 5b and 5c. However, metallic chromium was observed along with Cr^{3+} and Cr^{6+} at increasing depth profiles to 10 and 20 minutes.

It has been reported that the peak corresponding to the binding energy value of 576.7 eV indicates the presence of Cr^{3+} in the form of CrOOH[22-24]. From the above cases, it is inferred that the Cr^{3+} and Cr^{6+} are in the form of Cr$_2$O$_3$, CrOOH and CrO$_3$, respectively and agree with the previous reports.

3.3.3 Mo 3d$_{5/2}$ Spectra. For the type 316L SS, Mo in the form of Mo^{3+}, Mo^{4+}, Mo^{5+} and Mo^{6+} was detected in the outermost region with the corresponding binding energy values of 228.3 eV, 229.2 eV, 230.7 eV and 232.2 eV respectively. At the depth profile of 1, 3 and 5 minutes, molybdenum in the form of Mo^4, Mo^{5+} and Mo^{6+} was detected. As the depth profiling continued to 10 minutes, only Mo^{4+} and Mo^{6+} were observed. Metallic Mo was detected at the binding energy value of 227.2 eV along with Mo^{4+}, Mo^{5+} and Mo^{6+} at the depth profile of 20 minutes as shown in Figure 6a.

The molybdenum spectra obtained in the case of alloy C–276 and alloy 59 at the outermost region resolved into five peaks corresponding to metallic state, Mo^{3+}, Mo^{4+}, Mo^{5+} and Mo^{6+}. Compared to 316L SS, the peaks were intensified in the case of alloy C–276 and alloy 59, due to the presence of 16% of molybdenum. Almost all peaks were observed as the depth profiling continued to 3, 5, 10 and 20 minutes. However, the peak corresponding to Mo^{3+} completely disappeared after 1 minute as shown in Figures 6(b) and 6(c). The predominance of higher oxidation states at the depth profiles implies that molybdenum aids in the passivation phenomena of the alloy C-276 and alloy 59. Comparing the observed results with that of the earlier, works[26,27], molybdenum in the higher oxidation states namely Mo^{4+} and Mo^{6+} occurs mainly in the form of MoO$_2$ and MoO$_4{}^{2-}$.

3.3.4 O/Fe Peak Area Ratio. The O/Fe peak area ratio of type 316L SS, alloy C-276 and alloy 59 at the outermost, 1, 3, 5, 10 and 20 minutes of etching are shown in Figure 7. It was observed for all the alloys that the outermost region consisted of higher value of O/Fe ratio.

After 1 minute depth profile of the passive film, the O/Fe peak area ratio decreased and continued as the depth profiling increased. In the case of alloy C-276 and alloy 59, the ratio was calculated only after the depth profiling of 5 minutes, because of the lower content of Fe present.

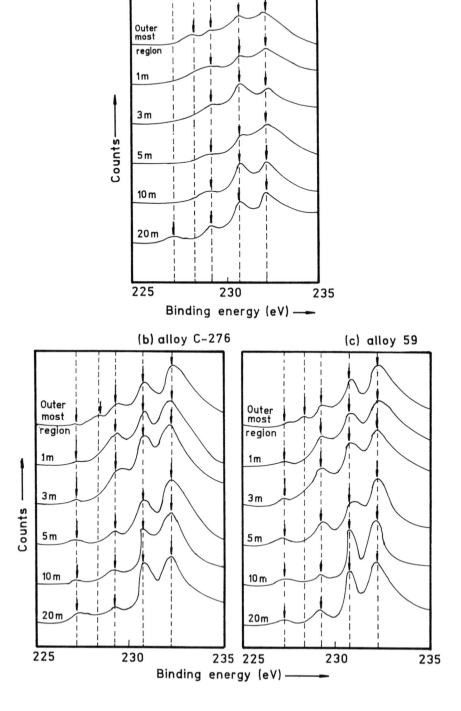

Figure 6 *High resolution spectra of molybdenum present in the passive films*

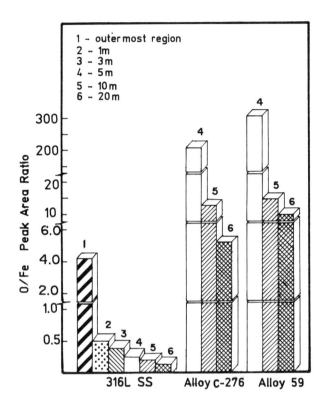

Figure 7 *O/Fe peak area ratio of type 316L SS, alloy C-276 and alloy 59*

The O/Fe peak area was used to indicate the nature of the passive film as hydrated, polymeric or amorphous and the ideal value of O/Fe peak area ratio greater than 2 indicates the hydrated oxide film.

In the present study, all the specimens showed a higher O/Fe ratio greater than 2 indicating that the above specimens were in hydrated state. As the amount of alloying elements namely Mo and Ni increases, the O/Fe peak aria ratio increases to very high value than 316L SS, which indicates the presence of oxygen in the form of bound water or hydroxyl ions (OH⁻) in the passive film and these ions will increase the stability of the passive film against the pitting attack.

4 CONCLUSIONS

From the above investigations, we can conclude that:

1. The shift in the critical pitting, pit-protection and crevice potentials of alloy C-276 and alloy 59 towards more noble direction compared to 316L SS is due to the beneficial effect of alloying elements.

2. The XPS analyses of the passive film formed on high Ni-Cr-Mo alloys revealed the presence of chromium enriched as CrOOH in the passive film. In contrast, the CrOOH was not detected in the reference type 316L SS. Hence, the presence of CrOOH in the high Ni-Cr Mo alloys imparted higher corrosion resistance.

3. The beneficial effect of molybdenum in the form of MoO_2, MoO_4^{2-} and MoO_3 in the passive film of the alloy C–276 and alloy 59 stabilizes the oxyhydroxides of chromium, thereby promoting higher corrosion resistance.

4. Thus, from the electrochemical and XPS investigations it can be concluded that the presence of higher amounts of alloying elements mainly chromium and molybdenum in the alloy C–276 and alloy 59 imparts higher resistance towards localized attack, compared to that of 316L SS.

Acknowledgement

One of the authors Mr.N.Rajendran is grateful to CSIR, New Delhi,
India for providing financial assistance.

References

1. O. Egra, *Chemical Engineering and Technology*, 1988, **11**, 402.
2. M. Garding and G. Shedberg, *Journal of the Air Pollution Control Association*, 1988, **38**, 1275.
3. S. Royse, *Process Engineering* (London), 1989, **70**, 43.
4. B. Weinstein, *Industrial and Engineering Chemistry Research*, 1989, 9 **28**, 256.
5. C. Tullln and E. Ljungstrom, *Energy and Fuels*, 1989, **3**, 284.
6. G. H. Koch, N. G. Thompson and J. L. Means, 'Proceedings of the Symposium on Materials Evaluation and Environmental Effects on Corrosion in Flue Gas Desulphurization Systems', Corrosion/84, NACE, 1984, p. 273.
7. M. G. Fontana, 'Corrosion Engineering', 3rd Edn., McGraw-Hill Book Company, New York, 1987, p. 416.
8. C. A. Johnson, 'Proceedings of the Symposium on Performance of Construction Materials in Flue Gas Desulphurization Systems', NACE, Houston, 1984, p. 29.
9. N. Rajendran and S. Rajeswari, *Journal of Materials Engineering and Performance* 1996, **5**, 46.
10. N. Rajendran, K. Ravichandran and S. Rajeswari, *Bull. of Electrochem.*, 1993, **9**, 4.
11. N. Rajendran, K. Ravichandran and S.Rajeswari, *Anti-Corrosion Methods and Materials*, 1995, **42**, 8.
12. N. Rajendran and S.Rajeswari, *Anti-Corrost'on Methods and Materials*, 1995, **42**, 13.
13. R. K. Dayal, N. Parvathavarthini and J. B. Gnanamoorthy, *Br. Corros. J.*, 1983, **187** 184.
14. R. J. Brigham, *Corrosion*, 1972, **28**, 177.
15. M. B. Rockel, *Corrosion*, 1973, **29**, 393.
16. M. J. Streicher, *Corrosion*, 1974, **30**, 77.
17. W. Yang, R. Chang and H. Z. Hua, *Corros. Sci.*, 1984, **24**, 691.
18. B. P. Lochell and H.H.Strehblow, *J. Electrochem. Soc.*, 1984, **131**, 713.

19. G. E. Therlault, T.L. Barry and M. J. B. Thomas, *Anal. Chem.*, 1975, **47,** 1492.
20. P. Marcus and J. M. Herbelin, *Corros. Sci.*, 1993, **34,** 1123.
21. N. S. McIntyre, D. G. Zofaruk and D. Owen, *J. Electrochem. Soc.*, 1979, **126,** 750.
22. J. A. L.Dobbelaar and J. H. W. de Wit, *J. Electrochem. Soc.*, 1990, **137,** 2038.
23. C. R. Clayton and Y. C. Lu, *J. Electrochem. Soc.*, 1986, **133,** 2465.
24. A. R. Brooks, C. R. Clayton, K. Doss and Y. C. Lu, *J. Electrochem. Soc.*, 1986, **133,** 2459.
25. P. Marcus and M. E. Bussell, *Appl. Surf Soc.*, 1992, **59,** 7.
26. W.C.Moshier, G. D. Davis and G. D. Cote, *J. Electrochem. Soc.*, 1989,**136,** 356.
27. W.C.Moshier, G. D. Davis, J. S. Aheran and H. F. Hough, *J. Electrochem. Soc.*, 1987, **134,** 266.
28. C. R. Clayton and Y. C. Lu, *Surf. and Inter.* Anal., 1989, **14,** 66.
29. M. Che, M. Fournier and J. P. Launay, *J. Chem.Phys.*, 1979, **71,** 15.
30. J. O. M. Bockris, *Corros. Scl.*, 1989, **29,** 291.
31. T. E. Poe, O. J. Murphy, V. Young, J. O. M. Bockris and L.L. Tongsen, *J. Electrochem. Soc.*, 1984, **131,** 1243.

3.5.2

Surface Engineering of Composite and Graded Coatings for Resistance to Solid Particle Erosion at Elevated Temperatures

M. M. Stack and D. Pena

CORROSION AND PROTECTION CENTRE, UMIST, P.O. BOX 88, SACKVILLE ST., MANCHESTER, UK

1 INTRODUCTION

Although there has been some work carried out on the erosive wear of metal matrix composites[1–5] at room temperatures, there has been little work carried out at elevated temperatures. This is despite the fact that such materials have many applications in such conditions. Additionally, gradation of hard reinforcement particulates or fibres through the material provides a means of minimizing adverse thermal mismatch effects. In particular, the use of such materials as surface treatments provides a means of ensuring adequate erosion and corrosion resistance, in addition to satisfying the load bearing requirements of the substrate material (generally steels).

In studies of abrasion of MMCs, there have been conflicting reports on the extent to which the reinforcement can reduce the wastage rate of the material. For the abrasive wear of an Al based MMC containing 20% volume fraction of SiC[3], a continuous decrease in the relative wear resistance, with increasing abrasive particle size, was recorded. In a study of an Al based MMC(reinforced with Alumina particulates[1]), the relative wear resistance reached a maximum value with increasing Alumina volume fraction, for exposure to 60 μm SiC particles, whereupon it commenced to decrease with further increases in % Alumina. However, when the abrasive particle size was decreased to 20 μm, the behaviour changed. In this case, a continuous increase in the relative wear resistance, with increasing Alumina volume fraction, was recorded. In research on the abrasive wear of an Al-Si based MMC containing 100 μm Zircon particles[4], an increase in the relative wear resistance with increasing Zircon volume fraction was recorded for "low stress" abrasion. For "high stress" abrasion, the relative wear resistance appeared independent of particulate volume fraction.

In further work, on the erosion of the Al/Alumina based MMC above[1], particle angularity and impact angle were shown to have a significant effect on the erosion behaviour. For exposure to 100 μm silicon carbide particles, the erosion was continually higher for the MMC containing 30% volume fraction Alumina than for the base material, as the impact angle was increased from 20° to 90°. For erosion with less angular 100 μm silica particles, changes in the ranking order of erosion resistance for the various MMCs were observed, as the impact angle was increased, i.e. the highest erosion rate at 20° was reported for the matrix material whereas at 90° the maximum erosion recorded was for the 30% Alumina material. For erosion of an Al based MMC by 250 μm silica in slurry conditions[5], a continuous decrease in the erosion rate,

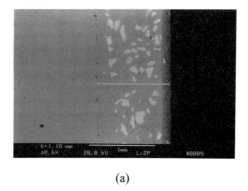

(a)

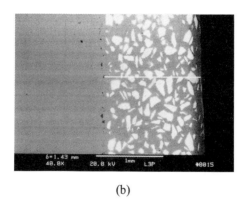

(b)

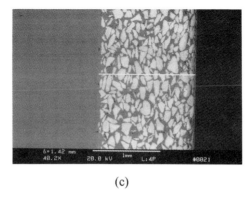

(c)

Figure 1 *Scanning electronmicrographs of the Ni-Cr/Wc MMCs (a) 9% WC, (b) 19% WC and (c) 34% WC*

with increasing fibre volume fraction, was recorded at 15°. However, this decrease was less pronounced for erosion at 90°.

Such conflicting results, particularly those on the effect of increasing reinforcement volume fraction, were attributed to the variation of erosion mechanism for MMCs in the different exposure conditions[2]. It is thought that additions of hard reinforcement fibres or particulates are only beneficial as long as fracture of the reinforcement can be avoided. Once fracture of the reinforcement occurs, then the performance of MMCs may, in some cases, be inferior to that of the base material.

At present, many MMC coatings are available commercially with variable volume fractions of reinforcement particles or fibres. However, on the basis of the above observations, there is good evidence that the composition required depends on properties of the impacting particle and the corrosive environment. Hence, there is a need to develop a rationale for selection of the appropriate coating composition for the exposure conditions.

One of the methods by which process or materials parameters can be optimized, in erosion and wear environments, is by the use of engineering "maps"[6-10]. Such maps can be used to identify the mechanism of damage and the level of wastage rate. At present, there is little published work on erosion mapping of composite materials; the objective of this work has been to address this issue.

Hence, the aim of this work has been to carry out a study of the erosion of the composite layers of a functionally graded Ni-Cr/WC material suitable for coating applications. The effects of increasing volume fraction of the WC volume fraction were evaluated at a range of velocities and temperatures. It was shown how erosion mechanism and materials performance maps could be constructed, on the basis of the results, and how such maps could be used as an aid to surface engineering of composite materials for resistance to elevated temperature erosion.

2 EXPERIMENTAL DETAILS

The erosion-corrosion apparatus[11], consisted of a fluidized bed of erodent particles. The velocity was controlled by a tachometer system which allowed the specimens to rotate at a range of angular velocities vertically through a gently bubbling bed of particles. The specimen assembly consisted of a cross piece on a rotating axle which was attached to the end of a tubular spindle supported inside self-aligned bearings.

The experimental run duration was 20 h for the tests reported below, and the relative velocity between the specimens and the particles was between 1 and 6 ms^{-1}. The specimens were rectangular (thickness 4mm, width 6mm and length 10mm) and were secured flush to the surface of the cross piece to give an impact angle of 90° between the particles and the main face. The erodent for these tests was alumina, and the average particle diameter of the erodent studied was 150 μm. Thickness loss measurements were made using a digital micrometer, with a precision of 1μm, and represent an average of 10 points over the surface. The specimens were subsequently examined by Scanning Electron Microscopy.

The MMCs consisted of WC particles (150 μm average diameter) in a Ni-Cr matrix(Ni-13% Cr-B-Si), and were supplied by Castolin S.A. Lausanne. The materials were flame sprayed. Scanning electron micrographs of the materials containing 9%, 19% and 34% volume fraction of WC particles, are shown on Figure 1.

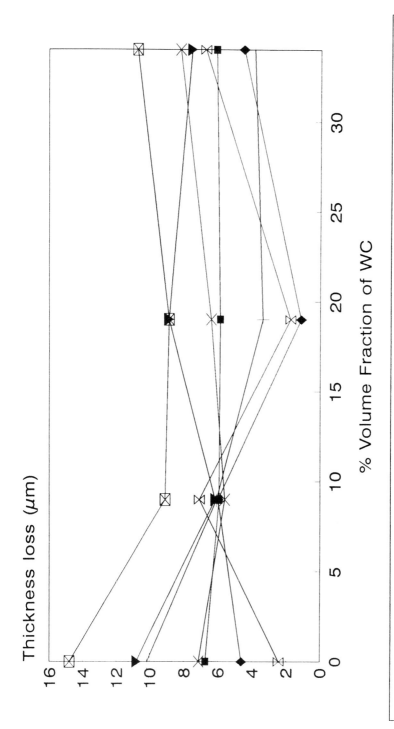

Figure 2(a) *The effect of velocity on the wastage rates of the MMCs and the matrix material at 25°C*

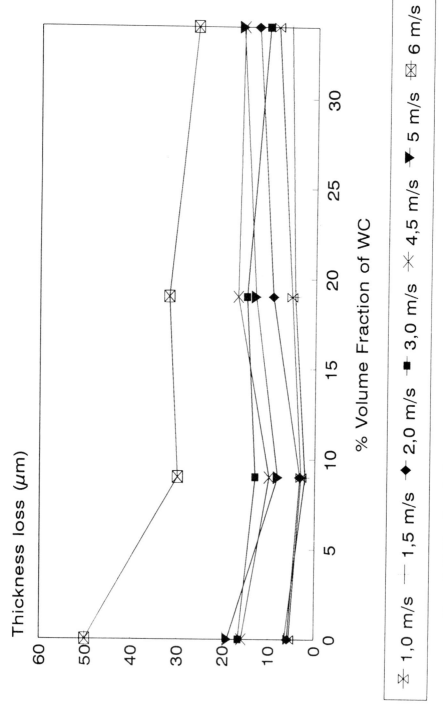

Figure 2 (b) *The effect of velocity on the wastage rates of the MMCs and the matrix material at 600°C*

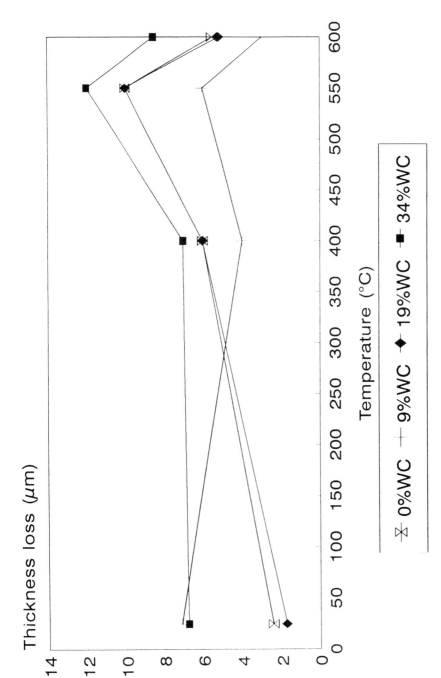

Figure 3 (a) *Effect of temperature on the wastage rates of the MMCs and the matrix material at 1 ms⁻¹*

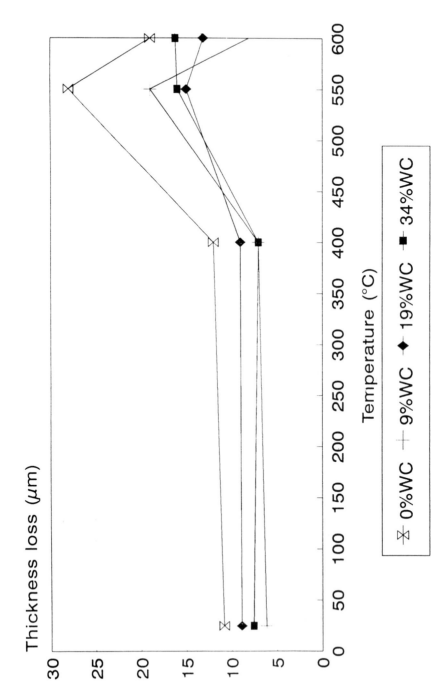

Figure 3 (b) *Effect of temperature on the wastage rates of the MMCs and the matrix material at 5 ms⁻¹*

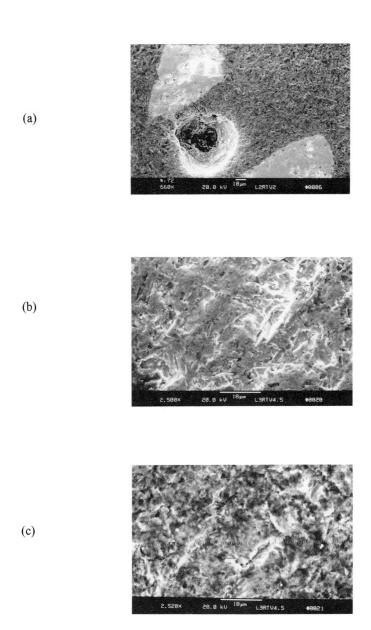

Figure 4 *Scanning electron micrographs of the Ni-Cr/WC MMCs at 25°C (a) 9% WC at 2 ms⁻¹ showing location where WC particle was removed from surface, (b) 19% WC at 4.5 ms⁻¹ showing high magnification of surface WC particles and (c) 19% WC at 4.5 ms⁻¹ showing surface of matrix material*

(a)

(b)

(c)

(d)

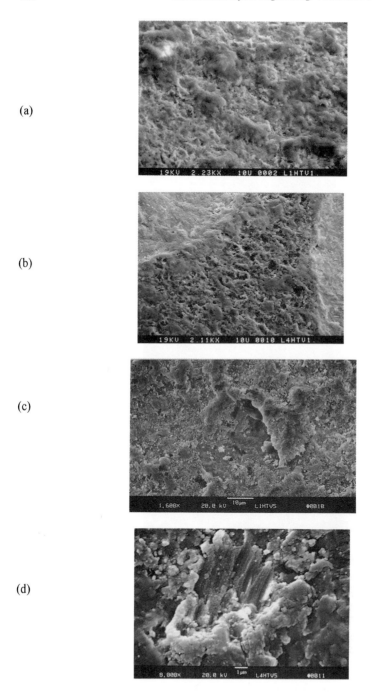

Figure 5 *Scanning electron micrographs of the Ni–Cr/WC MMCs at 600°C (a) matrix materials at 1 ms⁻¹, (b) 34% WC at 1 ms⁻¹, (c) matrix materials at 5 ms⁻¹ and (d) 34% WC at 5 ms⁻¹*

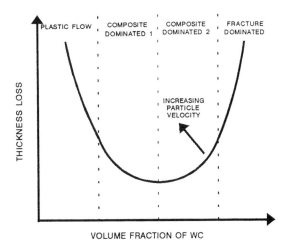

Figure 6 *Schematic diagram of the erosion regimes for the Ni–Cr/WC MMCs on the thickness loss versus %WC curve*

3 RESULTS

3.1 Volume Fraction Effects From 1 to 6 ms⁻¹

The general trend on the effect of volume fraction, Figure 2(a), was similar to that observed in previous work[12]. A minimum was observed in the thickness loss values recorded at intermediate volume fractions. This minimum in the curve appeared to shift to lower volume fractions with increasing impact velocity. For example, at 1.5 ms⁻¹, the minimum in the curve occurred at 19% WC; at 5 ms⁻¹ the minimum shifted to 9%

It was interesting that the effect of velocity varied for the various volume fractions. For example at 9% WC, there was little difference between the recorded thickness losses up to 5 ms⁻¹ whereas for the 19% WC material, the differences between the recorded thickness losses were more pronounced. (It should be noted that there was some scatter in the results, particularly at the lower temperatures and velocities where the recorded thickness losses were relatively low).

At 600°C, Figure 2(b), the effect of volume fraction was similar to that observed at room temperature i.e. a minimum in the curve was recorded. However, there was less evidence of this minimum shifting to lower volume fractions of WC with increasing velocity, as had been the case at the lower temperature of 25°C, Figure 2(a). The magnitudes of the wastage rates were also greater than at the lower temperature.

3.2 Effect of Temperature

The effect of temperature is shown for the various volume fractions at 1 ms⁻¹, Figure 3(a).

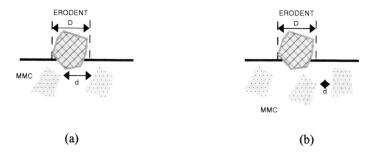

(a)   (b)

Figure 7 *Scematic diagram of regimes of behavoiur for erosion of MMCs (a) size of contact zone approaches distance between the WC particles and (b) size of contact zone is greater than distance between the WC particles*

The general pattern showed that the thickness loss increased with increasing temperature up to a critical value i.e. 550°C, whereupon it decreased with further increases in temperature. It was interesting that the ranking order of erosion resistance of the various MMCs changed as a function of temperature i.e. when the temperature was increased from 25°C to 550°C, the WC % at which the highest erosion rate was recorded changed from 9% to 34%.

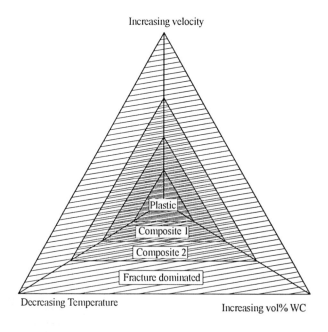

Figure 8 *Schematic diagram of three dimensional erosion mechanism map for MMCs showing the erosion transitions as a function of temperature, particle velocity and %WC*

At 5 ms⁻¹, Figure 3(b), the general effect of temperature was similar to that at 1 ms⁻¹. However, for the 34% WC containing material, no reduction in the thickness loss occurred at the higher temperature, 600°C. There was also less evidence of the ranking order of the various MMCs changing as a function of temperature(as compared to that observed at the lower velocities).

3.3 Microscopy of the Eroded Surfaces

Scanning electron microscopy showed evidence of brittle fracture of the reinforcement particulates, particularly at the higher velocities. At 2 ms⁻¹ there was some indication of removal of reinforcement particulates, Figure 4(a). At 4.5 ms⁻¹, the surface of the WC particulates and the matrix material were compared, Figure 4(b–c). Clearly, there was evidence of brittle chipping of the WC, whereas the matrix material appeared to have undergone a ductile cutting mechanism.

At 600°C, Figure 5(a), there appeared to have been oxidation of the surface at the low velocities i.e. 1 ms⁻¹. EDX analysis(not shown) also confirmed the presence of embedded alumina particles. At such low velocities, there was little evidence of erosion of the reinforcement particulates, Figure 5(b).

At 5 ms⁻¹, and at 600°C, the surface of the matrix material appeared to be covered by flakes of material, which were possibly a combination of extruded metal and oxide. There was also some evidence of deposition of erodent material, Figure 5(c). For the reinforcement particulates in such conditions, extensive fracture was observed, Figure 5(d).

4 DISCUSSION

4.1 Velocity, Temperature and Volume Fraction Effects on the Erosion of MMCs

The results show that the general effect of temperature, Figure 3, is similar to that observed for that of Fe based alloys, in fluidized bed conditions[11,13], with a peak in the thickness loss observed at intermediate temperatures. Various theories on the reasons for this peak have been proposed; one such theory[14] is that it is associated with the formation of a "critical oxide thickness", which provides resistance to erosion above the so-called critical temperature. Because the corrosion rate of the Ni–13%Cr matrix material is lower than that for Fe based alloys[15], the results obtained in this study might suggest that the peak is attributable to other reasons. The fact that the peak is not observed for the MMC containing higher % WC, above 5 ms⁻¹, might suggest that it represents a genuine "brittle" to "ductile" transition because this material would be more likely to be dominated by brittle erosion mechanisms, up to a higher temperature, than the lower % WC materials. However, further work needs to be carried out to address this issue.

It is interesting that the ranking order of material performance changes as a function of temperature, Figure 3. This indicates a variation in the erosion mechanism for the materials as a function of temperature; such changes in the ranking order of erosion-corrosion resistance for mild steel and stainless steel, as a function of temperature, have been observed in previous work[13,16]. The effect of increasing WC volume fraction is consistent with the results from the literature, on the erosion and abrasion of MMCs[1]. The minimum in the curve has been associated with the transition in the erosion behaviour from plastic flow of the material, to brittle fracture[2].

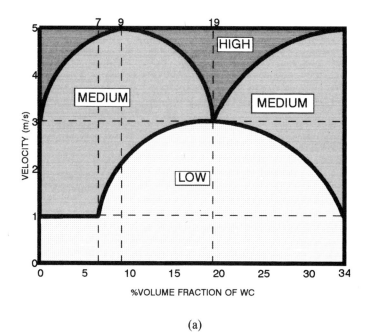

(a)

(b)

Figure 9 *Erosion materials performance map for MMCs showing the variation of wastage levels as a function of velocity and %WC for (a) 25°C and (b) 600°C*

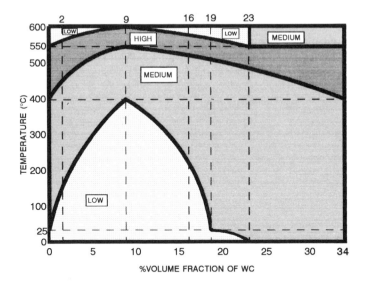

(a)

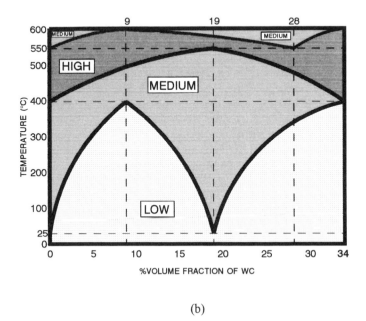

(b)

Figure 10 *Erosion materials performance map for MMCs showing the variation of wastage levels as a function of temperature and %WC for (a) 1 ms^{-1} and (b) 5 ms^{-1}*

This is addressed in more detail below[2].

The effect of velocity on the thickness loss versus WC volume fraction curve, Figure 2, has been to shift the minimum to lower % WC and higher wastage rates as outlined schematically, Figure 6. Again, this transition is associated with a change in erosion mechanism, because brittle erosion mechanisms, for MMCs, are favoured at higher velocities. The effect of temperature on the curve has shown that the region where WC additions are beneficial, Figure 6, expands to higher volume fractions with increasing temperature, an observation attributed to enhanced plasticity at higher temperatures.

4.2 Regimes of Damage for Composite and Graded Materials

The results have shown that regimes of damage can be identified for composite and graded materials, Figure 6. An attempt (in previous work[12]) has been made to identify four regimes of degradation for MMCs:

i.e.   (i)   plasticity dominated
        (ii)   composite dominated 1
        (iii) composite dominated 2
        (iv)   fracture dominated

The predominance of such regimes depends on the:

(a)   target i.e. volume fraction and size of reinforcement, oxidation resistance, hardness, and toughness of the MMC constituents.
(b)   particle i.e. size, shape, velocity, impact angle, toughness, hardness, flux
(c)   environment i.e. temperature, gas composition

For example, regimes of interaction can be defined as in Figure 7 on the basis of the distance between the reinforcement particles, and the size of the contact zone defined by the particle impact energy. Two regimes are identified in this case; one in which the distance between the reinforcement particles is less than the size of the contact zone, and the other where it is much greater. In the latter case, the erosion mechanism is be expected to be dominated by ductile erosion mechanisms i.e. "plasticity dominated" behaviour. In the former case, the mechanism is anticipated to be dominated by brittle fracture although this, in turn, would be affected by particle shape and temperature. Using such an analysis, Figure 8, a three dimensional erosion map can be constructed for MMCs showing the regime transitions, as a function of temperature, volume fraction of WC and impact velocity. The "fracture-dominated" regime is clearly identified with high velocities, low temperatures and high volume fractions of WC. The "plasticity-dominated" regime is favoured at high temperatures, low velocities, and low volume fractions of WC. Increasing particle flux, for a given % WC, shifts the "fracture-dominated" regime to higher temperatures and to lower particle velocities. The effect of increasing particle angularity is expected to have a similar effect on the transition boundaries. This map can also describe the behaviour of a graded material where the volume fraction of the reinforcement varies through the material.

It is important to note that the size of the contact zone versus the size of the reinforcement particle will also influence the erosion regime. If the size of the contact zone is significantly greater than the size of the reinforcement, then the response is said to be homogenous and the reinforcement is thought to have little effect on the erosion mechanism[2]. If the size of the contact zone approaches that of the reinforcement, then the erosion response is said to be heterogenous and this characterizes the situation in which the reinforcement may increase the erosion resistance under specific conditions. For the conditions in this study, the sizes of the erodent and WC particulates are equal i.e. 150 μm, which suggests a heterogenous response. However, for the erosion mechanism map above to have a more generic application, it is likely that the effect of reinforcement size also needs to be considered on the transition boundaries.

4.3 Surface Engineering of Composites and Graded Coatings Using Erosion Maps

An erosion mechanism map can be used as in the case of Figure 8, to identify the predominant mode of damage for the material. Knowledge of the degradation mechanism can be useful in materials selection decisions. However, identification of the actual wastage rate as a function of the main erosion and environmental variables is also important in order to assist in life prediction and process parameter optimization.

Figure 9(a) is an erosion materials performance map, based on the above results, showing the wastage rate ranging from "low" to "high" levels.(These levels are purely subjective limits and are dependent on the requirements of the process.) The map shows the various combinations of impact velocities and % WC over which the erosion is minimized for the conditions shown. At the higher temperature of 600°C, Figure 9(b), the boundaries show significant differences to those at room temperature, Figure 9(a). The fact that the boundaries are dependent on temperature means that surface engineering of such materials for erosion resistance must be approached with caution.

Alternatively, the boundaries on the map can be generated as a function of temperature and volume fraction of WC, Figure 10(a). The "low" region of wastage at the lower velocity of 1 ms⁻¹, at 600°C, may be attributed to a general transition to "ductile" behaviour, or to the beneficial effects of oxidation of the matrix material at higher temperatures. It is interesting that, at the higher velocity of 5 ms⁻¹, this region disappears, although there is still evidence of a reduction in the wastage at the higher temperatures, Figure 10(b), from "high" to "medium" levels.

Clearly such maps can be extremely useful in optimizing coatings and materials parameters for composite materials. They can be used to describe the performance of graded coatings, where the volume fraction of reinforcement particles decreases continually from the external layer to the coating-substrate interface, and thus may be an important tool in the future surface engineering of such materials.

5 CONCLUSIONS

i. A study has been undertaken of the combined effects of velocity, temperature and volume fraction for a range of Ni-Cr/WC metal matrix composites.

ii. Attempts have been made to identify regimes of damage for such materials.

iii. Erosion mechanism and materials performance maps have been constructed using the results from the study.

iv. Such maps may be a significant future aid in the surface engineering of MMCs and functionally graded coatings.

References

1. I. M. Hutchings, and A. Wang, Proc. Int. Nat. Conf. New. Mat. and their App., Univ. Warwick, Pub. Instit. Phys. Conf.Ser., 1990, Vol. 111, p. 91.
2. I. M. Hutchings, Proc. Euromat 91, Cambridge, Pub. Instit. Mat., 1991, p. 56.
3. S. Wilson and A. Ball, 'Tribology of Composite Materials', 1990, ASM, p. 103.
4. S. V. Prasad, P. K. Rohatgi and T. H. Kosel, *Mat. Sci. Eng.*, 1986, **80**, 213.
5. J. Masounave, S. Turenne, C. LeDore and G. Gagnon, Proc., ECF7, Budapest, 1990, p. 1255.
6. G. Sundararajan, *Wear*, 1991, **45**, 251.
7. D. J. Stephenson and J. R. Nicholls, *Wear*, 1995, **186-187**, 284.
8. M. M. Stack and L. Bray, *Wear*, 1995, **186-187**, 273.
9. S.C. Lim, M.F. Ashby, J.H. Brunton, *Acta. Metall.* 1987, **3**, 1343.
10. I. M. Hutchings, *J. Phys D. App. Phys.*, 1992, **25**, A212.
11. M. M. Stack, F.H. Stott and G.C. Wood, *Mat. Sci. Tech*, 1991, **7**, 1128.
12. M.M. Stack, J. Chacon-Nava and M.P. Jordan, *Mat. Sci. Tech.* 1996 **12**, 171.
13. A. J. Ninham, I. M. Hutchings and J. A. Little, *CORROSION*, 1990, **46**, 296.
14. V. Sethi and I. G. Wright, in 'Corrosion and Particle Erosion at High Temperatures', (ed. V. Srinivasan and K. Vedula), Pub. Minerals, Metals and Materials Society, Warrandale PA, 1989, p. 245.
15. N. Birks and G. H. Meier, 'An Introduction to High Temperature Oxidation of Metals', Pub. Edward Arnold, London, 1983.
16. M. M. Stack, F. H. Stott and G. C. Wood, *Wear*, 1993, **162-164(B)**, 706.

3.5.3

A Study of the Corrosion Resistance and Microstructure of a Laser Treated Layer on Duplex and Austenitic Stainless Steels

A. Neville[1] and T. Hodgkiess[2]

[1]DEPARTMENT OF MECHANICAL AND CHEMICAL ENGINEERING. HERIOT-WATT UNIVERSITY, EDINBURGH. SCOTLAND

[2]DEPARTMENT OF MECHANICAL ENGINEERING. GLASGOW UNIVERSITY. GLASGOW, SCOTLAND

1 INTRODUCTION

The demands made on engineering components today are often so exacting that it is not possible to meet them with conventional alloys. Materials' development has progressed significantly in recent years in the direction of surface engineering which describes a wide field of material processing, aimed solely at enhancing the properties (physical, mechanical or metallurgical) of a surface to closer meet with design requirements.

Surface engineering was born out of a need to address the limitations of traditional materials with respect to wear resistance, corrosion, fatigue strength and fretting resistance. The use of the more exotic state-of-the-art materials such as ceramics or metal-matrix composites is often unsuitable due to economic constraints or the inability to mechanically form the material. Surface engineering processes have the potential advantage of cost reduction by limiting the treatment to only the most susceptible areas. Also in some applications, the performance of less expensive substrate materials can be enhanced to give performance comparable with higher grade materials.

1.1 Laser Surface Modification

Since its development in the 1960s, the Laser has found application in a range of materials processing environments. Significant developments have been made in the laser cutting of components, to obtain a near-cut area free from heat affected zone effects[1,2]. Laser techniques are currently being developed to cut ceramic materials without detrimental brittle damage in the cut region. Laser welding has been extensively used in replacement of traditional techniques[3] in a number of industrial sectors[4] for example in the car industry because of its suitability for automation and the potential to increase production output.

In the field of surface engineering, there is much activity in industry and academia focusing on techniques such as laser cladding, surface melting and alloying and coating consolidation[5,6].

Laser cladding involves supplying sufficient heat input to a powder which then forms a metallurgical bond between the solidified material and the base metal surface[4,7]. Laser cladding is often used to apply hard facing (e.g. cobalt-base alloys) to a material surface to resist wear degradation[8,9]. Laser application uses a multi-pass technique to apply the material in powder form which has reduced the major problem of dilution from the substrate which was often associated with the TIG welding process[10].

Laser surface melting has mainly been performed with the intention of increasing the surface hardness of the material. Under an inert shrouding gas, surface hardening by laser treatment has been widely demonstrated; for instance on cast iron[11] and on Al-Si alloys[12]. Laser surface melting, when conducted in a gaseous atmosphere can also be termed laser surface alloying if the gas can diffuse sufficiently into the substrate. Laser surface alloying of titanium has successfully been achieved in a nitrogen atmosphere to produce a hard, wear resistant layer of TiN[13]. Further applications for the laser carbo-nitriding of titanium are being investigated.

Application of surface coatings consisting of a ceramic-metal matrix or pure ceramic, often results in gross porosity and a lack of adherence at the coating/substrate interface. Techniques under the description of laser consolidation have been attempted[14,15] to reduce detrimental porosity without causing a HAZ effect in the substrate. The porosity of Cr_2O_3 plasma sprayed coating could be significantly reduced by laser post treatment[14].

Laser surface irradiation of Fe-based alloys, including stainless steels has been studied in recent years by several groups[16,17], the bulk of work concentrating on austenitic stainless steels, in an attempt to improve the surface properties with respect to corrosion resistance. The resulting rate of cooling determines the microstructural changes during the phase transformation (i.e. the formation of precipitates etc.).

The beneficial effects of nitrogen as an alloying element in stainless steels have been recognised by several authors. Nitrogen has been reported to increase the resistance of stainless steels to stress corrosion cracking (SCC)[18], to decrease the risk of intermetallic phase formation in duplex stainless steels[19] and to increase the pitting potential of austenitic stainless steels[18]. There have been several methods attempted to increase the nitrogen content of the surface of a material, the most common of these being plasma nitriding and ion nitriding[20]. However the use of laser surface melting in a nitrogen atmosphere on stainless steels represents a logical method of increasing locally the level of nitrogen and hence the mechanical and corrosion properties. This paper represents an experimental study to investigate the physical and mechanical properties which occur during laser surface treatment of stainless steels under a nitrogen shielding gas.

2 EXPERIMENTAL METHODS

2.1 Laser Treatment

Surface irradiation was carried out using a CO_2 laser with nitrogen as the shrouding gas. The gas pressure was kept constant at 5 bar. Figure 1 is a schematic representation of the laser treatment equipment. From previous work, it had been established that defocusing the laser beam gave a better penetration of the surface and so in all tests, the workpiece was positioned 0.3mm above the focal point as shown in Figure 1.

A constant sample feed rate of 1000mm/min was used for all the surface treatments, again determined from a series of tests done prior to this work. Preliminary tests were performed using continuous power but excessive surface melting resulted in distortion of the metal sample and hence the remaining work was done on a pulsed power cycle of 2×10^{-3} seconds on and 1×10^{-3} seconds off. The power of the laser was 1 kW. The degree of transverse overlap was varied according to the results of microscopical examination and corrosion tests. Primarily laser treatment was concentrated on duplex stainless steels (UNS S32760 and UNS S31803)

Table 1 *Compositions of alloys as received*

	Cr	MO	Ni	N	C	Cu
UNS S32760	24.61	4.04	8.42	0.22	0.025	0.88
UNS S31803	21.56	3.5	5.5	0.17	0.026	
UNS S31603	17.04	2.76	11.61		0.02	
UNS S31254	19.68	6.41	17.93		0.02	0.74

but to conclude the study, two austenitic stainless steels (UNS 31603 and UNS S31254) were treated. Compositions of the alloys are given in Table 1.

2.2 Testing and Analysis

After laser treatment the specimens were cut in cross section in a line parallel to the run direction and also transverse to it. In cross section, a microhardness profile was obtained, the results quoted in Vickers hardness. The specimens were abraded and then polished to a 1

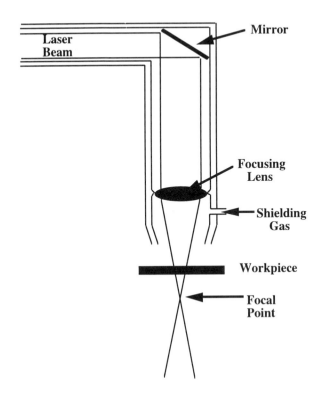

Figure 1 CO_2 *laser configuration for surface treatment of stainless steels*

Figure 2 *Untreated regions between laser runs on UNS S32760*

micron diamond finish. They were then etched electrolytically in solutions of either chromic and oxalic acid or in 40% potassium hydroxide depending on the metallurgical features of interest. In 40% KOH, the ferrite phase is darkened. The cross section was subsequently examined using light and scanning electron microscopy (SEM) to determine the thickness of the laser layer, the uniformity of the layer and the microstructure compared to the substrate material.

Corrosion tests on each laser treated material were carried out in seawater using DC-anodic polarisation techniques at ambient temperature (18°C) and at 50°C. The specimen, encapsulated in a non-conducting resin, consisted of a 1 cm^2 surface area, around which a lacquer was painted to seal the interface between the sample and the resin. The electrode potential, measured with reference to a saturated calomel electrode (SCE) was increased from the free corrosion potential E_{corr} at a rate of 15mV/min, until a sudden increase in the cell current was detected. Once an anodic current of 500μA was attained the electrode potential was scanned at the same rate back to E_{corr}. Comparisons were made between the performance of the laser treated layer and the substrate based on electrochemical behaviour and on the corrosion mechanisms observed via microscopy. Each polarisation test was repeated three times on each material.

Chemical analysis techniques were utilised to attempt to detect compositional changes (and in particular an increase in nitrogen concentration within the laser layer) and included the electron probe microanalysis (EPMA) facility on the SEM, XRD and a laser ionisation mass analyser (LIMA) process. LIMA is an analytical, mass spectrometry instrument capable of providing elemental, isotopic microanalysis.

200μm

Figure 3 *Plan view of the laser treated surface showing rough regions between runs*

3 RESULTS

3.1 Preliminary Tests and Laser Parameter Determination

Performance testing of the laser treated specimens began by concentrating on the duplex stainless steels UNS S32760 and UNS S31803. The degree of overlap was not established

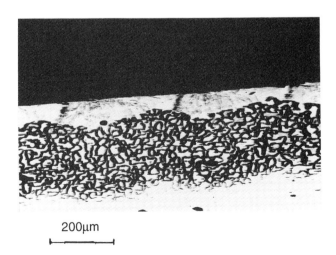

200μm

Figure 4 *Three layers of the UNS S32760 laser treated surface*

Table 2 *EPMA analyses on duplex UNS S32760 and laser treated layer*

	Cr	Mo	Ni
Austenite	23.8	2.5	8.6
Ferrite	28.4	4.1	5.6
Laser treated layer	27.0	2.6	7.5

and, as such, the procedure for testing followed an iterative process whereby the sample was treated, examined in plan, etched and examined in cross section and tested via electrochemical techniques in seawater. The results from all the tests were used to assess the uniformity of the laser treated layer and the parameters varied accordingly to yield optimum laser treatment.

Initially superduplex alloys (UNS S32760 and UNS S31608) were laser treated using a 0.3mm step between successive runs. Examined in cross-section, parallel to a single run, the treatment yielded a uniform layer but when examined perpendicular to a series of runs, it became apparent that there were regions, intermediate between runs, which had been treated only to a very small depth or in the worst cases not at all (Figure 2). In plan, the untreated regions appeared as in Figure 3 where, between the runs, the surface was rough and partial melting was evident. Taking a cross sectional view transverse to one laser run, the treated layer was examined using the SEM and EPMA. The cross section consisted of three distinct regions, shown in the SEM backscattered image, Figure 4. There was an upper layer (in the cross section of 75–100µm thick) the structure of which was not comparable with the untreated duplex structure and appeared to be primarily a single phase. The composition of this layer was between that of the austenite and the ferrite phases of the base material as shown in Table 2, where the numbers quoted have been corrected for atomic number, fluorescence and absorption effects.

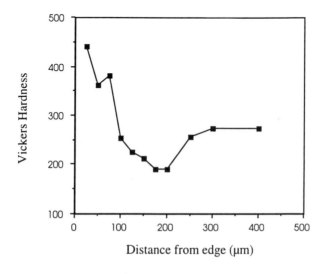

Figure 5 *Microhardness profile across the laser treated layer*

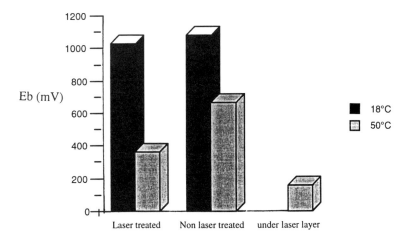

Figure 6 *Fall in E$_b$ rate on UNS S32760 due to laser surface treatment using a transverse shift of 0.3 mm between adjacent laser runs*

Under this layer was a second region, about two times the thickness of the upper layer, in which the duplex structure and composition was apparent but in which there were precipitates at the grain boundaries. Chemical analysis showed an enrichment of Cr at the grain boundary. An unaffected superduplex stainless steel, equal in composition and microstructure to the untreated material, was found under the two layers.

A microhardness profile was built up across the complete cross section and showed a significant hardening in the outer layer compared to the substrate. However, there was a loss of hardness across the layer immediately below (Figure 5).

3.2 Corrosion Tests

Corrosion tests at 18°C and 50°C in static conditions were undertaken on the aforementioned specimens of UNS S31803 and UNS S32760. In anodic polarisation tests on both materials at each temperature (repeated three times) the laser treated layer consistently showed resistance to passivity breakdown inferior to the untreated material as manifested in the lower potential, E$_b$, at which the current in the three-electrode electrochemical cell began to rise significantly. At 18°C on UNS S32760, the difference was only of the order of 60–70mV and there was therefore not a significant detrimental effect but at 50°C the magnitude of the difference in E$_b$ was accentuated, thus enhancing the detrimental effect of the laser treatment (Figure 6). The layer containing the grain boundary precipitates was tested after the surface was abraded to remove the initially treated layer and it was found to possess inferior resistance to passivity breakdown than the upper layer.

The lower Cr, UNS S31803 at 18°C showed a significant decrease in the E$_b$ value (450mV) after laser treatment and a further decrease when the temperature was increased to 50°C.

As may have been anticipated the susceptible areas to localised attack were shown to be the areas in between the adjacent runs where incomplete and inconsistent laser irradiation occurred, resulting in a non-homogeneous laser layer.

200µm

Figure 7 *Plan view of surface of UNS S31803 after laser treatment increasing the amount of overlap*

3.3 Modified Laser Treatment

3.3.1 Microstructure and Microhardness. In the second batch of laser treatment, the parameter governing the amount of overlap between adjacent laser runs was changed. The amount of shift perpendicular to the run just completed was reduced from 0.3mm to 0.175mm. In plan there was an obvious improvement in the homogeneity of the treated surface, in that there were no areas between the runs which appeared to be insufficiently melted. Figure 7 the surface in plan in which good continuity between laser layers is apparent.

The overlap proved to be sufficient to produce a layer of uniform thickness both along the weld and, more importantly, across several welds (Figure 8).

Since the parameters had then been established which would consistently produce a homogenous layer, the austenitic stainless steels were treated using the same conditions, for comparative purposes. It was found that the degree of overlap gave a layer uniform when sectioned in both directions.

Microhardness profiles for the four stainless steels are shown in Figure 9. The two duplex stainless steel showed a significant increase in hardness, in agreement with the results using the previous treatment parameters. There was no evidence of an intermediate layer between the substrate and the laser treated layer and as such there was no layer of significantly reduced hardness. There was no such increase in hardness on the austenitic stainless steels, even though there was a considerable change in the surface microstructure. A brief examination showed that laser treatment of the duplex stainless steels in an argon shrouding gas (but with all the parameters identical to those employed using nitrogen) yielded only very minor increases in surface hardness compared to those shown for nitrogen.

The microstructure of the duplex laser treated layer on both UNS S31803 and UNS S32760

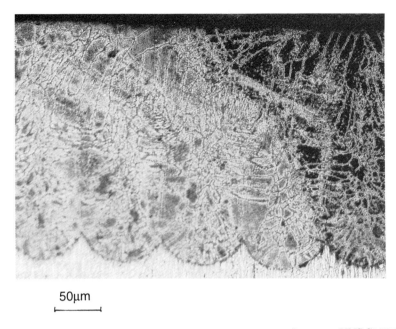

50μm

Figure 8 *Uniform thickness of laser treatment across several runs on UNS S32760 on increasing the overlap*

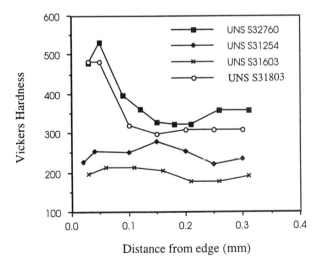

Figure 9 *Measured microhardness on the duplex and austenitic stainless steels after laser treatment in a nitrogen atmosphere*

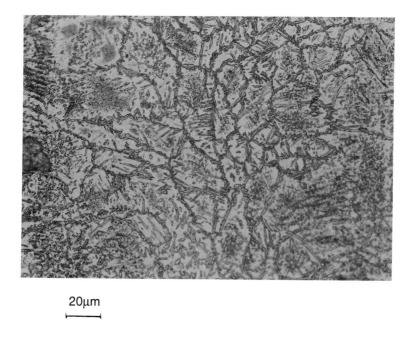

20μm

Figure 10 *Microstructure within the laser layer on UNS S32760*

differed significantly from a duplex stainless steel structure. Inside the laser layer there was a clear formation of precipitates or a secondary phase (Figure 10). The laser treated surface of the austenitic stainless steels appeared in the form of rods or dendrites, forming a complex multi-directional array (Figure 11). This is in agreement with microstructural features observed on austenitic stainless steel (types 304 and 316L)[21]. The overall structure appeared to be porous since spaces between the individual dendrites were identified.

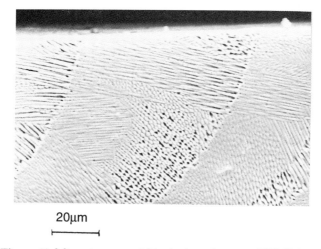

20μm

Figure 11 *Microstructure within the laser layer on UNS S31254*

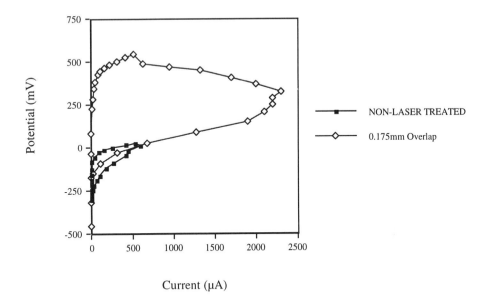

Figure 12 *Increased E_b on UNS S31803 after laser treatment in seawater at 50°C*

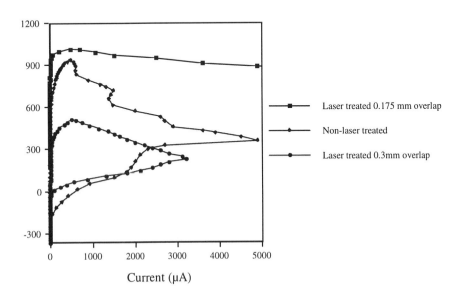

Figure 13 *Increased E_b on UNS S32760 after laser treatment in seawater*

200μm

Figure 14 *Corrosion between laser runs on UNS S32760 after anodic
polarisation at 50ºC*

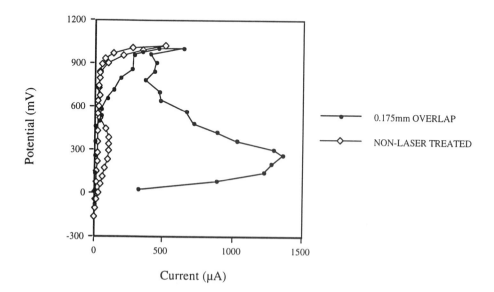

Figure 15 *Decreased resistance to the onset of localised corrosion in seawater at 50ºC on
laser treated UNS S31254*

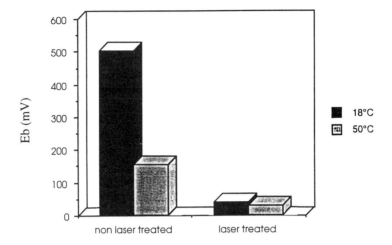

Figure 16 *Detrimental effect of laser treatment on austenitic stainless steels UNS S31603 at 18°C and at 50°C in seawater*

Since the shielding gas was nitrogen and this may well promote increased corrosion resistance and hardness (say by nitride formation in the laser treated layer), detection of nitrogen in the duplex stainless steels was attempted by three methods. These were XRD, EPMA (with a light element attachment) and LIMA.

However, no nitrides were found using XRD and nitrogen was not detected with EPMA. The LIMA Equipment detected nitrogen in the laser treated layer of UNS S32760 but at levels no higher than found in the substrate of this nitrogen containing steel. In addition, argon was successfully detected in the surface layer of the argon-treated duplex stainless steel.

3.3.2 Corrosion Performance. Interesting results were obtained, on the specimens subjected to the modified laser treatment, which showed that duplex stainless steels are amenable to an increase in resistance to corrosion initiation when irradiated under a nitrogen atmosphere. UNS S31803 exhibited the most beneficial effect of the laser treatment in that, at 50°C, the E_b was substantially ennobled by approximately 400mV (Figure 12). At 18°C, UNS S31803 already possesses good resistance to localised corrosion initiation and E_b is in the vicinity of +1000mV. This good resistance was maintained after laser treatment which had not been the case in the preliminary studies when areas of the surface between the runs were left untreated. As with UNS S31803, the superduplex UNS S32760 retained its passive potential range of in excess of 1200mV, in seawater at 18°C, after the modified laser treatment. Where UNS S32760 is particularly susceptible to localised corrosion is at elevated temperature[22] and the effect of the modified laser treatment under nitrogen has been to increase E_b, and hence retard the corrosion initiation (Figure 13). At elevated temperature, the current I_{max}, (defined as the current attained after reversal of the anodic polarisation scan), on the laser treated surface exceeded that on the untreated UNS S32760. On examination, it was found that as under the previous parameters, the susceptible area of the surface was the line between the runs and in this case pitting corrosion had been sustained (Figure 14).

The benefits observed on duplex stainless steels, manifested in increased hardness and corrosion resistance were not conferred to austenitic stainless steels when subjected to the same surface treatment. At ambient temperature, there was no detrimental effect on the initiation potential E_b on UNS S31254 but at 50°C the laser treated surface showed an increased susceptibility (Figure15). Corrosion attack occurred primarily at the interface between the sealing lacquer and the metal in the form of severe crevice attack. On UNS S31603, at both temperatures the E_b was lowered by a significant amount on laser treating as shown in Figure 16.

4 DISCUSSION

Laser irradiation of a surface involves consideration of numerous parameters, each of which can alter the heat input rate, the degree of overlap and other critical process conditions. This study has focused on three main aspects relating to the global process of laser irradiation of stainless steel surface, namely

- structural laser layer uniformity
- laser treated layer microhardness
- corrosion resistance

The assessment of laser treatment in this work has therefore relied on an iterative procedure of treatment, followed by examination of the treated surface and then feedback to the laser parameters. Initial tests attempted to optimise laser parameters which would consistently produce a uniform laser treated layer with improved mechanical hardness and corrosion resistance. Critical parameters for effective laser treatment were identified as pulsed power cycles, defocused laser beam and precise control of the overlapping of laser runs to ensure consistent laser treatment. Very small differences in step distance between parallel runs of 125 mm significantly changed the microstructural and mechanical characteristics of the laser treated layer, which were manifested in much reduced corrosion resistance. An important aspect of surface treatment is the ability to reproduce a uniform surface layer and laser treatment using the laser parameters in this study has been shown to do this. The laser layer appeared relatively smooth which is important from the point of view of mechanical properties where a significantly roughened surface could introduce stress concentration points and problems with design tolerances. This work has shown a double benefit on the duplex stainless steels of increased hardness and corrosion resistance.

The increased hardness is obviously linked to the metallurgy and composition of the laser treated layer and there are three possibilities. Increased hardness could be due to alloying with nitrogen which, on rapid cooling, precipitates a hard phase of perhaps CrN or could be primarily due to a metallurgical transformation on rapid cooling, independent of the shielding gas. Thirdly the nitrogen could be retained in solid solution.

The significant increase in hardness of the duplex stainless steel surfaces under a nitrogen atmosphere was not conferred to the duplex stainless steels under an argon atmosphere. This indicates that the notion that the effect of rapid solidification alone, independent of shielding gas, confers the increased hardness and corrosion resistance is not valid.

Precipitation of Cr_2N has been reported during rapid cooling or quenching from annealing temperatures. Fine precipitates of Cr_2N have been detected on rapid cooling of duplex stainless steels by Herbsieb and Schwaab[23]. Since nitrogen is less soluble in ferrite than in austenite, the Cr_2N formed preferentially in the ferrite phase. In the present study, Cr_2N has not been detected and therefore another explanation for the precipitated phase is sought.

Welding in a nitrogen atmosphere has been studied by several workers[24,25] and in this work rapid cooling is considered to try and draw analogies to laser welding under a nitrogen shielding gas. The work by Fourie et al.[26] states that the Cr and Ni equivalent ratio (from the Schaeffler diagram) can give information on the solidification of the alloy. For $C_{req}/N_{ieq} > 1.95$ (in the case for UNS S32760 $C_{req}/N_{ieq} = 4.62$) the weldment solidifies as a fully ferritic single phase structure. The austenite is later formed in the solid state by a Widmanstatten mechanism. In the work by Fourie et al., laser welding of UNS S31803 was shown to produce an almost fully ferritic structure with acicular austenite at the grain boundaries. No mention of N additions was made but the author did observe copious Cr_2N precipitation (the N coming from the alloy) within the ferrite grains with a precipitate free region near the reformed austenite. The microstructure of the laser layer in the present study was found to be similar to that reported by Fourie et al. In Figure 11 it can be seen that there appears to be a base ferritic structure with perhaps what could be austenite at the grain boundaries. Within the ferrite, fine precipitates are evident which, due to the evidence obtained during examination under dark field illumination, could perhaps be assumed to be the nitrides mentioned by Fourie et al. However, since nitrides were not detected using XRD, the precise nature of these precipitates is not known. It could be that their fine nature renders them undetectable in the bulk analysis XRD trace.

In terms of corrosion behaviour, this investigation has shown that the resistance to the initiation of localised corrosion of the duplex stainless steels is improved, particularly at elevated temperature. The greatest effect was observed in the lower alloyed UNS S31803 which is in agreement with Jonsson et al.[24] regarding the effect of nitrogen in welding of duplex stainless steels. Quite apart from the significant hardness increase, this represents an important finding from both a practical and a fundamental aspect since it is at elevated temperature that the superduplex shows limitations in resisting localised (especially crevice) attack[22]. There was some evidence to suggest that propagation might be enhanced. This could have been due to the form of the surface which because of its increased roughness proved to be more amenable to corrosion propagation through formation of local electrochemical cells and, as such, refinement of the finishing process could be important to provide a more uniform surface and eradicate the potential sites for corrosion initiation. This has been found to be the case on austenitic stainless steels by Stewart et al[16].

Nevertheless the indication of increased resistance to initiation of corrosion, especially at elevated temperature where the duplex stainless steels are susceptible, is a promising situation. The implications are that by further refinement of the parameters and processing techniques, there are additional benefits to be gained.

5 CONCLUSIONS

Laser irradiation under a nitrogen shielding gas has been shown to be effective as a method of improving the surface hardness of duplex stainless steels. Additionailly, evidence has been obtained to suggest that improvements in resistance to localised corrosion initiation can be obtained, particularly at elevated seawater temperatures. Points of uncertainty pertain to the precise metallurgy of the laser treated layer but it is postulated that the formation of nitride precipitates accounts for the hardening phase in the laser layer. On austenitic stainless steels, the laser parameters used in this study did not yield any significant surface hardening effect and indeed the surface treatment was shown to be detrimental to the seawater corrosion resistance.

Acknowledgements

The work described in this paper was made possible by means of an EPSRC research studentship awarded through the Marine Technology Directorate to A.N. Thanks are also due to Professor B. F. Scott, Head of Mechanical Engineering, University of Glasgow and Professor J. E. L. Simmons, Head of Mechanical and Chemical Engineering, Heriot Watt University, Edinburgh for the provision of laboratory facilities. The help of Dr P. John, of Edinburgh Surface Analysis Technology, Department of Chemistry, Heriot Watt University in providing analysis using LIMA equipment is acknowledged.

References

1 . J. Huber, W. Marx, 'Production laser cutting, in Applications of Lasers in Materials Processing', ed. E. A. Metzbower, American Society For Metals, 1979.

2. A. C. Lingenfeiter, C. D. Anglin, C. N. Westrich, J. R. Murchie, 'Precision cutting and drilling with the ND:YAG laser', in The Changing Frontiers of Laser Materials Processing, eds. C. M. Banas, G. L. Whitney, 1986.

3. E. A. Metzbower, D. W. Moon, 'Mechanical properties, fracture toughnesses and microstructures of laser welds in high strength alloys', in Applications of Lasers in Materials Processing, ed. E. A. Metzbower, American Society For Metals, 1979.

4. U. I. Chan, K. W. Casey, 'Laser welding of exhaust gas oxygen sensor', in Applications of Lasers in Materials Processing, ed. E. A. Metzbower, American Society For Metals, 1979

5. B. L. Mordike, H. W. Bergmann, *'Rapidly solidified amorphous and crystalline alloys'*, ed. Kear, Giessen, Cohen, North Holland Publishing Co. No. 8, 19823 463.

6. W. Gruhi, B. Grzemba, G. lbe, W. Hiller, *Metal*, 1978, **32**, 549.

7. S. Kosuge, M. Ono, K. Nakada, I. Watanabe, "Advanced Laser materials processing", in 'The Changing Frontiers of Laser Materials Processing', eds. C. M.   Banas, G. L. Whitney, 1986

8. B. L. Mordike, "State of the art of surface engineering with high energy beams", in 'Surface Engineering With High Energy Beams', eds. A. P. Loureiro, O. Conde, L. Guerra-Rosa, R. Vilar, Trans Tech Publications, 1990.

9. H. W. Bergmann, E. Schubert, "Modification of metal surfaces by excimer laser treatment - state of the art", in 'Surface Engineering With High Energy Beams', eds. A. P. Loureiro, 0. Conde, L. Guerra-Rosa, R. Vilar, Trans Tech Publications, 1990.

10. O. W. M. Steen, "Laser surface cladding", in 'Laser Surface Treatment of Metals', eds. C. W. Draper, P. Mazzoidi, NATO ASI series 1986.

11. 1. Z. D. Chen, D. R. F. West, W. M. Steen, "Laser melting of alloy cast iron", in 'The Changing Frontiers of Laser Materials Processing', eds. C. M. Banas, G. L. Whitney, 1986.

12. G. Coquerelle, J. L. Fachinetti, "Friction and wear of laser treated aluminium- silicon alloys", in 'The Changing Frontiers of Laser Materials Processing', eds. C. M. Banas, G. L. Whitney, 1986.

13. J. Folkes, D. R. F. West, W. M. Steen, "Laser surface melting and alloying of titanium", in 'Laser Surface Treatment of Metals', eds. C. W. Draper, P. Mazzoidi, NATO ASI series 1986.

14. L. N. Moskowitz, 'Application of HVOF thermal sprayng to solve corrosion problems in the petroleum industry', Proc. of the Int. Thermal Spray Conference and Exposition, Oriando, Florida, USA, 1992.

15. F. Uchiyama, K. Tsukamoto, Y. Kaga, T. Okuo, S. Yamaoka, Y. Ohno, Y. Takahagi, T. Kamoshida, 'Ceramic coating technique using laser spray process for solid oxide fuel cells', Proc. of the Int. Thermal Spray Conference and Exposition, Oriando, Florida, USA, 1992

16. J. Stewart and D. E. Williams, *Corrosion Science,* 1992, **3**, 457.

17. U. Kamachi Mundali, R. K. Dayal, G. L. Goswami, *Surface Engineering,* 1995, **11**, 331.

18. A. J. Sedriks, 'Effects of alloy composition and microstructure on the passivity of stainless steels', *Corrosion*, 1986, **42**, 71.

19. M. Vemeau, B. Bonnefois, 'Effects of alloying elements on the pitting corrosion resistance of duplex welded joints', Proc. of Duplex Stainless Steels'91 Int. Conf., Vol 11, eds. J. Charles, S. Bernhardsson, Bourgogne, 1991.

20. B. de Bendetti, E. Angelini, P. Bianco, F. Rosalbino, 'ion nitriding of duplex stainless steels : preferential dissolution phenomena in chloride media', Proc. of Duplex Stainless Steels'91 Int. Conf., Vol 11, eds. J. Charles, S. Bernhardsson, Bourgogne, 1991.

21. U. Mudali, R. K. Dayal, G. L. Goswami, *Surface Engineering*, 1995, **11**,4.

22. A. Neville and T. Hodgkiess, *Corrosion Science,* **38**, 61 1996.

23. G. Herbsieb, P. Schwaab, in R. A. Luia (ed), 'Duplex Stainless Steels', ASM, 1983 p. 15.

24. O. Jonsson, M. Liijas, P. Stenvall, 'The role of nitrogen in longitudinal welding of tubing in duplex stainless steels', Avesta Corrosion Management, 1994, No. 2.

25. R. Doyen, R. Niset, 'Welding of duplex and superduplex stainless steels', Proc. of Duplex Stainless Steels'91 Int. Conf., Vol 11, eds. J. Charles, S. Bernhardsson, Bourgogne, 1991.

26. J. W. Fourie, F. P. A. Robinson, *Journal of the South African Institute of Mining and Metallurgy,* March, 1990, 59.

3.5.4

The Use of Thermal Spray Coatings in the South African Power Industry

R. G. Wellman

ESKOM, TECHNOLOGY RESEARCH AND INVESTIGATIONS, PRIVATE BAG 40175, CLEVELAND 2022, RSA

1 INTRODUCTION

Unplanned downtime due to tube leaks caused by fly ash erosion, high temperature oxidation and fireside corrosion is costly due to lost generation. This should be reduced using all possible means. Often it is not possible to treat the root causes, viz. localised high velocities and localised turbulence, and it is necessary to address the symptoms. This is where metal spray coatings come to the fore. Depending on the composition used sprayed metal coatings offer protection of boiler tubes against erosion, high temperature oxidation and fireside corrosion.

Fly ash erosion of boiler tubes is a major problem due to the low quality coal used in SA power plants. The ash content of the coal varies from 20–45% with most power plants seeing an average ash content of 26%. Alumina–silicate species account for 60–90% of the ash. These are extremely abrasive particles and are the main cause of erosion problems experienced in the boilers. Fireside corrosion has also been identified as a problem on a number of power plants and metal spray coatings are currently being evaluated at Duvha Power Station to determine their effectiveness as a means of reducing tube wastage.

A number of different coatings are being used on SA power plants as protection against erosion and are increasingly being used for protection against high temperature oxidation. The choice of coating depends on the specific problems that need to be addressed. Stations that use coatings have reported a major reduction in tube leaks due to erosion since the application of metal sprayed coatings.

Four different coatings are currently being evaluated for use on turbine blades and high pressure and intermediate pressure inlet nozzles for protection against solid particle and liquid droplet erosion.

2 COATINGS FOR EROSION PROTECTION

Localised erosion, responsible for a number of tube leaks annually, is normally caused by localised high velocities and turbulence due to sudden changes in geometry in the boiler. In a number of instances this erosion can be reduced by modifying the flow characteristics within a specific region of the boiler by the use of screens. Cold air velocity distribution tests are used to establish the aerodynamic flow within a boiler and the results are then used to determine the positioning of screens to normalise the flow[1]. However there are instances where it is not

Table 1 *Coatings used to reduce erosion*

	Cr (%)	Fe (%)	B (%)	Mn (%)	Mo (%)	Al (%)	Ti (%)	Si (%)
Coating A	24–29	Base	3	-	-	-		-
Coating B	24–29	Base	-	2.5–4	-	4.7–7	-	-
Coating C	29	Base	3	0.7	-	-	-	-
Coating D	6.6	Base	-	2.4	0.8	-	-	-
Coating E	15	Base	-	-	-	-	4	1–2

possible to use screens or where the screens do not supply adequate protection. It is in these regions of the boiler that the metal spray coatings are used.

A number of different coatings is used by the different power stations for protection against erosion, these are listed in Table 1.

All of the coatings are applied using the arc spray method. This method is used due to its high deposition rate (9–12kg/hr) and the fact that it is easy to use for in situ applications, since most of the spraying is conducted inside the boiler.

Hendrina Power Station which comprises of ten 200 MW units uses metal spray coatings to protect the tubes in high wear areas in five of the boilers. No tube leaks have occurred in the areas that have been coated. Most of the metal spraying at Hendrina power station has been done using coating A. There has been a large reduction in the number of tube leaks experienced at Hendrina Power Station. The total number of tube leaks has been reduced from 24 per year to 17 in 1995. This reduction in the number of tube leaks is a direct result of the metal spray coatings that have been applied to the high wear areas of the boilers.

Lethabo, a relatively new station, runs six 660 MW units and uses metal spray coatings extensively throughout the boilers to reduce erosion damage. Between 1986 and 1990 Lethabo experienced 15 tube leaks due to fly ash erosion. During 1990 an extensive programme of metal spray coating was initiated to protect those areas most prone to fly ash erosion, which subsequently resulted in a great reduction in tube leaks due to erosion. Since 1991 Lethabo has experienced the following record for fly ash erosion related tube leaks: zero in 1991, three in 1992, one in 1993, one in 1994 and two in 1995. This represents a reduction in tube leaks of more than 50%. However, it must be noted that expanded metal screens were used in conjunction with the metal spray coatings. The screens normalise the flow in the boiler and reduce localised turbulence.

Matimba Power Station which also runs six 660 MW units uses metal spray coatings in high wear areas as well as for protection against high temperature oxidation. Matimba conducted some in situ testing of three different coatings during 1993/1995. A number of different tubes was coated and placed in a high wear area of the boiler. The tubes were in service from 16th October 1993 to 25th August 1995 during which time the boiler was operational for 13,019 hours. These tubes were situated in the economiser in the region of the hangers which, due to localised turbulence, is a high wear region of the boiler.

Thickness measurements were taken before and after the above mentioned service period. Table 2 gives the thicknesses as measured.

As can be seen from Table 2, coating C gave the most protection to the boiler tubes. An uncoated tube in this region of the boiler usually looses 0.6 mm per 12,500 steaming hours due to erosion. By applying a suitable coating it is possible to reduce the erosion rate quite

Table 2 *Coating thickness before and after 13,079 hours in service*

	Thickness before (μm)	Thickness after (μm)	Loss (μm)
Coating B	750	410	340
Coating C	540	420	120
Coating E	800	440	260

considerably, thus extending the life of the tube and reducing the number of tubes that need to be replaced during each outage.

3 HIGH TEMPERATURE OXIDATION

Tube wastage due to a combination of erosion and high temperature oxidation can also cause tube leaks and unplanned downtime. For this application a totally different type of metal spray coating is used. A 50/50 Ni/Cr has been found to be highly effective and more recently chrome-carbide/Cr/Ni coatings are being used. The 50/50 Ni/Cr is an arc sprayed coating while the chrome-carbide/Cr/Ni is a plasma sprayed coating that can only be applied in workshop conditions. Matimba Power Station has started to use these coatings in areas which experience high temperature oxidation.

Some of the areas which need to be protected are not accessible for in situ spraying. In these areas half shells are coated in a workshop and then fixed over the appropriate areas. This has the advantage that the half shells can be replaced when necessary without affecting the integrity of the plant, i.e. no welding of pressure parts. Since the application of these coatings at Matimba Power Station there have been no tube leaks due to metal wastage caused by high temperature oxidation, see Table 3.

4 FIRESIDE CORROSION

Metal wastage due to fireside corrosion in the waterwall of boilers in the region of fireball can result in tube leaks. During an outage at Duvha Power Station in August 1995 a number of tubes in the waterwall were found to be below minimum wall thickness and had to be replaced. Other tubes in this area, where metal wastage due to fireside corrosion had occurred, were coated with different types of coatings. These coatings will be monitored during future outages to determine the most suitable coating.

Three of the coatings (H, J & K) were arc sprayed while coating I was a powder applied by the high velocity oxy-fuel (HVOF) method, see Table 4.

Table 3 *Coatings used to prevent high temperature oxidation*

	Ni	Cr	CrC	Ti
Coating F	15	5	80	–
Coating G	51	45	–	4

Table 4 *Coatings used for protection against fireside corrosion*

	Ni (%)	Cr (%)	Ti (%)	Al (%)	Cu (%)
Coating H	51	45	4	-	-
Coating I	49	51	-	-	-
Coating J	58	41	1	-	-
Coating K	-	-	-	8	92

Under normal, slightly oxidizing conditions and a fireside tube metal temperature of 450°C, a protective oxide layer is formed on mild steel[2]. Due to the dynamic nature of the combustion process localised reducing conditions can be created at the tube surface that result in a complex range of non-equilibrium reactions. These in turn yield a range of different and relatively non-protective corrosion scales[3]. During combustion the alkali elements sodium and potassium form oxides that react with SO_3 to form pyrosulphates (e.g. $Na_2S_2O_7$). These pyrosulphates then react either with iron oxides that have formed on the tube surface or directly with the iron, where exposed to the fluegas. The temperatures at which the precipitation of the alkali pyrosulphates takes place vary from 370°C to 430°C. Although this is below the nominal steam temperature, it is the tube wall temperature that is of importance.

The reactions that take place at the tube wall have been reported to be the following: (note that potassium behaves in the same way as sodium)[4].

Sodium is oxidized in the flame:

$$4Na + O_2 = 2Na_2O$$

Sulphur oxidizes in the flame and deposits on the tube surfaces:

$$S + O_2 = SO_2$$
$$2SO_2 + O_2 = 2SO_3$$

Pyrosulphates form in the ash deposit:

$$Na_2O + 2SO_3 = Na_2S_2O_7$$

Carbon deposits in the ash either as unburned coal or by reduction:

$$2CO = CO_2 + C$$

Pyrosulphates are transported to the tube surface, dissociation takes place and SO_3 is released. Tube wastage then takes place by either of the following reactions:

$$SO_3 + 3C + Fe = FeS + 3CO$$
$$2SO_3 + 9C + Fe_2O_3 = 2FeS + 9CO$$

In the Ni/Cr coating, the chrome oxides, which microscopically form a dense lattice in the coating, inhibit the diffusion of sulphur species thus reducing the formation of iron oxides and sulphides. The nickel provides the coating with a thermal expansion coefficient that is compatible to carbon steel as well as a high bond strength, while titanium reduces the internal oxidation of the coating.

5 TURBINES

Two different types of erosion can occur in turbines. These are solid particle erosion (SPE) of high pressure turbine inlet nozzles and first stage blades and liquid droplet erosion (LDE) of the last stages of the LP turbines. Chrome carbide coatings with a 20 % binder phase are currently being investigated for protection against both SPE and LDE. Three different binder chemistries are being investigated Co-Ni-Cr-W, Fe-Cr-Al-Y and Ni-Cr-Mo[5]. The base material is a powder which can be applied by either HVOF or plasma spraying. It is anticipated that in situ testing of the coatings will be implemented early in 1996.

6 QUALITY CONTROL

Since all the coatings are applied to Eskom power plants by outside contractors it is necessary to maintain rigid quality control (QC) procedures in order to ensure a coating of adequate quality. This QC is performed by Eskom's Technology Research and Investigations section and involves the continuous monitoring of the application, as well as, the daily analysis of sprayed samples.

The QC involves daily checks on the spraying and grit blasting equipment, chemical analyses of the consumable being used and microscopic evaluation of the coating. Typically four to six samples are sprayed a day for microscopic evaluation. These samples are analysed, using an optical microscope coupled to an image analysis system, according to Eskom's internal specification for metal sprayed coatings[6].

7 CONCLUSIONS

The results obtained to date from coatings that have been in service at the various power stations are very promising. It appears that metal spray coatings are a cost effective means of combating the excessive erosion and corrosion experienced in certain regions of the boiler. It is however important that the correct type of coating is used for a particular application, and that the coating is compatible with the substrate onto which it is applied.

References

1.   D. Gibson, 'A Summary of the Investigations Conducted to Minimise Fly Ash Erosion within the Hendrina Power Station, El Paso', Design Boilers, Eskom TRR Report P94/099, December 1994.
2.   K. Rodseth, 'Fluegas Attack on the Fireside of Boilers', Eskom TRR Report S94/199, December 1994.

3. A. J. B. Cutler, T. Flatley and K. A. Hay, 'Fireside Corrosion in Power Station Boilers', CEGB Research, October 1987.
4. D. N. French, 'Metallurgical Failures in Fossil Fired Boilers', Second Edition, John Wiley and Sons Inc.
5. 'Erosion Resistant Coatings for Steam Turbines'; EPRI CS-5415, September, 1987.
6. 'Thermal Spraying of Boiler Tubing', Eskom Specification GGSS 0331.

Section 3.6 Marine

3.6.1

Properties of the Passive Films on SuperAustenitic Stainless Steels in Sea Water

G. Latha, N. Rajendran and S. Rajeswari

DEPARTMENT OF ANALYTICAL CHEMISTRY, UNIVERSITY OF MADRAS, GUINDY CAMPUS, MADRAS 600 025, INDIA

1 INTRODUCTION

Cooling water system is one of the major auxiliary supporting service systems for the operation of a thermal and nuclear power plant. The performance of this system has direct bearing on the output and efficiency of the plant. Among the various systems, the once through cooling systems are well known and widely used all over the world. Mostly in once through cooling systems where a large quantity of water is required, sea water forms the ultimate source for the heat sink[1-5]. Though the sea water satisfies mostly all the conditions of a coolant, the problems associated with it, mainly due to corrosion, need extensive attention. For the machinery components employed in such applications, due precautions have to be undertaken to their reliability in addition to cost optimization[6,7]. Since any modification of the selected material at a later stage would result in severe economic losses nearing to the initial cost of installation resulting to the total shutdown of the plants. It is preferable to adopt the conventional equipment design by judicious selection of materials for this critical system[8,9].

Over the past three decades great progress has been made in the selection of highly resistant materials to serve the need of the ever increasing corrosion problems in marine condition[8]. Even though the standard austenitic stainless steels, type 316L SS have been and continue to be the work horse in the marine related components, their vulnerability to localized corrosion failures mainly due to the presence of aggressive chloride ions has been a major problem, which necessitates the use of high resistant materials[10,11]. At this juncture, there is a need to employ materials with high corrosion performance compared to the currently used construction materials, which can be achieved by modifying the existing material with the addition of alloying elements like nitrogen, molybdenum and chromium. Such a material is the superaustenitic stainless steel, which is basically an austenitic stainless steel with the addition of higher amounts of nitrogen, molybdenum and chromium.

The objective of the present work is to study the beneficial affect of nitrogen, chromium and molybdenum on the passivity and the localized corrosion behaviour of superaustenitic stainless steels namely alloy 926 and alloy 31 in sea water.

2 EXPERIMENTAL PROCEDURE

2.1 Electrochemical Cell Assembly

The electrochemical cell used in the present study consists of 500 ml capacity with a three electrode system. Saturated calomel electrode (SCE) was used as the reference electrode, platinum foil as the counter electrode and the test material as the working electrode. In order to maintain the ideal condition employed in industrial units, natural sea water collected from the coastal area of Madras, India was used as the electrolyte under continuous stirring.

2.2 Electrode Preparation

The composition of the materials (reference type 316L SS, alloy 926 and alloy 31) used are given in Table 1.

The material was cut into 1×1×0.3 cm in size for electrochemical studies. Each piece was attached with a brass rod using silver paste for electrical contact. Then the samples were mounted in an epoxy resin in such a way that only one side with 1cm² surface area was exposed. The mounted samples were polished successively up to 800 grit emery paper and final polishing was done using diamond paste in order to get scratch free mirror finish. The electrodes were ultrasonically cleaned with soap solution, degreased using acetone and thoroughly rinsed in distilled water and dried. This served as the working electrode.

2.3 Pitting Corrosion Studies

The potentiodynamic anodic cyclic polarization studies were conducted in order to evaluate the pitting corrosion behaviour of the alloys. The working electrode was depolarised to − 1100 mV for 1 minute and allowed to stabilise for an hour until the constant potential was reached which is referred as the corrosion potential (E_{corr}). The potential was applied from the corrosion potential in the noble direction at a scan rate of 1 mV/sec until the breakdown potential (E_b) was attained, where the alloy entered the transpassive region or pitting. The sweep direction was then reversed after reaching an anodic current density of 3 mA/cm² until the potential where the reverse scan current density equals the upscan current density. The potential at which the reverse anodic scan meets the passive region is termed as repassivation potential or pit-protection potential (E_p).

The parameters of interest recorded during cyclic potentiodynamic polarization studies were (i) the corrosion potential (E_{corr}), (ii) the pitting potential (E_b) and (iii) the repassivation potential (E_p).

Table 1 *Chemical Composition of the alloys (wt. %)*

Alloys	Ni	Cr	N	Mo	C	S	Mn	Cu
316L SS	12.7	17.2	0.02	2.4	0.03	0.003	1.95	-
Alloy 926	25.0	21.0	0.20	6.5	0.03	0.002	0.70	0.96
Alloy 31	31.0	27.0	0.20	6.5	0.03	0.002	0.68	0.90

2.4 Crevice Corrosion Studies

The crevice corrosion studies were carried out on the working electrode by creating a crevice using a specially designed glass assembly described by Dayal et al.[12] The tip of the glass rod was brought into close contact with the electrode surface using a nut and threaded rod arrangement. To determine the critical crevice potential (E_{cc}), the potential was increased from E_{corr} in the noble direction at a scan rate of 1 mV/sec until the current increased rapidly.

2.5 X-ray Photoelectron Spectroscopic Study

The surface characterization of the passivated superaustenitic stainless steels (alloy 926 and alloy 31) was carried out by X-ray photoelectron spectroscopic study. Similar studies were also carried out on reference type 316L SS for comparison, to reveal the exact role of alloying elements on aiding the stability of the passive films.

The samples with 1 cm² surface area were anodically polarised at 200 mV in sea water for a period of 1 hour for the growth of the passive film. Then, the samples were removed from the polarization cell assembly and rinsed with distilled water, dried in a stream of flowing argon and kept inside a desiccator, until they were transferred to the evacuated sample chamber of the ESCA unit. The samples were irradiated with AlKα radiation with a mean kinetic energy of 1486.6 eV. All the experiments were carried out at a vacuum of 10^{-8} to 10^{-9} torrs. The survey spectrum of the material was first carried out followed by the individual spectra for C 1s, O 1s, N 1s Cr 2p, Fe 2p and Mo 3d electrons. The depth profiles of the passivated samples were performed for 7 minutes using argon ion. The output of the photoelectron analyses was obtained as binding energy versus intensity counts.

The binding energy of the alloying elements was measured from the spectra obtained for the respective elements and then the values were corrected with respect to the reference C 1s binding energy value. The area under the peaks obtained for iron and chromium was taken into account to calculate the peak area ratios.

3 RESULTS AND DISCUSSION

3.1 Effect of Chromium, Molybdenum and Nitrogen on the Pitting Potential

The potentiodynamic anodic cyclic polarization curves obtained for the alloys under study are depicted in Figure 1. The mean value of the critical pitting potential was taken by performing an average of five experiments. The E_b observed for type 316L SS was found to be 365 mV whereas for alloy 926 and alloy 31, the observed values corresponded to 1125 mV and 1182 mV respectively.

Irrespective of the various factors influencing the pitting potential, the amount of passivating elements namely chromium, molybdenum and nitrogen play a definite role. The extreme shift in the pitting potential in 926 and 31 towards the noble direction compared to 316L SS is due to the presence of 6.5% molybdenum and 0.2% of nitrogen for the superaustenitics.

The localized attack in an austenitic phase can be explained as follows. During the pit growth, the conditions prevailing at that site have been reported to be similar to those of the active dissolution state. During dissolution, elements like nickel, chromium dissolve whereas nitrogen enriches at the surface. Such an enrichment of the passive film in the alloys inhibits

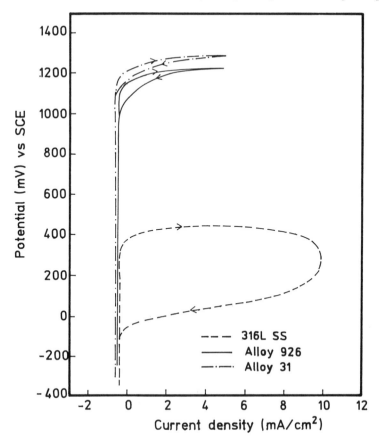

Figure 1 *Pitting corrosion behaviour of 316L SS, alloy 926 and alloy 31 in seawater*

the anodic dissolution by two orders of magnitude, presumably through the formation of iron nitride which inhibits the auto-catalytic process of pit formation and increases the tendency for any pit to heal[13].

The pit morphology observed from optical microscopy for the alloys under study is shown in Figure 2 (a–c). Figure 2a shows the pitted region of 316L SS. Roughly spherical pits of larger size were found to be present. This indicates the higher susceptibility of the material towards pitting attack. In the case of alloy 926 and alloy 31, (Figure 2b and 2c) very small pits, fewer in number, were found to occur compared to those of the reference type 316L SS. The increase in the contents of passivating elements such as molybdenum and nitrogen in the alloy 926 and alloy 31 decreases the aggressiveness of the environment at the prevailing site and hence prevents the growth of the pits.

Earlier studies have reported that the addition of small amounts of nitrogen can lead to an enhanced pitting resistance and passivating characteristics[14,15]. Newman et al.[16], have noticed an enrichment of molybdenum and nitrogen in the passive film and the enrichment of nitrogen is at least at a level of seven times the original concentration of nitrogen present in the alloy. This enrichment has proved to be the predominant factor for preventing further dissolution of substrate, consequent to the destruction of the passive, film. The synergistic influence of

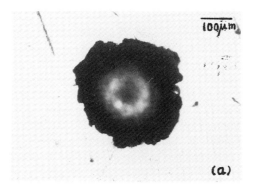

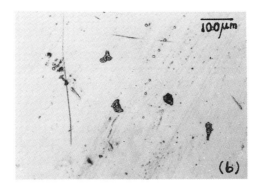

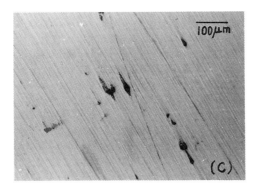

Figure 2 *Pit morphology of (a) 316L SS, (b) alloy 926 and (c) alloy 31*

nitrogen and molybdenum on the pitting corrosion resistance was also reported by Truman et al.[14] and Sedriks[17].

The repassivation potential (E_p) determined for the above alloys are shown in the Figure 1. The mean value of the repassivation potential increased in the noble direction for the alloy 926 and alloy 31. The observation can be interpreted by showing that new pits cannot be initiated above these potentials. The difference in the value of E_p from 316L SS and superaustenitics is basically due to the pit environment, which greatly influences E_p and varies as the pitting process proceeds. The largest active pit on the type 316L SS is likely to have developed the most aggressive environment (highly acidic and higher chloride ion content) within the pit sites; which is likely to control the repassivation potential. However, the increased amount of chromium, molybdenum and nitrogen, present in the superaustenitics hinders the development of new pits and also slows down the kinetics of growing pits thereby shifting the repassivation potential towards more noble direction.

3.2 Effect of Chromium, Molybdenum and Nitrogen on the Crevice Corrosion Potential

The environment prevailing at the actual operating condition of the cooling circuit involves

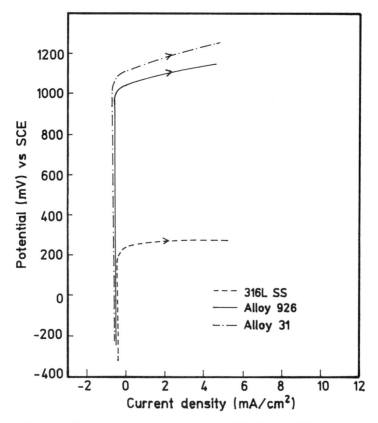

Figure 3 *Crevice corrosion behaviour of 316 SS, alloy 926 and alloy 31*

joints and nuts connected to the material which result in the crevice corrosion. Oxygen present in the crevice is consumed quite rapidly during passivation. Due to this, depletion of oxygen takes place leading to the formation of concentration cell which in turn accelerates the crevice corrosion.

The critical crevice corrosion potential values of the alloys under study were determined by the anodic polarization method and are shown in Figure 3.

The presence of 0.2% nitrogen and 6.5% molybdenum in the alloy 926 and alloy 31, shifted the E_{cc}, values towards more noble direction. Thus, the presence of nitrogen and molybdenum improves the crevice corrosion resistance of the superaustenitic stainless steel.

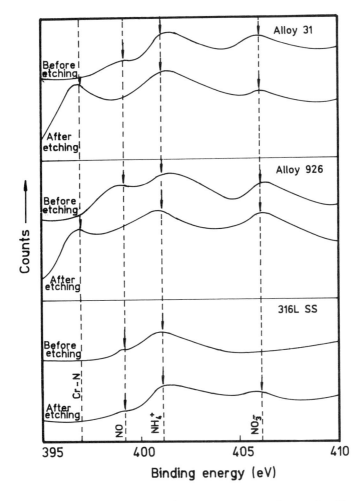

Figure 4 *XPS high resolution spectra of nitrogen present in the passive films of type 316L SS, alloy 926 and alloy 31*

3.3 XPS Study of the Passive Film on Alloy 926 and Alloy 31

The localized corrosion behaviour of the material is closely related to the composition of the passive film. Hence, the analysis of the passive film by XPS gives detailed information about the chemical environment of the elements in the passive film.

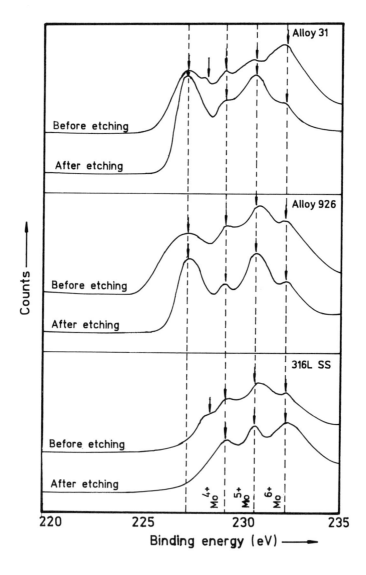

Figure 5 *XPS high resolution spectra of molybdenum present in the passive films of type 316L SS, alloy 926 and alloy 31*

3.3.1 N 1s Spectra. The N 1s spectra in the passive film on 316L SS revealed the peaks at 399.2 eV and 401.1 eV before sputtering. However, after sputtering, the peaks were observed at 399.2 eV, 401.1 eV and 406.2 eV and are represented in Figure 4.

The high resolution spectra of N 1s in the passive film on the alloy 926 and alloy 31 revealed the peaks at 399.2 eV, 401.1 eV and 406.2 eV at the surface of the passive film. After sputtering, the peak at 397.0 eV was observed along with the peaks at 401.2 eV and 406.2 eV.

On the basis of previously reported results[18], the peak located at low binding energy of 397.0 eV is due to nitride incorporation in the passive film. Nitrogen present in the form of NO^-, NH_4^+ and NO_3^- with their corresponding binding energy values of 399.2 eV, 402.0 eV and 406.2 eV showed a similar binding energy value when present in the passive film[19-23].

Thus, it can be suggested that an enrichment of nitrogen and in the form of NO^-, NH_4^+ appeared in the outer layer of alloy 926 and alloy 31. However, the inner layer consists of CrN, NO_3^- and NH_4^+ ions. The enrichment of CrN in the case of alloy 926 and alloy 31 would have impeded the release of metal ions through the passive film and increased the pitting resistance. However, CrN was not observed in the reference type 316L SS.

3.3.2 Mo $3d_{5/2}$ Spectra. The presence of molybdenum in the passive film of reference type 316L SS, alloy 926 and alloy 31 was analysed by XPS in the binding energy ranging from 225.0 to 240.00 eV.

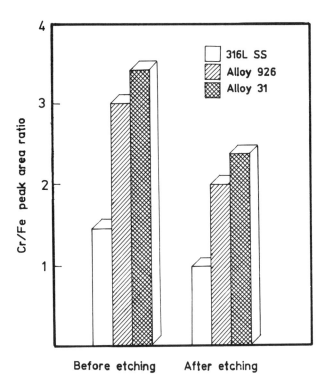

Figure 6 *Cr/Fe peak area ratio of type 316L SS, alloy 926 and alloy 31 before and after etching of the passive films*

From the spectra obtained for the reference type 316L SS, four peaks were deconvulted at binding energy values of 228.3 eV, 229.2 eV, 230.7 eV and 232.2 eV at the outermost region of the passive film corresponding to Mo^{3+}, Mo^{4+}, Mo^{5+} and Mo^{6+}. There was no significant change regarding the position or intensification of the peaks after sputtering.

Peaks at 227.2 eV, 228.3 eV, 229.2 eV, 230.7 eV and 232.2 eV corresponding to metallic Mo, Mo^{3+}, Mo^{4+}, Mo^{5+} and Mo^{6+} were observed at the outermost region of alloy 926 and alloy 31. However, the peaks were intensified in the case of alloy 926 and alloy 31. After sputtering, the peak at 228.3 eV disappeared, while the other peaks were observed (Figure 5).

Correlating the observed results with that of the previous reports, the molybdenum present in these alloys as Mo^{3+}, Mo^{4+}, Mo^{5+} and Mo^{6+} is mainly enriched as MoO_2 and MoO_4^{2-} in the passive oxide layer.

3.3.3 Cr/Fe Peak Area Ratio. The Cr/Fe ratio is a vital parameter in deciding the stability of the passive film against leaching. The ability of an alloy to retard the leaching of chromium from the passive film indicates the material's resistance towards localized attack. Previous study indicated that higher the Cr/Fe peak area ratio, higher is the corrosion resistance[29]. The Cr/Fe peak area ratio of the type 316L SS, alloy 926 and alloy 31 before and after sputtering are depicted in Figure 6.

From the results in Figure 6, it was observed that Cr/Fe peak area ratio was very high in the outer layer for all the alloys studied. After sputtering, the ratio was decreased.

Alloy 926 and alloy 31 showed a very high value of Cr/Fe peak area ratio when compared to the reference type 316L SS. From the above observations, the overall content of chromium in the passive film of superaustenitics was higher than 316L SS.

3.3.4 Role of Nitrogen in the Passive Film. The XPS investigation of the passive film formed on the superaustenitic stainless steels indicated the enrichment of nitrogen in the form of CrN, NH_4^+ and NO_3^- at the metal/film interface. This stabilizes the chromium ions present in the passive film which in turn impedes the ingress of halide ions by stabilizing the defects in the passive film. Hence, the higher nitrogen content in the superaustenitic stainless steel results in the enrichment of chromium.

Correlating the results from electrochemical and XPS investigation, the presence of CrN, NH_4^+ and NO_3^- decreases the acidity within the pits, which in turn accounts for the higher corrosion resistance of superaustenitic stainless steels.

4 CONCLUSIONS

1. The localized corrosion resistance of superaustenitic stainless steels was found to be higher than that of type 316L SS. This shows the beneficial effect of alloying elements present.

2. The passive films of superaustenitic stainless steels revealed an enrichment of nitrogen in the form of NH or NO, NH_4^+ and NO_3^-. The inner layer was mainly enriched with metal-nitride bond, especially CrN. This co-segregation of chromium and nitrogen hindered the release of metal ions through the passive film and accounts for the enhanced pitting and crevice corrosion resistance.

3. The Cr/Fe peak area ratio indicated that the superaustenitic stainless steels possessed a higher amount of chromium in the passive film than 316L SS.

Acknowlegement

The Authors would like to thank Di. S. V. Naraslmhan and Dr. Santhanu Bera, Scientific Officers, WSCL, IGCAR, Kalpakkam, India, for carrying out ESCA investigations. One of the authors, Ms. G. Latha is thankful to the University Grants Commission, New Delhi, India for financial assistance.

References

1. L. H. Boulton and A. J. Betts, *Br. Corros. J.*, 1991, **26**, 287.
2. F. C. Walsh and S. A. Campbeli, *Bull. of Electrochem.*, 1992, **8**, 311.
3. S. N. Manohar, R. Sundarajan and A. K.Bansal, 'Proceedings of the national seminar on cooling towers', 18–20 Jan 1990, New Delhi, NTPC, TS 111/49–56.
4. B. A. Shaw, P. J. Moran and P. Gartland, 'Proceedings of the 12th International Corrosion Congress', NACE, Houston, Vol. 3B, 1993, p. 1915.
5. F. P. Ijsseling, *Br. Corros. J.*, 1989, **24**, 55.
6. B. Wallen, S. Hendrikson, *Werkst.* und *Korros.*, 1989, **40**, 602.
7. F. A.Garner, H. R. Brager and R. J. Pulgh, *J. of Nucl. Mat.*, 1995, **133 & 134**, 535.
8. G. Latha and S. Rajeswari, 'Proceedings of National Seminar on Electrochemistry in Marine Environment', Madras, India, 1995, p. 140.
9. G. Latha and S. Rajeswari, *Anti-Corroslon Methods and Materials*, 1996, **443**, 13.
10. A. J. Sedrlks, *Corrosion,* 1989, **45,** 510.
11. P. B. Lindsay, *Materials Performance*, 1986, **25**, 23.
12. R. K. Dayal, N. Parvathavarthini and J.B.Gnanamoorthy, *Br. Corros. J.*, 1983, **18**, 184.
13. Y. C. Lu, J. L. Luo and M. B. Ives, *Corrosion*, 1991, **47**, 835.
14. J. E. Truman, M. J. Coleman and K. R. Pirt, *Br. Corros. J.*, 1977, **12**, 236.
15. Y. C. Lu, M. B. Ives and C. R.Clayton, *Corros. Sci.*, 1993, **35**, 89.
16. R. C. Newman, Y. C. Lu, R. Bandy and C. R. Clayton,'Proceedings of the 9th International Congress on Metallic Corrosion', National Research Council, Toronto, Vol. 1, 1984, p. 394.
17. A. J. Sedrlks, *Inter. Mat. Rev.*, 1983, **28**, 306.
18. C. R. Clayton, Passivity mechanisms in stainless steels, Mo-N Synergism, Report No.N00014-85-K-0437, New York, 1986.
19. K. S. Kim and N.Winograd, *Surf Sci.*, 1974, **43**, 25.
20. A. Swift, A. G. Paul and J. C.Vickerman, *Surf Interface Anal.*, 1993, **20, 27**.
21. I. Olefjord and B. O. Elfstrom, *Corrosion*, 1982, **38,** 46.
22. P. Marcus and M. E. Bussell, *Appl.Surf Sci.*, 1992, **59,** 7.
23. A. Sadough Vanini, J.P. Audourd and P. Marcus, *Corros. Sci.*, 1994, **36**, 1825.
24. K. Asami, M. Naka, H. Hasimoto and T. Masumoto, *J. Electrochem. Soc.*, 1980, **127**, 2130.
25. M.Sakashita and N.Sato, *Corros. Sci.*, 1977, **17,** p493.
26. W. C. Moshier, G. D. Davis, J. S. Ahern and H. F. Hough, *J. Electrochem. Soc.*, 1987, **134**, 266.
27. K. Sugimoto and Y. Swada, *Corrosion*, 1976, **32**, 347.

28. W. C. Moshier, G. D. Davis, J. S. Ahern and H. F. Hough, *J. Electrochem. Soc.*, 1986, **133,** 1063.

29. J. E. Fulghum, G. E. Mcquire, I. H. Musselman, R. J. Nemanich, J. M. White and O. R. Chopra, *Anal. Chem.*, 1989, **61,** 243R.

3.6.2

Scheme of Coating and Surface Treatment for Improving Corrosion Resistance with Application to Marine Structures

M. M. El-Gammal

NAVAL ARCHITECTURE AND MARINE DEPARTMENT, FACULTY OF ENGINEERING, UNIVERSITY OF ALEXANDRIA, ALEXANDRIA, EGYPT

1 INTRODUCTION

Corrosion causes many problems in marine structures. Figure 1 illustrates this point. The ballast and deep tanks in ships show major amounts of corrosive areas. The splash areas of an offshore structure suffer to a great extent from corrosion. The corrosion process of steel in sea water is electrochemical in nature[1]. Corrosion of steel is thus a predominant phenomenon due to the formation of microcells, and thus loss of electrons is considered an inevitable process[2] – Figure 2. This may be attributed to the fact that the anode, which is the steel or metal, goes into solution as iron ions and electrons are lost. However, at the cathode, oxygen combines with the water; the result is what is known as hydroxyl ion formation, Figure 3. The next step in the corrosion process is the formation of a secondary reaction, termed as rust. The rust is a result of the reaction of the iron ions with the surrounding environment. The formation of soluble iron oxides and hydroxyls will cause the formation of rust, Figure 4.

The velocity of the reaction with the steel structure will be controlled by the extent of aggression of the surrounding environment[3]. Thus the corrosion rate and the extent of the corrosion cells both in number and areas will be affected by the nature, aggression and severity of the environment. The corrosion rate is determined by a reaction barrier, called polarization. There, exists many types of polarization and for corrosion of steel in sea water, the rate is usually controlled by the cathodic reaction. In other words, the oxygen plays a major role in speeding up the corrosion operation. Thus it is preferred to reduce the oxygen, if the rate of corrosion is to be kept to a minimum.

2 TYPES OF CORROSION

There are several types of corrosion known to the material engineers – general corrosion, pitting corrosion, galvanic corrosion and finally bacterial corrosion. First, the general corrosion is formed by the presence of oxygen in sea water and the salinity and temperature of the surrounding environment plays an importnt role in speeding up this kind of corrosion. Seconds, the galvanic corrosion results from the action of the anode/cathode electro-chemical solution potential phenomenon and two metals come into contact in the presence of electrolyte. Thus steel in sea water will suffer from such a type of corrosion. As is the case for most offshore structures, both fixed and movable, it is of great importance to study the corrosion rate for

(a)   (b)

(c)   (d)

(e)   (f)

Figure 1 *Typical cases of corrosion in ships: (a) extent of corrosion on outer hull of a passenger ship, (b) all types of corrosion as exhibited on the outer hull of a ship, (c) badly corroded ship at side shell plating of a cargo liner, (d) corroded hatch covers have lead to the leakage of water into the cargo hold, (e) a capsized ship due to loss of stability as the internal   partitions are found to be badly corroded and (f) a foundered ship due to water leakage into the double hull areas due to corrosion*

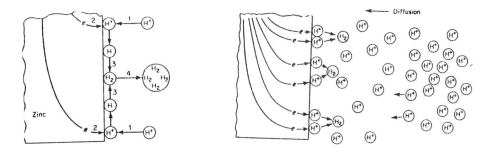

Figure 2 *Corrosion process considered as loss of electrons and hence a loss in weight and dimensions*

such structures. Figure 5 shows the different zones of corrosion as accrued in the case of fixed offshore structures. From this figure it may be concluded that the splash zone is the most exposed zone. The third type, pitting corrosion, accrues due to heterogeneity and inhomogeneity in the steel surface. This type of corrosion could be due to the mill scale presence or may be due to sputter if in the case of improper welding process. It depends on the nature of local attack causing pitting. The corrosion rate within the pit can be 10 to 50 times that of the general attacks. Thus the expected local severity of anodic steel pits and the surrounding cathodic surfices, penetration of pitting through the structure will occur. The result is leakage, and/or shorter life of the exposed members of the structure. Pitting corrosion may not represent a strengthwise problem. Nevertheless, it jeopardizes the tightness resulting in serious problems[4]. Specifically if a leak happens in an oil pipeline, this may in turn cause pollution. For instance a steel piling into the sea bed will be subjected to galvanic corrosion due to the effect of two different environments acting upon the pile[5,6]. As is explained in Figure 5 that the highest corrosion area is the splash zone followed by the submerged water area, while the lowest is the mud. The same effect and the same nature of corrosion are observed as in narrow hooks, in crevices and in welds[7]. The fused area of the heat affected zone may act as an anode and thus suffers from preferential galvanic and pitting corrosion. This may be due to the small anodic area combined with great cathodic areas, which yields the most dangerous situation. The fourth, the bacterial corrosion, can occur once the conditions prevailing matching together,

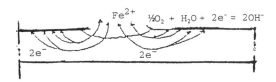

$$Fe^{2+} \quad \tfrac{1}{2}O_2 + H_2O + 2e^- = 2OH^-$$

$$2e^- \qquad 2e^-$$

Figure 3 *Formation of hydroxyl*

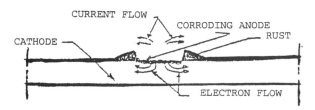

Figure 4 *Formation of rust*

such as the existence of sulphate reducing agents and in zones of estuaries. Sulphate and organic materials are necessary for the creation of this type of corrosion[8]. Needless to say that both of those constituents are found in sewage areas and in areas of harbours. This type of corrosion will be found in the sea and hence will occur in close offshore systems and for systems buried in the sea bed[9].

The movable will suffer from differences in metal potentials due to the presence of the propeller, usually fabricated from a noble material and the stern hull of the offshore rig made of steel. The action will result in a galvanic cell. Nevertheless, corrosion could take place even in the protected area due to the movement of the hull and the exposed collided areas with other structures. The worse the surface finish, the more will be the action of corrosion; thus a combination of more than one type of corrosion can be expected to occur. Special care must be given to buried structures such as siphons, pilings, sewage pipelines and oil pipelines. With this picture in mind, measures to safeguard engineering and strategic projects from engineering are required. This is the main aim of the present investigation. This paper will attempt to safeguard offshore structures against corrosion.

3 CRITICS AND VIEWS ON THE CURRENT CATHODIC PROTECTION SYSTEM

At the present time, current cathodic protection systems cannot provide complete protection against the corrosion phenomena stated above. There is no means to protect simultaneously against the four categories of corrosion. Nontheless, there are individual methods in promoting separate means of protection against each type of corrosion. The current cathodic protection system of metallic structures involves either sacrificial anodes or impressed current methods or a combination of both. Figure 6 gives a schematic representation of the theoretical applications of sacrificial anodes and that of the impressed current. Simply, the idea is to drive a flow of electrons to neutralize the effect of the formation of the hydroxyl ions as shown in Figure 6. However, there are certain factors which need to be considered in choosing the appropriate system and in the choice of the components of the cathodic system itself. The choice of the type of anode and anode material reflects in itself a problem. This may be due to the fact that some types of anodes are more active than others. The chemical constituents of anodic material and the electrolyte, the activity of the anode consumption and anode capacitance are thus the important parameters which need to be estimated. Figure 7 shows the principles of cathodic protection systems.

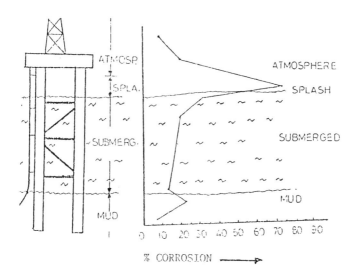

% CORROSION ⟶

Figure 5 *Different corrosion zones in an offshore structure*

4 THE PROPOSED APPROACH

The present approach attempts to promote a method to cover all the corrosion sources and to give a thorough protection at a reasonable cost. Thus both engineering and economical aspects are to be emphasized and achieved. Keeping in mind the above stated aspects it is rather not advisable to safegaurd structures against corrosion by deriving a complicated technique. These effects have been discussed previously[10-13]. Figure 8 summarizes the steps involveded in the calculations and the determination of the proper alternatives, from Figure 8 the following main steps can be calculated:

1. Calculate the total area subjected to corrosion.
2. Carry out the measurements of the environmental resistivity. This is to be done at various times with different. parameters, e.g.,environmental temperatures and at different Beaufort Scales. i.e, different set conditions.
3. Take the worst of the values determined in the previous step.
4. Calculate the current density and the total protection current.
5. Determine the alloying elements to overcome the aggression of the environment. Hence determine the optimum thickness of the alloying element to counteract the total current and develop the protection of the external hull of the marine structure.
6. Determine whether the steel is to be clad or sprayed with the coating as predetermined above. The copper-based alloy is to be either in the form of sheets clad with steel plates during milling operation or may be sprayed onto the external surface of the hull. The hull is locally heated to a temperature less than the eutectoid temperature that could be estimated from Sepherian's formula for pre-heating of steel as in the following equation:

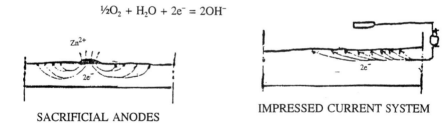

$$\tfrac{1}{2}O_2 + H_2O + 2e^- = 2OH^-$$

SACRIFICIAL ANODES   IMPRESSED CURRENT SYSTEM

Figure 6 *Schematic representation of protection systems involving sacrificial anodes and impressed current systems*

$$T = 350\sqrt{[C]'} - 0.25 \qquad \{1\}$$

where $[C]'$ is the total carbon and thickness equivalents given by:

$$[C]' = [C]c + [C]t \qquad \{2\}$$

$[C]c$ is the carbon equivalent, deduced from:

$$[C]c = C + \frac{1}{9}(Mn + Cr) + \frac{1}{18}Ni + \frac{7}{90}Mo \qquad \{3\}$$

$[C]t$ is the thickness equivalent, estimated from:

$$[C]t = 0.005.h.[C]c \qquad \{4\}$$

and h is the thickness of the plate.

7. If the clad steel sheet is used, it is justifiable to use the proper welding process and to choose the best type of electrode suited for a proper welding seqence. This is to be done to achieve the best recommended practice of welding of clad steel.

8. If the second alternative is applied, i.e. if the steel is sprayed with a brass coating it is important to sand blast the external surface of the hull to 0–3. Of course the sand blasting is to be carried out after removing reinforcement height at welded joints. Seams and butts are to be flushed to the surface, then the brass-based alloy, in molten condition, is to be sprayed to coat the heated portion of the hull. Thus the process of spraying is to be repeated until the required thickness is achieved.

9. Of course the required thickness of coating is to be determined by gaining experience of spraying adjustments and thus the number of layers to produce a certain thickness is to be known before carrying out the spray operation. Also, the spray pressure and the conditions of the surface play an important role in the determination of the final thickness.

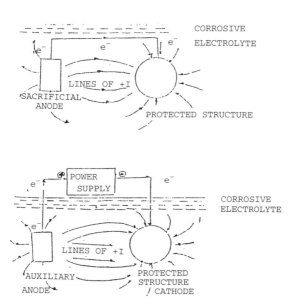

Figure 7 *Principles of cathodic protection systems*

5 DISCUSSION AND APPLICATIONS

The present recommended practice based on the idea of using coated clad steel or sheathed alloying elements is not a new technique. What is new is that the type of alloying elements sheathing the base metal is to be covered whatever be its condition, and after the completion of the welding, cutting, shaping and fabrication process.

The conditions for the success of operation of coat-cladding or sheathing to protect the steels against corrosion require that these are to be maintained by the necessity of heating the surface to the temperature as determined in equation {1}. Thus, in order to promote this technology, it is rather important to subject the steel to a reasonably high temperature. Thus the brass based alloy had to be poured or sprayed as shown in Figure 9, onto the surface of the heated steel giving a flush surface. The flush surface has a good reflection of the Sun's rays, besides its strength quality that was experimentally obtained by El-Gammal[11], to be 5–10 times that of ordinary bare steels. This is only achieved by, and attributed to, the good quality of coated-clad or the sheathed steel. Thus no wonder that this is the material of the future. This is mainly due to the good properties of the material, and its good working life. Also this material is found to have good mechanical properties both in static and dynamic conditions, besides the anti-fouling properties achieved by the sheathing material. Thus, as expected, the factor of safety is roughly 1/5th to 1/10th that of the normal bare steels, giving good notch resistance and prevailing good fatigue endurance limits. Consequently, this material will have an industrial impact, both on economical and engineering levels. Besides, the brass based

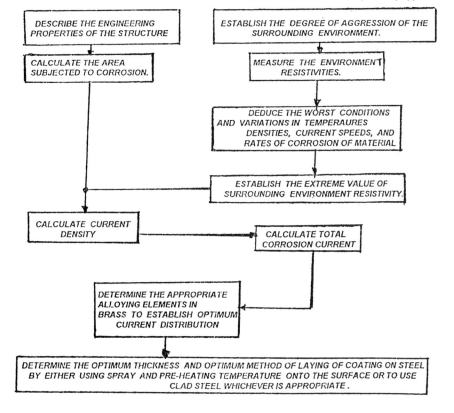

Figure 8 *Flow chart for the proposed scheme of cathodic protection*

alloy as a coating for steel, will cause the water flow to be smooth around the hull. This will lead to better propulsion performance and less water resistance and frictional resistance. The good anti-corrosive property of the clad or sheathed steel or of the coated sprayed steel is thus dependent on the tight contact between the coat-clad and steel. Air voids, air bubbles and air cavities can produce drastic effects on the coated or clad steel. Also, the welding operation plays a major role in the strength as well as the anti corrosive properties of the coated, sheathed or clad steels.

This is simply the idea adapted by the contemporary materials engineer[10–14]. The recent developments in both sciences of material and in the science of welding technologies play a major rôle in these applications. The first is achieved by giving soluble solutions and thus ionic bond is being resulted. The second gives metallurgical and mechanical bonds of such materials.

6 CONCLUSIONS

As has been stated within the text, the present coated, sheathed or clad copper-based alloys, will have good future use in engineering practice. It is gaining popularity due to economical

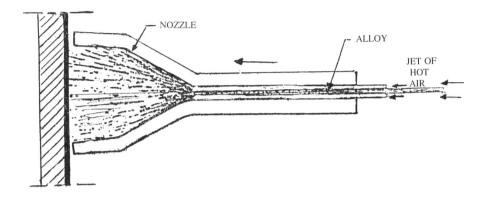

Figure 9 *Spraying method of cupro-alloys onto the pre-heated surface of steel. This process leads to better engineering component life*

and engineering features. The first is achieved by the advent of light alloyed clad-steel that will promote the technology of using plated ships instead of the conventional way used nowadays which is to use framed and shells to the current practice. The plated ships will not only have large spaces, but will also have an increase in speed. The technology simply could be applied by the use of sprayed hot-metal alloy, onto the surface of steels. Thus, the process could be either applied on local parts or could be applied globally. This could be achieved only if results in gaining of good strength and the better bond of the sprayed coated-steels were to be satisfied. The chemical constituents of the coat will of course be different from that of the clad. Thus, experimentation of different copper-based alloys with different grades of steel will determine the best material suitable for a certain aggressive environment. The aggression of the environment will play a major rôle and affect the selection of alloying elements in brass to promote certain chemical, mechanical, physical and environmental features. This would be achieved by giving the component better wear resistance and anti corrosive characteristics. Besides, the process could be developed to incorporate protection of moving parts such as pistons, piston pins, connecting rods and crankshafts against wear. Thus this development is aimed at incorporating the optimum percentage of the alloying elements. Therefore, the sheathing must be based on metallurgical as well as mechanical bonding of the sprayed coat or of the clad material to the surafce of the parent material.

References.

1. B. Wyatt , 'Design and Operation Guidance on Cathodic Protection of Offshore Structures', Sections 9 and 12, Subsea Installation and Pipeline, MTD Ltd., 1990.
2. B. Wyatt and D J Irvine, *Materials Performance*, 1987, **26**, 12.
3. NACE/ASTM, 'Pipeline Corrosion Control', 1990.
4. B. Wyatt,'Design and Operation Guidance on Cathodic Protection of Offshore Structures', Sub-Sea Installations and Pipelines, Section 12, Marine Technology Directorate Ltd., London, 1990.

5.   Det Norsk Veritas, 'Cathodic Protection Design', Recommended Practice RPB 401, March 1986.

6.   Department of Energy, 'Offshore Installations: Guidance on Design and Construction', HMSO, London, UK, 4th Edition, 1990.

7.   Lloyds Register of Shipping, 'Draft Rules and Regulations for the Classification of Fixed Offshore Installations', London, UK, 1987.

8.   Det Norsk Veritas, 'Monitoring Cathodic Protection Sytems', Recommended Practice RPB 403, March 1987.

9.   British Standards Institution, "Code of Practice for Cathodic Protection", BS 7361, 1991, Part 1.

10. M. M. El-Gammal, 'Prospects of Cathodic Protection Systems for Marine Units', SME&S, Arab Branch, Alexandria, Egypt, December 1991.

11. M. M. El-Gammal, 'Sandwich Course: Cathodic Protection Systems', Community Development Centre, CDC, Alexandria University, Alexandria, Egypt, 1995.

12. M. M. El-Gammal, Proceedings of the International Conference, Dublin Ireland, Edited by M. S. J. Hashmi, 1995, Vol. 1, p. 577.

13. M. M. El-Gammal, 'New Technological Anti-Corrosion Process with Application to Marine Structures', In press.

3.6.3
Enhancement of Tooling Life Using Surface Engineering Technologies

K. N. Strafford[1], T. P. Wilks[2], C. Subramanian[3] and W. McMillan[4]

[1]INSTITUTE OF MATERIALS, LONDON, UK

[2]CENTRE FOR MANUFACTURING, SOUTH AUSTRALIA

[3]GARTRELL SCHOOL OF MINING, METALLURGY AND APPLIED GEOLOGY, UNIVERSITY OF SOUTH AUSTRALIA

[4]IAN WARK RESEARCH INSTITUTE, UNIVERSITY OF SOUTH AUSTRALIA, LEVELS CAMPUS, SOUTH AUSTRALIA

1 INTRODUCTION

Wear of press forming tools and dies in the automotive industry has serious implications for both quality and productivity achievements when pressing sheet steel materials. McMillan[1] has provided a review. The ability to achieve Just-in-Time delivery and sustain lean manufacturing philosophies demands advanced process capabilities. The increased use of coated materials e.g. Zn alloy coated steel for corrosion protection[2] adds another demand on die and wear performance. McMillan[1] has considered problems associated with automotive panel manufacture, covering aspects such as material selection factors – the choice of coated, uncoated, cold rolled or hot rolled steel; forming factors, such as the severity of the forming operation and the work done in producing a panel – i.e. the forming operation as a whole: stretching, drawing, trimming etc. and, most importantly, the nature of the occurrence of scoring or other inherent wear problems associated with the die.

For press forming dies, the most prevalent problem is wear, characterised by severe scoring of the panel surfaces at critical locations, thereby reducing panel quality. Such scoring can be a combination of adhesive and abrasive wear. The die steel tends to pick up material from the panel it is forming, (a softer material), and this builds up on the die surface. With subsequent panels being produced, this adhered material will work harden, and abrasion of the panels results, creating scores and gouges[2].

Poor quality die surfaces, in addition to increasing friction at draw beads and critical radii, result in the "pick-up" of particulate debris from the pressed material on to die surfaces. Due to work hardening of the pressed material during the forming operation, the initial score marks may act as a file and cause further deterioration of die surfaces.

The significance of friction in wear has been noted by Eyre[3]. He has pointed out that friction and wear are often given different interpretations, depending upon the industry concerned. As with Ludema[4], Eyre[3] agrees that these parameters are not intrinsic material properties, but are a characteristic of the engineering system. There is no simple, universal relationship between friction and wear: often very large changes in wear result from only small changes in friction. However, in general, low friction will result in low wear, and changes in friction values are likely to indicate an important change in wear rate and/or wear mechanisms.

Surface engineering (SE) treatments applied to press forming dies reduce die wear and maintain part quality because they may be expected to:

i.   reduce friction between the die surface and the sheet steel being formed, with attendant reduction in interface forces and less heat generation;

ii. reduce the pick-up of the sheet steel and protective (corrosion) coatings (if present), and thus lower the potential for adhesive wear, and

iii. reduce abrasive wear and indentation damage (pimples) for skin panels due to the harder die surface.

Laboratory wear tests have been conducted to attempt to qualify various properties of surface engineering treatment such as the Toyota Diffusion Process, hard chromium plating, electroless nickel coatings with, and without PTFE and a number of thermal treatments, i.e. conventional hardening treatments. In addition to characterising and evaluating tribological performance of these rival SE systems, a detailed forensic assessment of wear mechanisms has been carried out using various optical, electron optical and other surface analysis techniques. In this way it was hoped to be able to understand more fully the wear mechanisms interacting, and to be able to provide useful information for the design engineer. The laboratory programme was to complement a parallel investigative programme in the automotive production environment, as reported elsewhere[1,5].

2 EXPERIMENTAL

2. 1 Materials and Treatments

2.1.1 Materials. The substrate materials used for laboratory evaluations were those most commonly utilised in the press shop at GMHA for dies and die inserts:

(a) D2 high carbon/high chromium tool steel (GMH D2), AS 1239/D2A.
     Composition: 1.5–1.6C, 0.3_{max}Si, 0.3_{max}Mn, 12.0Cr, 0.8–1.0Mo, 0.8-1.0V

(b) Flame hardenable grey cast iron (GM241), AS 1830/H-269.
     Composition: 2.8–3.2C, 1.8–2.25Si, 0.75–1.05Mn, 0.12_{max}P, 0.12_{max}S, 0.35–0.5Cr, 0.35–0.5Mo

(c) Flame hardenable cast steel. Composition of casting used was close to both of these standards:
     L2A: 0.45–0.55C, 0.75_{max}Si, 0.5–1.0Mn, 0.06_{max}P, 0.06_{max}S, 0.8–1.2Cr
     L2B: 0.55–0.65C, 0.75_{max}Si, 0.5–1.0Mn, 0.06_{max}P, 0.06_{max}S, 0.8–1.5Cr, 0.2–0.4Mo

(d) ASSAB specially formulated steels
     1.   ASSAB Vanadis 4 powder metallurgical cold work tool steel
          Composition: 1.5C, 1.0Si, 0.4Mn, 8.0Cr, 1.5Mo, 4.0V
     2.   ASSAB Calmax plastic mould and cold work tool steel
          Composition: 0.6C, 0.35Si. 0.8Mn, 4.5Cr, 0.5Mo, 0.2V

The specimens made from these materials were derived from bar stock, circular cross-section of 71 to 74 mm diameter, 5 mm thickness and containing a central hole of 9.5 mm diameter to accommodate the apparatus used for subsequent wear testing. Surface grinding to remove possible decarburisation layer and finishing to press shop requirements was then carried out. The disc specimens were then treated using a number of conventional hardening as well as coating techniques.

2.1.2 Treatments: Conventional Hardening Treatments

a. Fully hardened (59–61 HRC) D2 tool steel (Designation: Full hard-D2)

Heat treatment procedure:

Preheated at 600°C for 2 hours.

Austenitised at nominally 1010°C with a soaking time of 45 minutes to give a pre-tempering hardness of 64 HRC, in a vacuum furnace to protect material from the atmosphere. Quenched in oil.

Tempered twice at 300°C for 2 hours, with intermediate cooling to room temperature, to give final hardness of 59–61 HRC (VHN_{200}810).

b. Flame hardened (50 HRC) grey cast iron (Designation: Flame hard-CI)

Flame hardening procedure:

An oxyacetylene flame was used to rapidly heat the surface of the specimens above the transformation temperature (i.e. >650°C).

The surface was rapidly quenched using a jet of air to obtain a surface hardness of 50 HRC (VHN_{200}780).

c. Flame hardened (50 HRC) cast steel (Designation: Flame hard-CS)

Flame hardening procedure:

Similar to that of grey cast iron to give a surface hardness of 50 HRC (VHN_{200}560)

d. Fully hardened (59–61 HRC) ASSAB Vanadis 4 (Designation: Full hard-Van)

Heat treatment procedure:

Preheated at 600°C for 2 hours.

Austenitised at nominally 1010°C with a soaking time of 30 minutes to give a pre-tempering hardness of 60 HRC, in a vacuum furnace to protect material from the atmosphere. Quenched in oil.

Tempered twice at 300°C for 2 hours, with intermediate cooling to room temperature, to give final hardness of 59–61 HRC (VHN_{200}859).

e. Fully hardened (59–61 HRC) ASSAB Calmax (Designation: Full hard-Cal)

Heat treatment procedure:

Preheated at 700°C for 2 hours.

Austenitised at nominally 960°C with a soaking time of 30 minutes to give a pre-tempering hardness of 63 HRC, in a vacuum furnace to protect material from the atmosphere.

Quenched in martempering bath at 200°C followed by forced air cooling to 25°C to avoid shrinkage.

Tempered twice at 200°C for 2 hours, with intermediate cooling to room temperature, to give final hardness of 59-61 HRC (VHN_{200}812).

Surface engineering treatments

a. Toyota Diffusion (VC) treated D2 tool steel (Designation: TD-D2)

Treatment procedure: Because of commercial confidentiality and the proprietary nature of the process, only general details of the procedure can be given here. However the main processing steps to create a vanadium carbide coating included rigorous degreasing and cleaning of samples, a preheating step (~ 550°C), dipping in the molten salt bath (essentially a mixture comprising anhydrous borax, additive (V_2O_5), plus a reducing agent (B_4C) typically held at 800–1250°C, post-treatments (oil quenching at 200°C, followed by tempering). McMillan[1] has provided a critique of the total process. The vanadium carbide coating thickness produced was about 2–2.5μm of VHN_{200} maximum 1250. Bath composition: HEEFCHROME (High Efficiency Etch Free Chromium) solution containing 220–240 g/1 chromic acid, 3g/1 sulphate anion and 15–25 ml/l activated catalyst.

Anodes: 7% tin-lead insoluble alloy.

Bath temperature: Nominally 60°C.

Current: 17000-18000 A (approximately 70% of rated rectifier capacity).

Plating schedule: Initial chrome strike for 5 minutes at a current density of 6 A/in^2 followed by plating for 1–2 hours at a current density of 1–1.5 A/in^2.

The coating thickness was some 13μm thick with a VHN$_{200}$ of 350.

b.   Electroless nickel plated grey cast iron (Designation: Ni-CI)

Plating procedure:

Bath composition: Enthone Inc. Enplate NI-423 plating solution was used, containing ~9–11 % by weight P and a nickel concentration of nominally 5.8 g/l, in the concentration 200 ml/l with distilled water.

Bath temperature: 84–89°C, optimum of 87°C.

Bath pH: 5.0–5.2

Plating time: 2 hours (plating rate nominally 17 m/hr).

The coating produced was 35–38μm thickness, with a VHN$_{200}$ of ~560.

c.   Heat treated electroless nickel plated grey cast iron (Designation: H/Ni-CI)

Plating procedure:

Same plating procedure as electroless nickel plating with the addition of a heat treatment after plating, at 300°C for 2 hours. Hardness was VHN$_{200}$600.

d.   Electroless nickel with PTFE on grey cast iron (Designation: Ni/PT-CI)

Plating procedure:

Same plating procedure as for electroless nickel plating with the addition of 20–25% by volume PTFE. The VHN$_{200}$ was 320.

2.2 Wear Testing and SEM Evaluation

The wear testing of hardened and/or coated discs of die materials was conducted using a pin-on-disc machine to rank the treatments and coatings, relative to each other, by keeping test parameters constant.

In the pin-on-disc wear machine a zirconia ball counterface was slid against a disc specimen of the metal substrate with respective treatment (e.g. fully hardened, TD treated, hard chromium plated, etc.) under a normal load of 2 kg (19.62 N), at a chosen linear sliding speed of nominally 0.3 m/sec. The duration of each test was 2 hours to give a sliding distance of ~ 2160 m. The test was performed under unlubricated (dry) conditions to test the severity of the wear operation. For all tests a temperature of 20°C (i.e. ambient temperature) was used, to standardise the test conditions and avoid external heating effects.

The wear rig used in this study has been described elsewhere[1]. In operation, the disc was rotated anti-clockwise and the ball was offset from the disc centre, so that the friction force on the pin was directed away from the load arm pivot. A load cell was used to measure the tangential (frictional) force of the system. These digital output signals were logged to a microprocessor at a controlled rate (every second).

Both disc and ball were cleaned thoroughly in acetone and weighed to the nearest tenth of a milligram prior to testing. After completion of the test both the ball and the disc were cleaned and re-weighed to measure the weight loss, and allow calculation of wear rates. The disc dimensions allowed a maximum of three wear tests per surface and where possible both surfaces of a particular disc were tested.

A CamScan scanning electron microscope (SEM) was employed to assess the wear mode and general characteristics of selected wear test tracks of each treatment evaluated. The associated SEM EDX (energy dispersive analysis by X-ray) analytical apparatus was used to determine the presence of and characterise adherent counterface material, and to assess whether a particular coating was entirely removed from the substrate, or if cohesive failure was present.

3 RESULTS

The tribological performances of the various materials were sytematically assessed involving the determination of: (1) wear rate and weight loss data; (2) friction data; and (3) characterisation of morphological features of wear.

3.1 Wear and Weight Loss Data

Average (unlubricated) wear rate data for all of the twenty four different types of materials / coatings tested (identified in the Experimental) have been summarised in Figure 1.

The average wear rates shown were derived from weight losses experienced by the test discs, divided by both the standard applied load (2 kg) and a nominally fixed sliding distance of ~ 2200m. The minimum and maximum wear rates varied considerably with a spread of almost two orders of magnitude. It was possible to rank wear rates into five classifications from low to high, reflecting the broad response of the different materials / coatings as follows:

Wear rate ranking   **Treatment/material**

Low $(2 - 7 \times 10^{-4}$ mg/m/kg).
a.   Toyota Diffusion (VC) treated D2 tool steel (TD-D2)
b.   Electroless nickel with PTFE on grey cast iron (Ni/PT-CI)
Low to medium $(3 \times 10^{4} - 2 \times 10^{-3})$ mg/m.kg.)
c.   Fully hardened (59-61 HRC) D2 tool steel (Full-hard-D2)
d.   Fully hardened (59-61 HRC) ASSAB Vanadis 4 (Full hard-Van)
e.   Fully hardened (59-61 HRC) ASSAB Calmax (Full hard-Cal)
Medium $(6 \times 10^{-4} - 3 \times 10^{-3}$ mg/ni/kg)
f.   Heat treated electroless nickel plated grey cast iron (H/Ni-CI)
g.   Flame hardened (50 HRC) cast steel (Flame hard-CS)
h.   Flame hardened (50 HRC) cast iron (Flame hard-CI)
Medium to high   (1 x 10-3 - 1 x 10-2 mg/m.kg)
i.   Electroless nickel plated grey cast iron (Ni-CI)
High $(4 \times 10^{-3} - 1 \times 10^{-2}$ mg/m.kg)
j.   Hard chromium plated cast steel - Vendor 1 (Vl CR-CS)
k.   Hard chromium plated cast steel - Vendor 2 (V2 CR-CS)

In general, it was interesting to note that the majority of the coated samples exhibited better wear performance than the hardened steels, although the fully hardened tool steel was superior to the steel specimens which had been simply flame hardened. Clearly, samples which had received the Toyota Diffusion treatment to yield a VC "coating", showed the best overall behaviour, and also within the behaviour of all of the coated specimens. Also of particular interest was the excellent relative performance of cast iron samples coated with electroless Ni

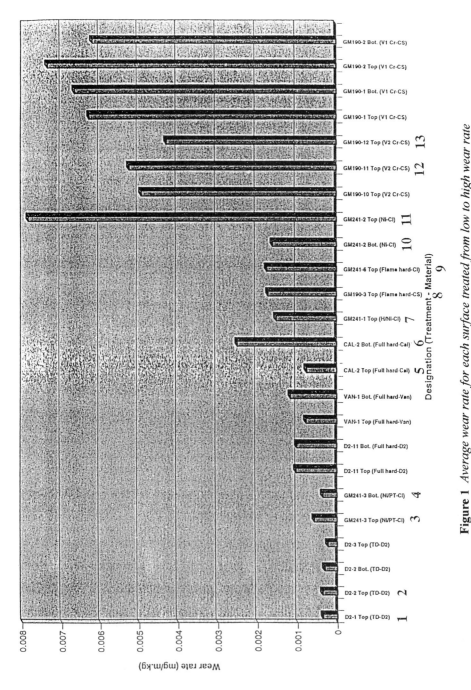

Figure 1 *Average wear rate for each surface treated from low to high wear rate*

bearing PTFE. It was perhaps surprising that the hard chromium plating on the grey iron substrate exhibited poor wear resistance.

Table 1 has been derived from information collected from the pin-on-disc tests involving just thirteen materials / coatings, chosen to provide examples illustrative of the tribological behaviour observed for the seven limiting different types of material / coating investigated - viz. fully hardened steel, flame-hardened steel, the vanadium carbide "coated" steel, and cast steel, coated with electrolytic chromium, or three types of electroless Ni-P coatings on grey iron, including the PTFE variant. These thirteen test samples, numbered 1-13, also to are the subject of detailed evaluation and discussion have been identified in Table 1 and Figure 1.

Linear speeds varied from a minimum of 0.287 to a maximum of 0.330 m/s, while elapsed running times ranged from 7,198 to 7,442s (i.e. nominally 2h). Sliding distances varied from 2,224 to 2,641m. Of more importance are the weight losses measured, disc by disc, which range from a minimum of 1.3mg (Specimen 1) to a maximum of 44.4mg for Specimen 11. This latter, unusually high, figure contrasts with the low value noted for the second grey iron sample coated with electroless nickel: such scatter was reflected in the variable morphologies of the wear patterns (see Sub-Section 3.3). In general, however, the weight losses exhibited by the discs mirror wear rates, specimen by specimen. Also of interest in Table 1 are the weight losses experienced by the zirconia ball (the "pin"). Generally, these are less than 0.5mg and a small percentage of the disc weight loss. However, in the total wear system, the possible roles of small amounts of zirconia particles setting up a three-body abrasion situation and/or in transferring to the disc, so enhancing adhesive wear, have to be recognised (see Sub-Section 3.3).

3.2 Frictional Force Data

Figure 2 summarises the mean maximum and final frictional force values observed for twenty two types of materials/coatings. In general, when comparing this Figure with Figure 1 (relating to wear rates), it is clear that the frictional forces, unlike the wear rates, although varying, do not exhibit such a broad range of values, and, also, do not systematically increase from the levels associated with the VC-coated samples (Specimens 1 and 2) through to those for chromium plated grey iron (Specimens 12 and 13).

Indeed, it could be said that friction values are, clearly unlike wear rates, relatively invariant, specimen by specimen, with a mean value of some 18-20N (see Table 2). In detail there are interesting and important variations. Thus the mean values at the completion of the wear tests after 120 minutes are generally lower, and sometimes substantially lower than the mean maximum observed values. Noteworthy, but not surprising are the lowest forces observed with the electroless nickel loaded with PTFE (Specimen 3) in total contrast to the exceptionally high values recorded for the flame hardened steel sample, Specimen 8. An overall conclusion, however, is that the frictional forces do not in general follow the wear rate patterns, material coating by material coating. Also coated samples numbered 1-4, 7, 10-11, 12 and 13 are not associated with especially low friction forces, although it is interesting to note that the nickel plated specimens (specimens 3, 4, 7, 10, 11), do not appear to exhibit lower than average (20.8 mean maximum, and 16.8 mean at 120 minutes) average values. Again, the VC coated samples, 1 and 2, while exhibiting the lowest wear rates, are not associated with unusually low frictional force values.

Typical graphs of frictional force versus running time are shown in Figures 3–7. Figures 3 and 4 contrast the low friction force characteristic of the flame hardened cast iron (Specimen

Table 1 *Parameters summarizing unlubricated wear testing of materials and surface treatments from low to high wear rates*

Spec. No.	Test I.D.	Substrate	Coating/Treatment	Linear speed m/s	Elapsed Time s	Sliding distance m	Weight loss mg Disc	Ball	Wear rate mg/mKg
1	Low wear rate D2-1 Top 2	Steel AS1239/D2A	VC (Aust) (Toyota D)	0.287	7220	2072	1.3	0.4	3.137 exp-4
2	D2-2 Top 1	Steel AS1239/D2A	VC (Aust) (Toyota D)	0.366	7225	2641	1.7	0.8	6.630 exp-4
3	GM241-3 Top 2	Iron AS1830/H-269	E/less Ni-P (PTFE)	0.309	7442	2359	2.3	0.0	4.875 exp-4
4	GM241-3 Bot 1	Iron AS1830/H-269	E/less Ni-P (PTFE)	0.328	7200	2362	2.1	0.2	4.446 exp-4
5	Low to medium wear rate CAL-2 Top 2	Steel ASSAB CALMAX	Full hard. 59-61 HRC	0.323	7200	2327	2.8	0.5	6.018 exp-4
6	CAL-2 Bot 1	Steel ASSAB CALMAX	Full hard. 59-61 HRC	0.320	7200	2304	11.5	0.4	2.495 exp-3
7	Medium wear rate GM241-1 Top 1	Iron AS1830/H-269	E/less Ni-P H.Treat.	0.309	7199	2224	5.8	0.1	1.304 exp-3
8	GM190-3 Top1	Steel AS2074/L2A, L2B	Flame hard. 50 HRC	0.303	7200	2184	11.7	1.5	2.679 exp-3
9	GM241-6 Top 1	Iron AS1830/H-269	Flame hard. 50 HRC	0.309	7219	2233	11.8	0.1	2.642 exp-3
10	Medium to high wear rate GM241-2 Bot 1	Iron AS1830/H-269	E/less Ni-P	0.320	7199	2304	6.3	0.1	1.367 exp-3
11	GM241-2 Top 1	Iron AS1830/H-269	E/less Ni-P	0.309	7198	2224	44.4	0.2	9.981 exp-3
12	High wear rate GM190-11 Top 3	Steel AS2074/L2A, L2B	Hard Cr (Ven^2)	0.314	7200	2258	22.1	1.2	4.894 exp-3
13	GM190-12 Top 3	Steel AS2074/L2A, L2B	Hard Cr (Ven^2)	0.330	7203	2376	21.9	1.2	4.608 exp-3

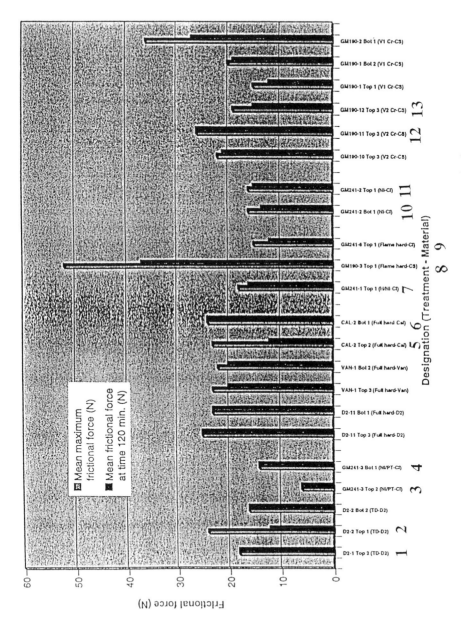

Figure 2 *Mean maximum and final friction force, from low to high wear rate*

9) with that for the flame hardened cast steel (Specimen 8): these materials however were associated with very similar wear rates of 2.642×10^{-3} and 2.679×10^{-3} mg/m kg, respectively. The average-to-high friction values for the Toyota Diffusion treated (VC "coatings") samples (Specimens 1 and 2) are evident in Figures 5 and 6, where the large variations over short periods of time, creating a broad trace pattern are also unusual. The lower final "steady rate" force values of Specimen 2 relative to Specimen 1, albeit after the high values in the earliest stages of wear, are also noteworthy: that is, scatter is observed with nominally similar materials. Figure 7, relating to grey cast iron coated with electroless Ni containing PTFE (Specimen 3) shows in complete contrast, after an initial period of spread, a remarkably "focused" trace: also, the very low values of the frictional force are evident.

3.3 Surface Damage – Morphology and Composition

All of the wear tracks on the test discs were systematically examined using SEM to appraise and identify the wear mechanisms - abrasive or adhesive wear, and with coatings to additionally assess coating integrity - coherence and adherence to the substrate.

Of particular interest was the nature and associated extent of material transfer between disc and the zirconia ball, bearing in mind the mass changes evident with both these entities following the completion of the tests (see Table 1). The EDX spectra generated from the worn specimen surface were particularly useful in this context. Selected observations are reported here, to illustrate the range of behaviour exhibited, and to provide evidence for

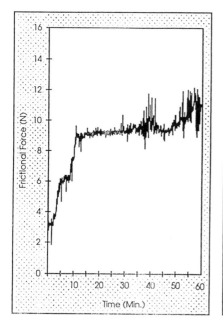
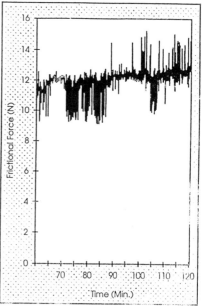

Figure 3 *Wear test output for flame hardened grey cast iron (Specimen 9)*

Table 2 *Maximum and final frictional forces associated with low to high wear rates*

Material ID	Specimen No.	Mean max. frictional force (N)	Mean frictional force at time 2h (N)
D2-1 Top 2	1	18	18
D2-2 Top 2	2	24	12
GM241-3 Top 2	3	6	6
GM241-3 Bot. 1	4	14	14
CAL-2 Top 2	5	23	12
Cal-2 Bot. 1	6	24	24
GM241-1 Top 1	7	18	16
GM190-3 Top 1	8	52	37
GM241-6 Top	9	15	12
GM241-2 Bot. 1	10	16	13.5
GM241-2 Top1	11	16	15
GM190-11 Top 3	12	26	26
GM190-12 Top 3	13	19	15
Averages		20.8	16.8

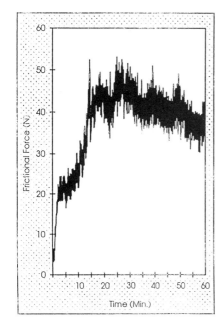
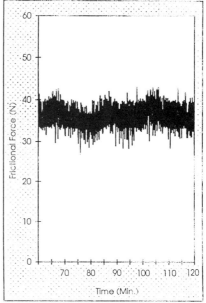

Figure 4 *Wear test output for flame hardened cast steel (Specimen 8)*

enunciation of likely wear damage mechanisms. Again, as in the previous sub-Sections 3.1 and 3.2, attention is focused on samples from among the Specimens numbered 1 to 13 – Figure 1 and Table 1 refer.

3.3.1 Surface Engineered (Coated) Materials. The Toyota Diffusion treated D2 tool steels (Specimens 1 and 2) exhibited the lowest wear rates.

Figure 8 (upper) is an SEM micrograph of wear track on Specimen 1 that shows a combination of intact "coating" and spallation of the "layer". It is possible that the formation of these spalled regions was the cause of the evident frictional force fluctuations seen in Figure 5, resulting in differential contact of the counterface with the surface of the disc. There appears to be no scoring of, or obvious adhesion of, wear debris to the track. Figure 8 (lower) gives detail of the wear track, and reveals evidence of plastic deformation of the coating within the spalled area, with no visible changes in the intact "coating" portions. Again, there is no evidence of scoring but the presence of plastically-deformed material indicates an adhesive wear mechanism has been at play, also involving polishing, as such, at residual, still-intact, areas.

Figure 9 (upper) is the analytical spectrum attributed to the intact portion of wear track of Specimen 1. It can be seen that it is virtually all "VC" with very little zirconium pick-up from the counterface. There was also a small amount of iron evident in this region, indicating possible areas of residual or developed porosity in the VC layer and/or contamination of either the treatment bath or the surface.

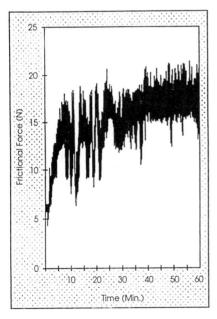
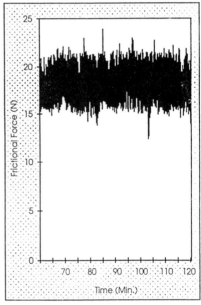

Figure 5 *Wear test output for Toyota Diffusion treated tool steel (Specimen 2)*

Figure 9 (lower) is the analytical spectrum derived from the spalled region of the same sample. Here there was considerable zirconium pick-up from the counterface compared with the intact region. A significant increase in the quantity of iron detected, coupled with the fact that there was significantly less vanadium present, indicates an almost total loss of coating in these areas. Either the layer was broken down almost completely or the layer was thin enough to allow for the detection of iron. However, Figure 9 shows a considerable amount of coating is still present, thereby favouring the latter explanation.

Also, in the low wear rate category, with essentially comparable performance to the VC-coated samples, were the electroless Ni coated grey cast irons where the coating contained some 20% of PTFE (Samples numbered 3 and 4). Figure 10 (upper), shows an SEM micrograph of the wear track from Specimen 3, electroless nickel coating containing PTFE on grey cast iron, and reveals very little pick-up or scoring. Essentially there is a polishing action only, flattening out the initial nodular surface associated with the addition of PTFE. In detail Figure 10 (lower), on the inside of the track, some areas of porosity can be seen, clearly associated directly with the surface nodularity. There appears to be minor pick-up from re-deposited wear debris, as well as light score marks from the counterface. Figure 11 shows the interface between the track and the unworn coating. There is no cracking, only evidence of the pressing of the nodules into flat formations, and subsequent plastic deformation, or smearing of the coating over the interface.

Figure 12 is the EDS spectrum for the worn area of Figure 11 and shows strong Ni and P signals only, indicating that there is a completely intact layer, with no pick-up from the counterface. This correlates well with the fact that there was no wear – weight loss – of the counterface – see Table 1.

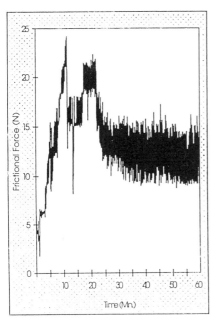
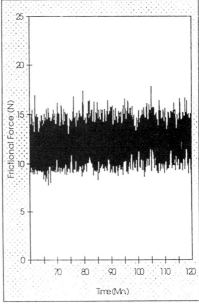

Figure 6

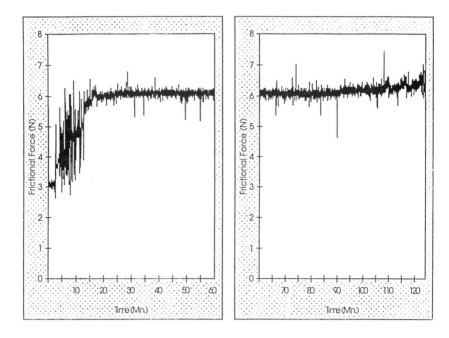

Figure 7 *Wear test output for electroless nickel with PTFE on grey cast iron (Specimen 3)*

Figure 13 (upper), shows an SEM micrograph of wear track GM241-3 Bottom 1, electroless nickel coating containing PTFE on grey cast iron, revealing a high degree of roughness in the track associated with both adhesion of re-deposited wear debris, as well as porosity associated with the flattening of the nodular surface. Scoring is also evident. All of these factors would account for the increase in frictional force observed during the wear test, particularly at the latter stages when surface contact would have been aggravated by increased temperatures and counterface material wear, see Figure 14. Figure 13 (lower), reveals the interface between the unworn surface and the wear track. Score marks can be clearly seen as well as the high degree of porosity and some cracking near the edge of the track. Also evident are smears of re-deposited wear debris, all acting to increase friction, though, as shown, not necessarily increasing wear rate.

The wear behaviour of two other types of electroless Ni coating on grey cast iron - that is, Ni-P coatings without, and with, heat treatment was also examined, Specimens 10 and 11, and 7, respectively (see Table 1). Specimens 10 and 11 were ranked in the medium to high wear rate category: Specimen 7 was placed in the medium wear rate class.

Microscopy revealed two areas in the wear track from Specimen 11, electroless nickel on AS 1830/H-269 grey cast iron, of differing morphologies. Figure 15 (upper), shows the delaminated nickel coating. There appears to be a combination of complete removal of the coating, as well as re-deposition and subsequent embedding of wear debris into the surface of the substrate. It can be seen at the left-hand edge that the layer is virtually separated from the substrate and there is evidently a tendency for the nickel to fail cohesively; in other words, some failure occurred within the layer itself to produce a conchoidal fracture. The edge cracking

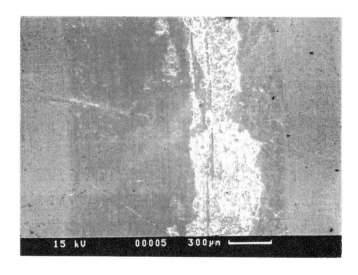

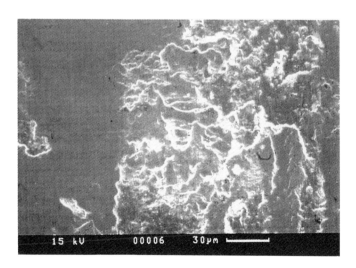

Figure 8 *Upper – SEM micrograph of wear track of specimen 1 showing a combination of intact VC layer and spallation/grinding of the layer. Lower – close up revealing plastic deformation of the VC layer within the spalled area with no damage (apart from polishing) at the intact portion*

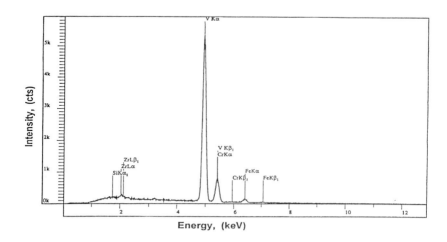

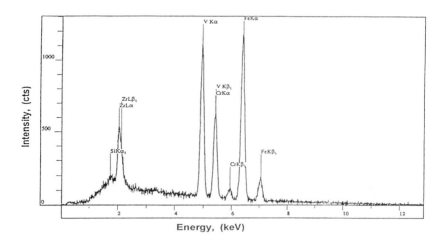

Figure 9 *Upper – analytical spectrum for the intact portion of the wear track of Figure 8, showing virtually complete VC layer, no pick-up from the counterface and only a negligible amount of iron due to porosity or contamination of the layer. Lower – analytical spectrum for the spalled region of the wear track of Figure 8, showing considerable pick-up of zirconium from the counterface. Significant quantities of iron and a reduction in vanadium content, indicate an almost total loss of VC layer in these areas*

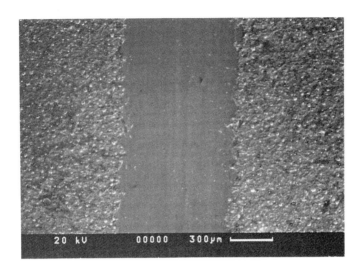

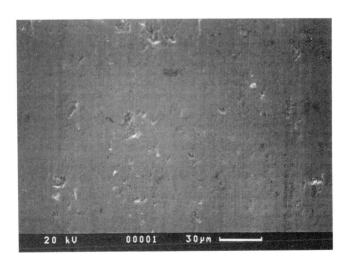

Figure 10 *Upper – SEM micrograph of wear track from Specimen 3, revealing very little pick-up or scoring. Instead, there was a polishing and "flattening" of the nodular surface associated with the addition of PTFE. Lower – close-up of wear track Specimen 3, showing more clearly some porosity associated with the nodular surface as well as minor pick-up of wear debris and light score marks*

did not appear to be as regular as for chromium plating and did not tend to "crumble" (i.e. there was a clean fracture with very little debris present). Figure 15, (lower), shows a similar failure pattern, with complete removal of the coating to reveal the graphite flakes in the grey cast iron substrate. There also appears to be re-deposition and embedding of wear debris in the track itself, as well as at the right-hand edge of the track, where it appears to predominate.

Figure 16 is the spectral analysis associated with the delaminated area shown in Figure 15, (upper), revealing the presence of re-deposited nickel-phosphorus, as well as the substrate material. Figure 17 (upper), is the same wear track at a region of intact coating. Here a polishing of the coating and smearing of wear debris along the track has occurred. There does not appear to be any sign of scoring. Figure 17, (lower), is the spectrum for this region of the wear track, revealing the coating to be virtually intact and comprised almost entirely of the coating constituents, with only a small quantity of iron present from the underlying substrate.

Figure 18, (top), illustrates a micrograph of the wear track from Specimen 7, heat treated electroless nickel on grey cast iron. It reveals a more cohesive structure compared with as-plated electroless nickel PTFE, with no cracking and spalling at the edge – cf Figure 15 (upper) and 17 (upper) and some re-deposited wear debris smeared along the track. There is no evidence whatsoever of delamination or decohesion, and only small-scale scoring is present. Figure 18 (bottom), clearly details the inside track features – wear debris and the score marks associated with the counterface.

The spectral analysis (not shown) associated with the wear track of Figure 18 (bottom) reveals the presence of an intact coating, with no substrate or counterface material present, thereby confirming the formation of a strong metallurgical bond between the coating and the substrate, as opposed to a weak mechanical bond, or adhesive bond, which is prone to easy delamination.

The chromium plated cast steels (Specimens 12 and 13) were at almost the other extreme of wear behaviour, exhibiting wear rates an order of magnitude greater than those for the VC-coated (TD) Specimens 1 and 2.

Figure 19 (upper), is an SEM micrograph of the wear track from Specimen 12, and reveals a high degree of deformation and slight scoring, particularly towards the centre of the track. Figure 19 (lower), details a section of track and shows that it contains what appears to be a portion of intact coating. Deformation of the coating is quite pronounced and scoring can be observed. Hence, here it appears a combination of adhesive and abrasive wear has taken place.

Figure 20 (upper), is the spectral analysis of the region of apparently intact Cr featured in Figure 19 (lower). There is an appreciable amount of Cr present, as well as the substrate material, this latter probably due to beam scatter or thinning of the deposit in these regions. Associated with this portion of the wear track is a high degree of Zr – bearing material pick-up from the counterface, indicating appreciable adhesion has occurred between the Cr layer and the counterface. Comparing this spectrum with Figure 20, (lower), derived from an area of gross spallation of the layer, it can be seen that very little of the chromium coating is still present, with the former substrate material revealed. Also of interest is the absence of the zirconium-bearing substance in this region, indicating that the counterface material, again, had adhered to the pockets of intact chromium coating and did not remain in the spalled areas of the track.

3.3.2 Conventionally-Hardened Materials. Wear tests were conducted (Table 1) on three types of conventionally-hardened materials – fully-hardened steel (Specimen 5), a flame hard-

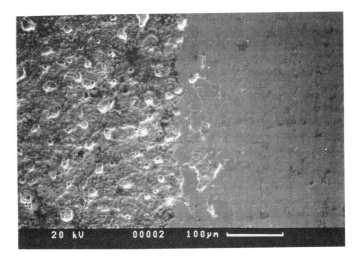

Figure 11 *Close-up of the edge of wear track from Specimen 3, showing the interface between the track and the unworn coating. There is no cracking, only a pressing of nodules and plastic deformation, or smearing, of the coating over the interface*

ened cast steel (Specimen 8), and a flame-hardened grey cast iron (Specimen 9). These materials generally exhibited medium wear rates, as recorded in sub-Section 3, similar to that measured with the electroless Ni heat treated coated grey iron (Specimen 7), whose characteristics have been reported above.

Figure 21 (upper) shows an SEM micrograph of wear track CAL-2 Top 2, fully hardened (59–61 HRC) steel ASSAB Calmax (Specimen 5) and exhibits smeared wear debris on the surface, as well as what appears to be scoring. However, at higher magnification, Figure 21, (lower), what appeared to be scoring is, in fact, also smeared wear debris. There is some deposition of wear debris, which appears to be zirconium-rich, from Figure 21, (bottom), the analytical spectrum. This presumably accounts for the lower frictional forces achieved during the latter stages of the test, as these adhered zirconium-rich particles are embedded in the surface of the track, acting as "wear plates", or preferential wear zones, protecting the surface from extensive wear and scoring. The wear mechanism, therefore, was pure adhesive wear.

Figure 22 (upper), is the SEM micrograph of the wear track from Specimen 8, flame hardened cast steel of grade AS 2074/L2A, L2B. The severity of the scoring that can be observed was far greater than for flame hardened grey cast iron (Figure 23), which was reflected in the much higher frictional forces reached in the cast steel (cf Figures 3 and 4). Figure 22 (lower), is a close-up of the wear track and reveals more clearly the extent of the scoring and also the degree of adhesive pick-up in the track.

Three spectral analyses (Figure 24), were produced for this track which give information as to what may have occurred in the wear track in the initial stages of wear. Figure 24 reveals iron, chromium and a little oxygen to be present. Figure 24 (middle), is an analysis of the bed of the wear track and does not show any appreciable difference to the analysis for the unworn surface except a very minute increase in the oxygen level. However the analysis in Figure 24 (bottom), of the adhered material (as seen in Figure 22 (lower)), confirms it to be a zirconiun oxide particle.

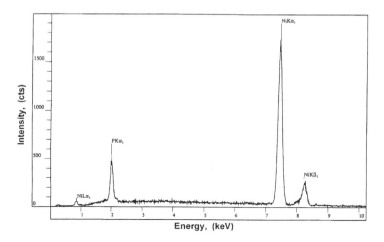

Figure 12 *Analytical spectrum associated with the unworn nickel/PTFE layer of Figure 11, revealing the coating constituents of nickel and phosphorus*

Finally, Figure 23 (upper), is an SEM micrograph of wear track from Specimen 9, flame hardened grey cast iron of grade AS 1830/H-269. Clearly seen are areas of adhesive pick-up, while there is also some scoring in the wear track. Graphite flakes can be seen in the unworn areas, less so in the wear track. This indicates either removal of graphite from the surface, or that the re-deposited, i.e. adhered material, has obscured the graphitic regions.

Figure 23 (lower), is an enlarged image of the same wear track, revealing light score marks and regions of adhesive pick-up. There appear to be two regions where graphite flakes have been torn from the substrate, leaving behind holes that have cracked at the edges. The darker areas are remnants of an oxide film, the mid-grey areas the underlying substrate, whilst the light areas peppered throughout the track and around these holes are regions of adhered zirconium derived from the counterface.

Figure 25 (upper), is the spectral analysis for the unworn surface of disc Specimen 9, revealing very little oxygen and some carbon, but predominantly iron and some silicon. Figure 25 (lower), is the spectral analysis of the wear track, associated with Figure 23 (lower). It can be seen that there was pick-up of zirconium-containing material from the counterface. Another interesting point is the fact that there is oxygen present in the track (but not the unworn surface), giving credence to the assumption that an oxide layer formed during the test and was subsequently worn away, leaving remnants behind. Visual observation showed the presence of such oxide layers in all of the wear tracks on this specimen.

Another feature of Figure 25, (lower), is the reduced presence of carbon compared to the unworn surface (Figure 25 (upper)), which again suggests the possibility of graphite removal from the disc surface during testing.

4 DISCUSSION

It is hard to compare, quantitatively, the results of the present study with those of other

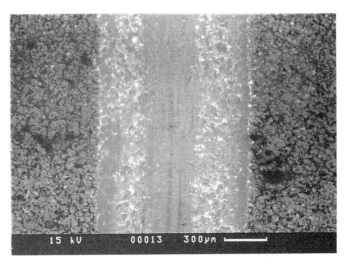

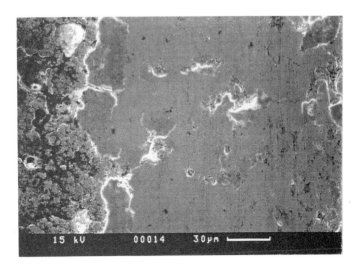

Figure 13 *Upper – SEM micrograph of wear track from Specimen 4, showing high degree of roughening due to adhesion of re-deposited wear debris and porosity from "flattening" of surface nodules. Scoring is also evident. Lower – close-up of wear track from Specimen 4, showing the interface between the unworn surface and the wear tracks. Score marks, porosity, plastic deformation and some cracking near the edge can be seen. Also evident are regions of re-deposited wear debris*

researchers due to the major differences encountered in terms of, *inter alia,* wear test apparatus and testing conditions. However, in qualitative terms, the present wear tests generally correlate well with overall findings in the literature. It has been shown that Toyota Diffusion treated D2 tool steel performed the best, with a high degree of wear resistance. This agrees with the findings of Child et al[6-9], Chatterjee-Fischer[10], Arai et al[11-15], Cocks and Fisher[16], Vaccari[17], Glaser[18], Plumb[19], Braza[20] and Habig[21] amongst others, who have all studied, in detail, the wear resistance of layers produced by the Toyota Diffusion Process.

Hard chromium plating in this study did not perform as well as the reported behaviour. Hard chromium can be compared to electroless nickel coatings in terms of poor wear resistance. However heat treatment of the electroless nickel elevated it to a higher degree of wear resistance. Gawne and Ma[22] found that even though heat treated electroless nickel is more effective than as-plated electroless nickel, hard chromium plating performs better than both of these coatings. This present study has found that heat treatment of electroless nickel coatings effects a transition to a higher wear resistance, even greater than that for hard chromium plating. However, Strafford, Datta and co-workers[23] made similar observations, including the merits of the electroless Ni-B variant, not considered in the present programme. One explanation for such differences in wear performance compared to reported behaviour could be associated with differences in the substrate materials.

In discussing the observed wear patterns reported in Section 3, it is useful to consider behaviour (1) in the context of surface engineered (coated) materials, and (2) conventionally hardened materials.

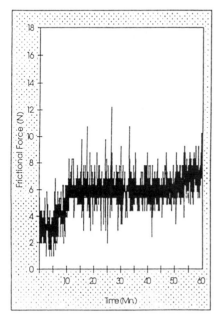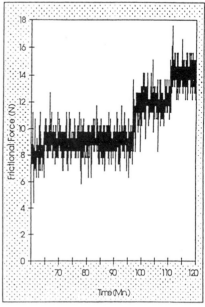

Figure 14 *Wear test output for electroless nickel with PTFE on grey cast iron*

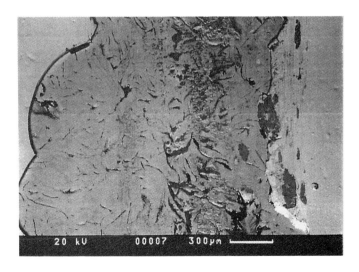

Figure 15 *Upper – SEM micrograph of wear track from Specimen 11, revealing complete delamination and cohesive failure of the nickel coating. Also present is embedded wear debris. Lower – SEM micrograph of wear track from Specimen 11, showing similar failure pattern as above, with complete removal of the nickel coating to reveal the graphite flakes in the grey cast iron substrate. Also shown are regions of re-deposited wear debris*

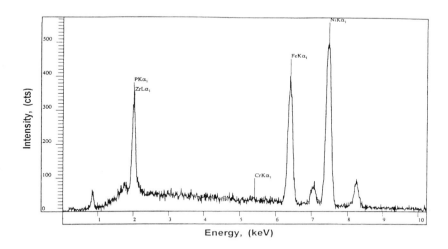

Figure 16 *Analytical spectrum associated with the delaminated areas of Figure 15, upper, revealing the presence of re-deposited nickel coating as well as the substrate material*

4.1.1 Surface Engineered (Coated) Materials. Toyota Diffusion (VC) treated AS1239/D2A tool steel (Specimens 1 and 2) performed best, in terms of wear resistance, out of all wear tests conducted. The excellent wear resistance may be, in part, attributed to the exceptionally high hardness (reported values of up to 3800 VHN[10–19] and even 4000 VHN[24], experimental values of up to 2355 VHN_{200}, which correlates with some findings[8]), and also due to the fact it is not a coating *per se,* but a diffusion layer, creating a hard wear resistant surface composed of vanadium carbide within a metal matrix, with no distinct interface, dependent upon adhesion to bond.

For Toyota Diffusion (VC) treated D2 tool steel, a large initial frictional force (20–25 N) was generally followed by a steady state grinding and polishing regime, after initial wear of the VC layer and subsequent re-deposition of wear debris from both the test disc and the counterface. An increase in frictional force near the end of the test presumably indicates further layer breakdown, with debris being swept away, to be re-deposited at a later stage.

From SEM analysis, Toyota Diffusion (VC) treated D2 tool steel generally showed two different wear morphologies. Firstly, there were regions of vanadium carbide layer spallation, caused by local fatigue failure from repeated contact stresses, thermal gradient or high friction sliding. In fact, frictional force fluctuations were generally evident on these specimens due to this macro-rough surface, creating differential contact between the counterface and the disc. The pattern of failure in this spalled region was a cleavage surface, with some plastic deformation of the coating. The second morphology was that of pure polishing, due to very fine abrasion, with no apparent destruction of the coating apart from the slight loss in thickness. This polishing

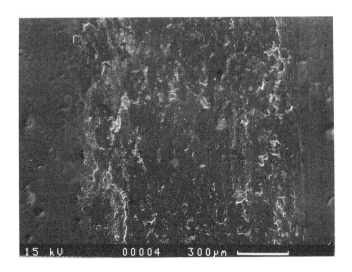

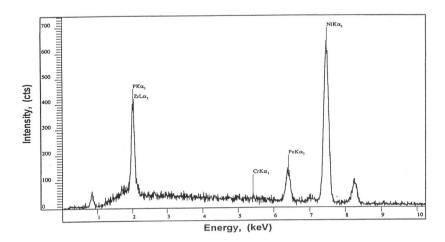

Figure 17 *Upper – SEM micrograph of an intact region of wear track from Specimen 11, showing polishing of the coating and smearing of wear debris along the track. Lower – analytical spectrum for Figure 17, upper, revealing an almost intact nickel coating with only a small quantity of substrate material present due to a number of small cracks in the coating*

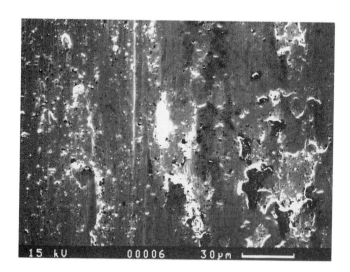

Figure 18 *Upper – SEM micrograph of wear track from Specimen 7, revealing a cohesive structure with no cracking at the edge or delamination. Also evident are regions of re-deposited wear debris. Lower – close-up of wear track from Specimen 7, clearly showing wear debris and score marks*

action resulted in both a micro- and macro-smooth wear track. As expected, EDAX analysis revealed virtually no zirconium-bearing material on the polished surface, together with a completely intact coating while, in the region of spallation there was significant quantities of zirconium-bearing material, as well as the presence of iron and chromium, indicating the detection of the underlying substrate. Concurrent with this, there was a decrease in vanadium (i.e. vanadium carbide) resultant from the loss of layer stability.

Interestingly, the spalled region of specimen D2-1 Top 2 (Specimen 1) contained both more zirconium-bearing material and vanadium carbide than specimen D2-2 Top 1, (Specimen 2). Therefore, for D2-1 Top 2, an increase in adhering wear debris had very little bearing on the subsequent removal of the vanadium carbide layer, i.e. it did not tend to aggravate the wear process.

Hard chromium plating on cast steel. Specimens 12 and 13, perhaps, surprisingly, were associated with high wear for both the discs and the counterface materials, indicating a mutual wearing mechanism, with high interfacial friction. From the results obtained, it would appear that hard chromium on cast steel would not be suitable for press forming dies. However, the trend in automotive manufacture in Europe, Japan and North America regarding hard chromium plating is to use this technology. Hard chromium plating, applied to large automotive draw dies, has resulted in major benefits[1,5]. However, it must be remembered that in the present laboratory work no lubricant was used and the test itself was severe.

Hard chromium plating generally exhibited high frictional forces, especially within the first few minutes of the test. This can be attributed to the severe breakdown of the chromium layer. Hard chromium plating, by its very nature, is prone to gross spallation. Its adhesive and wear properties are generally quite good. However, once failure is initiated, the chromium plating tends to fail severely, with gross spallation of the layer. This is presumably due to the inherent micro-cracks in the coating[22,25,26]

More generally, the present experimental work exhibited scatter in weight loss data, sample by sample, and this indicates substrate properties, as regards compositional variations, hardness and cleanliness to obtain reproducible behaviour. The hard chromium plating process is well known to be fickle and difficult to control, especially on surface of complex geometry.

Electroless nickel plated grey cast iron (GM241-2 Top), (Specimen 11), showed a high wear rate for the disc and a low wear rate for the counterface. The reverse side of this specimen (GM241-2 Bottom), (Specimen 10) however showed a relatively low wear rate for both disc and counterface. This leads to the conclusion that there was poor adhesion on the top side of the disc compared to the bottom side, as wear of the counterface was not appreciably different. Therefore, the wear tests served indirectly as a quality assessment , as well as a test method to obtain wear data, providing an indication of coating process conformity.

Electroless nickel-phosphorus on cast iron had a similar degree of wear resistance as hard chromium plating on cast steel (Specimens 12 and 13, see Table 1). However, electroless nickel exhibited a greater standard deviation in terms of wear rate, indicating less reproducible results, as well as a lack of consistency in the process. However, the results do indicate a potentially significant benefit from this surface engineering technique.

The frictional forces encountered during the wear of electroless nickel plated cast iron were quite irregular, with some large frictional fluctuations occurring, presumably due to a mixture of adhesive and cohesive failure of the coating, as observed from forensic analysis under the SEM. Frictional spiking was also evident on these samples, possibly due to continued fracture and delamination of the coating, as well as embedding and subsequent wearing of re-deposited debris. Also a contributory factor was continual edge cracking and flaking, with cohesive failure of the coating itself, producing conchoidal fracture patterns.

Microscopy revealed two distinct areas in wear track GM241-2 Top 1 (Specimen 11) of differing morphology. Thus there appeared to be a combination of complete removal of the coating as well as re-deposition and subsequent embedding of wear debris into the surface of the substrate. There was a tendency for the nickel coating to cohesively fail; in other words, some failure occurred within the layer itself to produce a conchoidal fracture. Edge cracking did not appear to be as regular as for chromium plating and did not appear to involve "crumbling" (i.e. there was a clean fracture with very little debris present). The removal of the coating allowed the observation of the grey cast iron substrate, detailing the graphite flakes.

The EDAX analysis associated with the delaminated area revealed the presence of re-deposited nickel-phosphorus as well as the substrate material. Within the same wear track, however, there were regions of intact coating. Here a polishing, due to fine abrasion, of the coating and smearing of wear debris along the track had occurred. There did not appear to be any sign of scoring or galling. The EDAX analysis for this region of the wear track revealed that the coating was virtually intact and comprised almost entirely of the coating constituents, with only a small quantity of iron present from the underlying substrate.

There was a significant improvement in wear resistance when electroless nickel on grey cast iron was subsequently heat treated at 300°C for 2 hours. The heat treatment effected a transition from a high wear regime (44.4 mg weight loss -9.98×10^{-3} mg/m.kg wear rate) to a low wear regime (5.8 mg weight loss -1.30×10^{-3} mg/m.kg wear rate) due to the presence of hard and coherent Ni_3P precipitates known to be produced by the heat treatment, as well as the development of a metallurgical bond between the coating and the substrate, through an interdiffusion mechanism[23]. Low frictional force values (Figure 6)were also observed for the HT electrolytic Ni due, chiefly, to the embedding of re-deposited wear debris in the track acting as preferential zones or "wear plates".

SEM analysis of heat treated electroless nickel on grey cast iron revealed a more cohesive structure compared with as-plated electroless nickel. There was generally no cracking or spalling at the edge, with only minor re-deposited wear debris smeared along the track. EDAX analysis indicated an intact coating with none of the wear debris deposited being from the counterface. The presence of an intact coating tends to confirm that the heat treatment creates both a harder layer, due to the precipitation of Ni_3P, as well as a metallurgical bond, by promoting interdiffusion between the substrate and the layer.

There do appear then to be significant benefits from heat treatment of electroless nickel coatings, based on the results compared to the as-plated condition, but the practical aspects of, firstly plating, then heat treating large, complex tools precludes its application at present. However, there may be scope in the future for using such coatings for use in the automotive industry through process scale-up.

Electroless nickel with added PTFE exhibited very high wear resistance, mainly due to the self-lubricating properties that the polymer imparts to the coating. It is evident then that friction is an important factor when considering the wear properties of a surface, though by no means the only consideration.

From visual inspection, polytetrafluoroethylene (Teflon) gives electroless nickel a self lubricating characteristic, with the surface being slippery to the touch. This caused a very significant reduction in frictional coefficient, resulting in the conversion from a high wear regime, associated with as-plated electroless nickel, to a low wear regime with the added PTFE.

In fact, the lowest frictional force values observed were encountered in theses specimens,

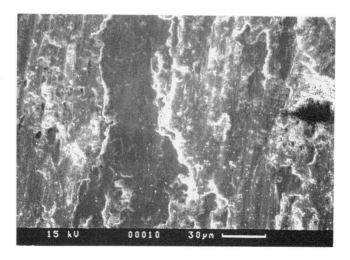

Figure 19 *Upper – SEM micrograph of wear track from Specimen 12, showing a high degree of plastic deformation and slight scoring towards the centre of the track. Lower – close-up of wear track from Specimen 12, containing a large portion of what appears to be intact coating. Deformation is quite pronounced and scoring can be observed*

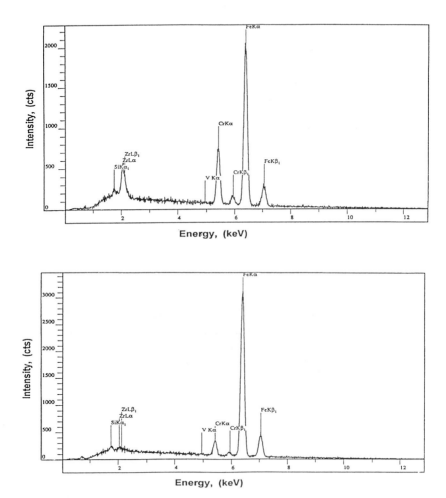

Figure 20 *Upper – analytical spectrum for the region of apparent intact chromium seen in Figure 19, lower. There is an appreciable amount of chromium, substrate material (iron and silicon) and adhesion of zirconium from the counterface. Lower – analytical spectrum associated with spalled area of Figure 19, however, showing very little chromium present (virtually all substrate). There was no zirconium pick-up in this region, indicating adhesion was to chromium pockets only*

due to the addition of PTFE, although there were differences in frictional characteristics between the two surfaces tested (cf Figures 7 and 18). Thus for GM241-3 Top 2 (Specimen 3), a linear increase in frictional force to 6 N occurred after the first 15 minutes of the test, where it stayed at steady state until the completion of the test. However, the reverse side of the specimen (GM241-3 Bottom 1)(Specimen 4) showed different frictional characteristics, whereby a steady state of 6 N was reached after 5 minutes where it remained until time 50 minutes, after which it steadily rose to a maximum of 14 N at the completion of the test.

This illustrates that even though higher frictional forces may be encountered, there is no specific indication that this affects the wear rate of the material. Thus virtually identical wear rates were encountered in both the electroless nickel with PTFE tests, even though frictional forces were very different.

SEM analysis for GM241-3 Bottom 1 (Specimen 4), reveals a texture in the track that suggests both adhesion of re-deposited wear debris, as well as porosity associated with the flattening of the nodular surface. There, indeed, appears to be a combination of adhesive and abrasive wear operating on this surface, often classed as "scuffing". This "roughness" may account for the slight increase in friction associated with this surface.

Therefore, in terms of frictional forces encountered, it is evident that the initial "wearing in" period of a coated component is most critical when considering the ramifications on the subsequent extent of wear for the useful life of that component. This initial wear period governs the frictional forces that a component or tool will undergo for the majority of its lifetime.

It has been observed that there was no strict correlation between friction and wear rate, with the maximum degree of friction occurring at 52 N for flame hardened cast steel, which, however, was only in the medium wear rate classification. However, the lowest degree of frictional force (6 N for electroless nickel with PTFE on cast iron) coincided with a low wear classification. The second highest frictional force (37 N for Vendor 1 hard chromium plating on cast steel) resulted in a high wear classification. Therefore, even though there are no strict correlations between wear rate and frictional force, some trends can be seen, together with some anomalies. This indicates that no one variable can adequately give an indication of wear resistance. There are too many interacting variables that provide a network of properties for the enhancement of wear performance.

4.1.2 Conventionally-Hardened Materials. Flame hardened grey cast iron (Specimen 9) has been shown to be in the medium wear rate classification. The microstructure of this material is essentially Type A graphite in a pearlite matrix for the core, while the case has a matrix of tempered martensite. This microstructure has given the material a reasonable degree of wear resistance.

When discussing frictional force with regards to flame hardened grey cast iron (Specimen 9), frictional variation, or fluctuation, becomes quite pronounced. Two theories of why this frictional variation occurs can be discussed. Firstly, this variation, at elapsed time range 71-86 minutes, may indicate the formation of a protective oxide film due to the intrinsic temperatures reached at this stage of the test, thus lowering the friction between the counterface and the cast iron. Beyond 86 minutes, this area of lower friction disappeared, presumably, the oxide layer was itself worn away. This cycle then tended to repeat itself, with formation and destruction of oxide layers (Figure 3).

The other possible explanation for this frictional variation could have been the removal of graphite form the grey cast iron surface during wear, depositing on to the wear track and creating a lubricating surface until the graphite itself was worn away. In general, the overall low frictional forces exhibited by the grey cast iron were probably due to this oxide film. It is unlikely that the graphite acted as a lubricant due to the nature of the flakes.

Flame hardened grey cast iron appears to wear by a combination of adhesive and abrasive mechanisms. The wear tracks on this surface can be described as scuffed, and, from the developed micro-smoothness of the track, is believed to be due to a progressive loss and reformation of surface films, such as an oxide, by fine abrasion, and/or tractive stresses, mutually imposed by adhesive or viscous interaction, with macro-roughness perpetuated by severe adhesion. The tracks appeared slightly shiny presumably due, again, to the presence of graphite flakes in the wear track, indicating either complete removal from the surface, or, alternatively, adhered material, such as re-deposited wear debris, may be obscuring these regions. On closer inspection, there are a number of places in the wear track where graphite has been "torn" from the surface of the substrate, leaving behind elongated, sharp, holes that have experienced cracking at the tips. This damage would act to aggravate the wear process, particularly the abrasive component.

Like the grey cast iron, flame hardened cast steel (Specimen 8) was also in the medium wear rate category. This is consistent with the microstructure of the material and its anticipated properties, which had a core of fine ferrite and spheroidised pearlite, and a case consisting of, again, tempered martensite from the flame treatment.

Flame hardened cast steel specimens displayed the highest frictional force values of all tests. This high friction is related to the counterface behaviour, which underwent a high degree of weight loss (1.5 mg, see Table 1). This high wear of the counterface would have thus created a high friction contact. It is impossible to say whether the friction was a result of counterface wear, or if counterface wear was a product of intrinsically high friction existing between the contacting surfaces.

From SEM analysis of the fully hardened cast steel (Specimens 5 and 6), the wear mechanism can apparently be defined as a combination of abrasive and adhesive wear, with abrasive wear being the more predominant mechanism. Overall, the description of the wear process is essentially the same as for grey cast iron. The analysis for the unworn surface reveals iron, chromium and a little oxygen to be present. The analysis of the bed of the wear track does not show any appreciable difference to the analysis for the unworn surface except a very minute increase in the oxygen level. However, the analysis of the deformed adherent material in the wear track indicates large quantities of zirconium and oxygen, in the form of zirconium/oxide particles.

These features fit well with the description of dry wear, or unlubricated sliding, linking this to the associated frictional behaviour of the cast steel, suggests that an oxide film formed and continually wore during the initial stages of the test. The oxide layer probably did not form until after the first 2 minutes or so, when the frictional force reached a mean of $20 \pm 3N$. This appeared to slow the rate of wear down as the layer itself was steadily worn, until severe wear of the oxide layer was achieved. Particles of re-deposited wear debris (containing counterface material and oxide "clumps") were deformed and smeared along the track, with associated scoring becoming predominant. A slight frictional increase then resulted as some of the wear debris particles embedded into the track, while other, non-adherent, particles produced a three-body abrasion mechanism situation. The gradual decrease in friction to a steady state occurred only after full embedding, creating a thin zirconium/oxide layer that gradually wore away. Very little evidence of oxide debris away from these discrete particles (i.e. in the bed of the wear track) was found, indicating that these combined zirconium/oxide particles were cohesive.

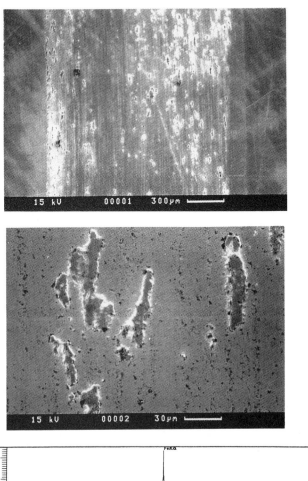

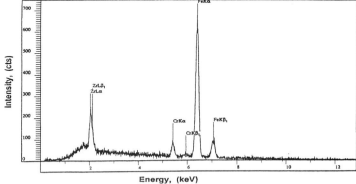

Figure 21 *Top – SEM micrograph of wear track from Specimen 5, showing adhered wear debris and what appears to be score marks. Middle – close up of wear track from Specimen 5, showing what appears to be score marks in Figure 21 top, but are in fact regions of adhered wear debris. Bottom – analytical spectrum associated with Figure 21, middle, showing that the wear debris adhering to the surface is predominantly zirconium from the counterface*

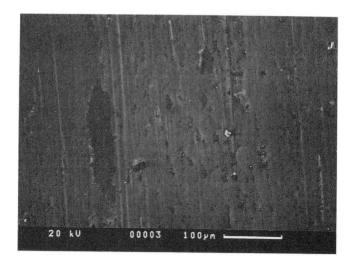

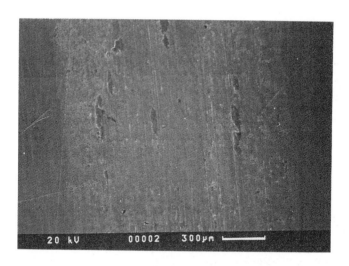

Figure 22 *Upper – SEM micrograph of wear track from Specimen 8, showing a high degree of scoring. Lower – close-up of wear track from Specimen 8, revealing a combination of abrasive and adhesive wear, with the former being the predominant wear mechanism*

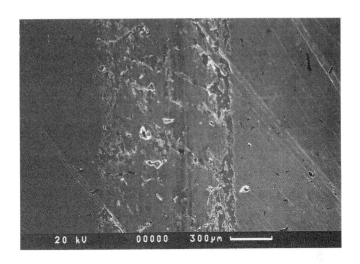

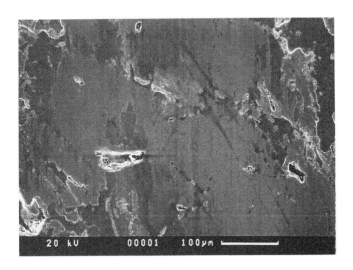

Figure 23 *Upper – SEM micrograph of wear track from Specimen 9, revealing areas of adhesive pick-up and scoring in the wear track. Graphite flakes are visible in the unworn areas. Lower – close-up of wear track from Specimen 9, detailing score marks and regions of adhesive pick-up. The darker areas are remnants of an oxide film, the mid-grey areas the underlying substrate and the light areas are adhered zirconium from the counterface*

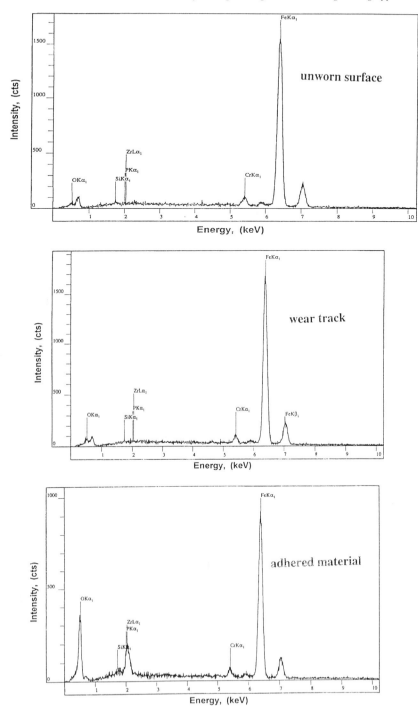

Figure 24 *Analytical spectra derived from Specimen 8 (cf. Figure 22): top – unworn as received surface; middle – wear track; and bottom – adhered material*

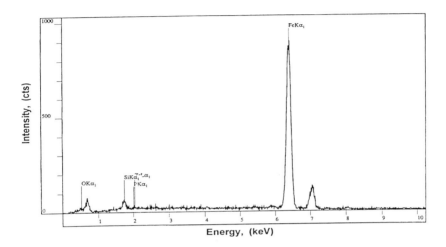

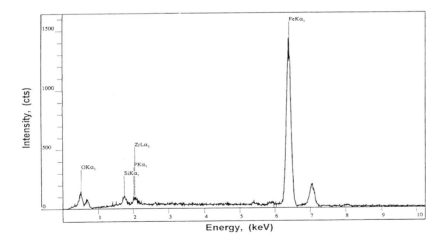

Figure 25 *Analytical spectra, derived from Specimen 9; upper – from unworn surface revealing very little oxygen and some carbon; and lower – from the wear track indicating zirconium pick-up and the presence of oxygen conducive with oxide layer formation*

5 CONCLUSIONS

- It has been shown that a number of surface coatings enhance the wear resistance in laboratory testing of steels and grey cast irons, relative to the corresponding performance of the same materials which have been conventionally surface engineered via full or flame hardening treatments.

- Toyota Diffusion (VC) treated steel showed the best overall resistance, followed by electroless nickel with added PTFE.

- The excellent wear behaviour of the electroless nickel PTFE variant is clearly associated with the observed low friction force. However, more generally, no systematic correlation between friction and wear was noted amongst the various coatings and treatments.

- Heat treatment of electroless Ni-P coatings on a grey iron has major benefit relative to the same coating as-plated.

- Hard chromium plating on steel exhibited a wear resistance similar to as-plated electroless Ni but inferior to heat treated electroless Ni. It is believed this may reflect, in part, the role of the substrate.

- Generally, the results lend substantial credence to the potential for these coatings to enhance wear resistance and tool/die life. However, appropriate choice of coating thickness and metallurgical condition, its hardness and the hardness of the substrate will be necessary within the overall design audit of a particular die/tool/fabrication operation for optimization and enhancement of manufacturing efficiency via the creation of the most favourable tribological conditions in the total wear system.

Acknowlegements

The authors are pleased to acknowledge the generous support of General Motors-Holden's Automotive Limited, Salisbury, S.A.

References

1. W. McMillan, 'Enhancement of Tooling Life in the Automotive Industry Using Surface Engineering Technologies', M. Eng. Thesis, University of South Australia, 1995.
2. K. N. Strafford, P. K. Datta and J. S. Gray, Editors, 'Surface Engineering' Practice: Processes, Fundamentals and Applications in Corrosion and Wear'. Idem., 1990, p. 397, Ellis Horwood, London.
3. T. S. Eyre, *Metals and Materials*, March 1991, 143.
4. K. C. Ludema, 'Selecting Material for Wear Resistance', *Wear of Materials*, March/April, 1981.
5. W. McMillan. Various Reports to General Motors Holdens Automotive, Adelaide, South Australia, 1993–1195.
6. H. C. Child, *Heat Treatment of Metals*, 1983, **39**, 77.
7. D. T. Jackson and H. C. Child, *Heat Treatment of Metals*, 1985, **5**, 92.
8. H. C. Child, S. A. Plumb and J. J. McDermott, in 'Heat Treatment '84', The Metals Society, London, 1984.

9. H. C. Child, S.A. Plumb and G. Reeves, Proc. 2nd Intl. Conf. on Heat Treatment of Metals, Florence, 1982, p. 89.

10. R. Chatterjee-Fischer, 'Boriding and Diffusion Metallising', in 'Surface Modification Technologies: An Engineers Guide', ed. T.S. Sudarshan, Marcel Dekker Inc., New York, 1989, p. 567.

11. T. Arai and N. Komatsu, *Metals Australasia*, 1982, 8.

12. Idem, Proc. 18th Int. Machine Tool Design and Research Conf., Ed. J. M. Alexander 1977, 225.

13. T. Arai, J. *Heat Treating*, 1979, **1**, 15.

14. T. Arai, 'Surface Modification Technologies III', ed. T. S. Susarshan and D.G. Bhat, TMS9 1990, 587.

15. T. Arai, *Wire*, 1981, **3**, 102.

16. G. J. Cocks and P. Fisher, 13th Biennial Conf. on Efficiency in Sheet Metal Forming, Melbourne, 1984, 515.

17. J. A. Vaccari, *American Machinist*, 1991, 45.

18. H. M. Glasser, *Worldclass Productivity*, 1991, **1**, 507.

19. S. A. Plumb, *Metallurgica*, Feb., 1995, 59.

20. J. F. Braza, 'Surface Modification Technologies V', Ed. T. S. Sudarshan and J. F. Braza, The Institute of Metals, London, 1992, 749.

21. K. H. Habig, *Tribology International*, 1989, **22**, 65.

22. D. T. Gawne and U. Ma, *Surface Eng.*, 1988, **4**, 239.

23. P. K. Datta, K. N. Strafford, A. Storey and A. O'Donnell, in 'Coatings and Surface Treatment for Corrosion and Wear Resistance'. Ed. K. N. Strafford, P. K. Datta and C. G. Googan, Ellis Horwood, Chichester, 1984, p.46.

24. E. D. Doyle and P. Jewsbury, *Materials Australasia*, June 1986, 8.

25. R. K. Guffie, *Products Finishing*, Jan., 1987, 62.

26. W. Allen, Ibid, March 1988, 66.

3.6.4
Progress in Coated Cutting Tool Science and Technology

K. N. Strafford[*]

IAN WARK RESEARCH INSTITUTE, UNIVERSITY OF SOUTH AUSTRALIA, AUSTRALIA

[*]PRESENT ADDRESS: INSTITUTE OF MATERIALS, LONDON, UK

1 INTRODUCTION

Over past decades, since the Industrial Revolution, various manufacturing industries have engaged a variety of metal processing techniques. Some of the major manufacturing processes may be classified as casting, metal forming and machining operations. Of over-riding significance and importance are the various machining processes involving cutting tools in one form or another. Cutting tools have been subject to changes over the years leading to improvement in their quality and design, always with the view to increase productivity and manufacturing efficiency and quality. Tool materials such as carbon steels, high speed steels and carbides have been in use for many years. More advanced tool materials include alumina, the sialons, and diamond and cubic boron nitride. Many of these tool materials have been found commercially viable for various manufacturing processes, appropriate selection however with respect to usage being critical. Figure 1[1] depicts the evolution of tool materials in the Twentieth Century in terms of the significantly reducing times taken to machine a standard component over the years.

Surface coatings – introduced over the past 25 years or so – have been used on various engineering components in order to enhance performance, especially in the area of cutting tools and dies where the wear of surfaces is obviously very critical. The use of ceramics as coating materials has played an important rôle. To improve tool materials themselves, properties have been improved by several hardening/strengthening techniques, including plasma nitriding. In this way the performance of tool and coating together – the coating system – may be optimised.

Ceramic materials have been used very effectively as hard coatings in providing wear resistant surfaces, particularly in the area of tooling and dies. A number of ceramics such as oxides, carbides, borides and nitrides has been used, or is of interest, in the context of creation of coated tools for enhanced tribological performance (Table 1)[2]. In recent years significant progress has been made both in respect of tooling itself – design and materials – and in the design and development of ceramic coatings, where the new process technologies of chemical vapour deposition (CVD) and especially physical vapour deposition (PVD) have played major roles.

Cemented carbides have become widely used tool materials. These materials are composed of the carbides of titanium, tungsten or tantalum sintered in a binder matrix involving processing by powder metallurgy techniques[3]. These tools are normally recognised for their high hardness,

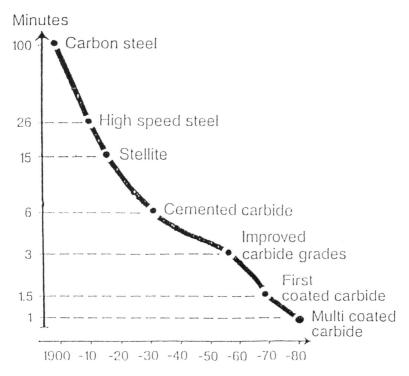

Figure 1 *Evolution of tool materials in terms of time taken to machine a standard component[1]*

compression strength and hot hardness. The production of thin ceramic films using novel processes has enabled a new generation of coated tool to emerge[4, 5]. These various coating process technologies and their advantages have been well addressed in the literature[6–10] – the advent of CVD and especially PVD technology has attracted enormous interest for the coating of tools and dies. In the recent years, PVD seems to have gained special recognition for its ability to coat materials at lower temperatures than CVD, thus enabling many substrate materials to be utilised, apart from its inherent flexibility with respect to the possible coating compositions and structures. These coating techniques have allowed the effective creation of novel tool/ die systems, exploiting the best and most useful range of properties of a number of alloys and ceramic coatings in combinations.

It has been possible to design complex systems, involving single layer, double layers or multilayers, often with multicomponent ceramics, on substrates having complicated geometries, with high accuracy.

The object of the present paper has been to review some of the background theory concerning the design of coated tools, and to appraise the reported performance of various coated tools. Particular attention has been focused on uncoated carbide and multilayer coated carbide cutting tools in the machine environment. Wear patterns and mechanisms for coated and uncoated tools have been analysed. A number of areas has been covered including aspects of the design of coating systems, where the coating architecture is critical, drawing upon basic concepts of materials science, as well as experience and intuition.

Table 1 *Compounds of carbides, borides, oxides and nitrides commonly used to enhance tribological performance[2].*

	Ti	Hf	Zr	Ta	V	Nb	Cr	Si	W	Mo	Al	Others
CARBIDES	TiC	HfC	ZrC	TaC	VC	NbC	Cr_3C_2, Cr_7C_3, $Cr_{23}C_6$	SiC	WC, W_2C	MoC	----	TiC–TiN; TiC–VC; Ti-SiC; $(Fe,Mn)_3C$
NITRIDES	TiN	HfN	ZrN	TaN	VN	NbN	CrN	Si_3N_4	----	----	AlN	TiN–TiC; $(Ti,V)N$; $(Si,Al)N$; Fe_4N
OXIDES	TiO_x	HfO_2	ZrO_2	Ta_2O_5	V_2O_3	----	Cr_2O_3	SiO_2	----	----	Al_2O_3	TiC–Ti_xO_y
BORIDES	TiB_2	HfB_2	ZrB_2	TaB_2	VB	----	----	----	WB	MoB		

Table 2 *Single layer coatings*

S.No	Coating Material	Processing Technology	Substrate	Reference
1.	TiN			
2.	(Ti, Zr)N	PVD	42C4Mo4V	11
3.	ZrN			
4.	TiC, TiN, Ti(C,N)	PVD	Cemented carbide	12
5.	TiN	PVD Sputtering	HSS	13, 14
6.	TiC, TiB$_2$	Sputtering	HSS	15
7.	ZrN	Sputtering Reactive sputtering	Cemented carbide M2 HSS 304 Stainless steel	16, 17, 18
8.	NbN$_x$, MoN$_x$	Sputtering	HSS	16
9.	TiN, TiC	Diode Sputtering	Cemented carbides	19
ION PLATING				
10.	TiN, ZrN, HfN	Cathodic ion plating	52100 steel	20
11.	TiN, ZrN, HfN	Arc ion plating	18Cr--8 Ni	21
12.	NbN, TiN, (Ti, Al)N ZrN	Triode ion plating	HSS	22
13.	ZrN TiN	Ion beam assisted	Gcr 15 steel --	23, 24
CVD				
14.	Ti(Cx, Ny), TiC,TiN, HfN, Al$_2$O$_3$	CVD	--	
15.	TiC, TiCxNy, TiN	CVD	TTR	25

2 TRIBOLOGICAL COATING MATERIALS FOR CUTTING TOOLS

2.1 Coatings – Rôle, Choice and Availability

The coatings used on tools play a very important part in enhancing their performance in many ways. Coatings can be considered as new surfaces on existing tools, enabling improved wear resistance and tool life without altering tool geometry by the possession of unique properties in the **combined** format with the substrate material. The thin film of coating is also required to accommodate the variables affecting the tool material, acting as a part of the tool material and yet without affecting the tool material itself, as shown in Figure 2. Here coating materials such as TiN, by virtue of possession of properties such as low coefficient of friction and high hardness, have the ability to withstand high temperatures and have been used widely as coating materials. Various coating materials have been used as a single layer, double layers and multilayer formats, as shown in Tables 2–4[11–35]. Many researchers[36–41] have reported the different wear mechanisms occurring in the tool materials; however there is a paucity of information, especially in the coating area. Some of the key factors necessary for the coated tool system to perform have been described below.

In the context of the tribology of hard coatings, materials[42] such as TiN, hard Cr, and

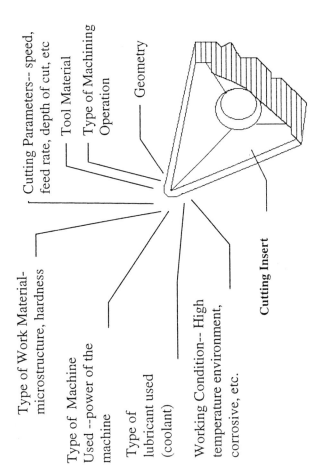

Cutting Parameters-- speed, feed rate, depth of cut, etc

Tool Material

Type of Machining Operation

Geometry

Type of Work Material- microstructure, hardness

Type of Machine Used --power of the machine

Type of lubricant used (coolant)

Working Condition-- High temperature environment, corrosive, etc.

Cutting Insert

Figure 2 *Variables influencing cutting tool performance*

Table 3 *Two layer coating systems*

	Coating Material	Process Technology	Substrate	Total Thickness	No. of layers	Ref.
System I	TiC/TiB₂	Sputtering " PVD	Cemented carbide HSS	5 µm 4 µm 5 µm	100 1000 10-each 0.5µm	26, 27 28
System II	TiN/TiB₂	Sputtering PVD	Cemented carbide --	5 µm 5 µm	100 10-each 0.5µm	27. 28
System III	TiC/TiN	Sputtering PVD CVD CVD CVD	Cemented carbide Cemented carbide Steel AISI440C Tool Materials Carbides	5 µm TiC-1µm TiN-3µm 5- 6 µm 10 µm 8 µm	1000	26 29. 30 31. 32
System IV	TiC/Al₂O₃	CVD CVD CVD CVD Sputtering	Cemented WC Tool materials Cemented carbides Carbide M10,P25,P30 Cemented carbideM15	10- 11µm 10- µm 10- µm 7- µm 5- µm	each8.3 µm 2 30	61. 31 32. 33 34
System V	Ti(C,N)/TiN	CVD CVD PVD	Carbide Tool Materials X155CrVMo121cd2	8 µm 10 µm 0.36 µm	each2.6 µm	31. 32
System VI	HfC/HfN	PVD	--	--	9	
System VII	Ti(C,N)/Al₂O₃	CVD	Carbide	10 µm		33

Table 4 *Three layer coating systems*

	Coating material	Process Technology	Substrate	Thickness	No. of Layers	Ref
System I	TiC/ Al₂O₃/ TiN	CVD CVD CVD	Cemented WC Tool Materials Carbides	10 µm 10 µm 8 µm	3 -- 3	31, 32, 34
System II	TiC/ Ti(C,N)/ TiN	CVD CVD CVD	Cemented carbides Steel Carbide K 20	7 µm 5 µm 6 µm		30, 34
System III	TiN/TiC/TiN	PVD CVD	Cemented carbide Carbide	5 µm 3 µm	1.1.3µm	29, 32
System IV	Ti(C,N)/Al₂O₃/TiN	CVD CVD CVD	-- Carbide Carbide M10, P25, P30	18 µm 9 µm 7 µm	7	32, 29, 61
System V	Ti(C,N)/Al₂O₃/TiN/ Al₂O₃/TiN/Al₂O₃/ Ti(C,N)/TiN	CVD	Carbide	Ti(C,N)-6, Al₂O₃-3, Ti(C,N)-1µm		32

elecroless nickel have also been considered in their rôles as "solid lubricants" between the substrate and the work material. In this context the coating may be considered as an intermediate region between the work material and the substrate, and should possess a low coefficient of friction - a critical property of any lubricant.

2.2 Some Factors to be Considered in the Design of a Coating Tool System

2.2.1 Introduction. The performance of a coated tool and its mode of failure will be determined, for a particular situation, by the properties of the coating(s) and the tool itself, where the interactive role of the coating and the substrate must also be recognised. Of fundamental importance is the adhesion of the coating to the substrate. Coherency in the coating itself is also critical. In examining some of the factors which determine coating performance, it is useful to consider the factors of a working coating system as described by Strafford et al[43], Figure 3. A most significant observation here is that the fundamental requirement of compatibility is quite different at the two working interfaces. For adequate coating adhesion to the substrate good compatibility is essential, achieved, *inter alia*, by factors such as matching elastic moduli and coefficients of thermal expansion. On the other hand to avoid adhesive/diffusion (wear) there should be poor compatibility between the coating and the work material.

Some of the factors to be considered in design are set out in Figure 4. Strafford et al[43] have discussed the various properties of the coating material and their perceived importance in determining coated tool performance/life.

It has been suggested[44] that the reasons for failure occurring between a coating and substrate or coating and a coating are not only as a result of differences in chemical composition but are also influenced by mechanical properties. Mechanical properties such as Young's modulus, yield strength, creep, fatigue, etc. are a crucial influence on coating behaviour. Mismatched coating/ substrate physical properties such as thermal conductivity and thermal expansion, if not recognised at the design stage, clearly could have detrimental effects in the performance of the coatings.

The geometry of the tool or die is important for two principal reasons:
1. the ability thereby for the substrate to withstand deflection under load and to provide support to the coating: it is of no use employing a coating material with an attractive portfolio of properties on a mechanically weak tool; and
2. in relation to the coating process efficiency, with associated implications for coating thickness, uniformity, etc.

Also to be considered, clearly, (Figure 3) must be the precise operating/environment conditions. Wear, as noted earlier, occurs by a variety of processes (such as adhesive, abrasive, diffusive wear, etc.) and it has been shown repeatedly that a good wear performance in one situation may not be exhibited in another – "fitness of purpose" must be the guideline. A good knowledge of the details of the progressive wear of a particular tool (geometry) in a given fabrication operation must be available in order to 'design' the necessary features in the tool coating, features which may differ on different parts of the tool working surfaces particularly in one of complex geometry. As detailed by[43] in describing the fundamental coating system (Figure 3), to accommodate suitably the variables in the coating system (between substrate-coating, coating-coating) it would be necessary to examine the individual and combined material properties such as thermal expansion coefficient, thermal conductivity, friction, miscibility (mutual solubility), etc., Table 5[45].

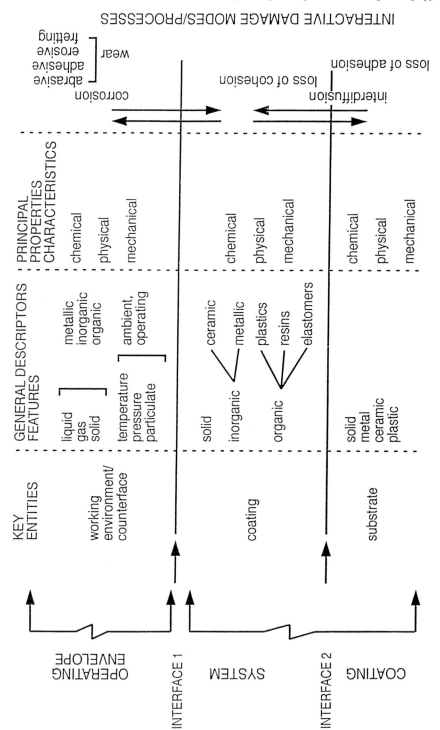

Figure 3 *Factors influencing a working coating system*[43]

Table 5 *Material solubility in coating materials[45]*

Carbides\Nitrides	TiC	ZrC	HfC	VC	NbC	TaC
TiN	●	●	●	●	●	●
ZrN	●	●	●	○	●	●
HfN	●	●	●	¦	●	●
VN	●	○	○	●	●	●
NbN	●	●	●	●	●	●

● = Fully miscible

○ = No or limited miscibility

Nitrides

	TiN	ZrN	HfN	VN
ZrN	●			
HfN	●	●		
VN	●	○	○	
NbN	●	●	●	●

2.2.2 Coating Adhesion Considerations. Adhesion, which is most important requirement in the "coating system", could be considered as a most decisive factor in the performance of the same. Many failures occurring in a coating system could be directed towards a poor adhesion mechanism and hence in design, concentration of attention in this area would seem to be of primary importance. As noted earlier the ideal requirement for a material as a coating would be that the material should have least adherence (less tendency to react) with the contacting work material and high adherence with the substrate or the immediate coating layer it is in contact with.

Complete or partial spallation of coatings in use could be indications of adhesion failures. Some of the causes of adhesion failures have been attributed to the variables in the coating processes[7], such as:

a. Substrate: substrate defects – for example, substrates with pre-existing cracks (macroscopic or microscopic) could act as initiation points for crack propagation that could result in failure of the coated system.

b. Coating materials: the purity of the coating materials - for example, some inclusions transferred and present in the coating materials could act as stress raisers and failure initiators.

c. Processing conditions: processing conditions such as process parameters (e.g. such as partial gas pressures in ion plating, operating temperatures, voltage, current density, etc) are known to affect coating properties[6] such as thickness, composition, structure, etc; cleanliness of the equipment (eg. some alien material present in the coating equipment) could affect the plasma, for instance in the PVD process, that could result in undesirable variations in deposition rate etc.

d. Interface characteristics such as deposit growth, uniformity in composition, etc.

e. Compatibility: this would be an important factor as failure could occur due to improper process parameters, mismatch of material properties or failure under operating conditions such as surface interactions in high temperature applications/ corrosive environments, etc.

f. Internal stress: high internal stress could break the bond strength between the coating and the substrate. It is known that stresses are developed in the coating during the deposition process and/ or there may also be substrate residual stresses[46]. Stresses acquired during the process, such as compressive stresses, could increase the risk of buckling in the system while minimising tensile failure, while tensile stresses could produce an opposite effect. Ultimately it is the net stress in the coating system which is crucial and it has been suggested[46] that developing coatings with low compressive stresses to reduce tendency to tensile cracking and failure without increasing the risk of buckling, is a desirable objective.

The adhesion "property" will be influenced to a large extent by interface characteristics. In particular, interdiffusion either during processing or in service will be of importance. It has been suggested[7] that intermediate layers could be used that would be soluble in both the substrate and the coating materials to effect good bonding and hence good adhesion. The miscibility between various materials has been shown in Table 5[45], but there is, in general, a paucity of data.

To obtain good adhesion and compatibility implies satisfying requirements of mutual solubility, bonding and interface characteristics between the coating/coating or coating/ substrate at the two interfaces 1 and 2 in Figure 3.

2.2.3 Significance of Hardness and Coating Performance. Many researchers have considered hardness as an important property in relation to adhesive and abrasive wear behaviour[7,47,48]. Although hardness values for some coating materials and some coated systems have been reported (Table 6), there is, in general, a lack of information in the open literature, detailing especially that in context of the sequence of hardness "required" for coating materials when used in multilayer formats.

While the hardness of a coated system, as a whole, has not been found in general to be equal to the sum of the hardness of the individual layers, there must be a contribution from each individual layer to the overall hardness. Differences in the overall hardness in a multilayered coating, as a whole, when compared with the cumulative sum of all the layers, could be partially associated with the variable resistance offered to indentation by the substrate immediately beneath the coating/substrate interface.

Within the concept of the design of a multilayer coating system, not much information is available in the open literature mentioning the "suitable" constituent materials to be chosen as the inner layer, intermediate layer(s) and outer layer. However it has been suggested[49] that a prime requisite of the inner layer would be to provide adhesion, with the intermediate layer(s) principally contributing to hardness and strength, while the outer layer should possess a low coefficient of thermal expansion, and reactivity (to reduce wear).

A possible optimised sequencing arrangement with respect to hardness would be for the inner layer to be the hardest, with an intermediate layer of lower hardness than the inner layer but with a higher value than that of the outer layer. Figure 4 illustrates this system schematically.

In the design scheme of Figure 4 (a), the object would be for the outer layer to wear first in a uniform manner, adequately supported by the inner layer. The frictional property of the outermost layer would also be an important variable.

In an alternative design, Figure 4 (b) wear resistance would be well provided for in the initial stages of operation, especially in machining applications. Further, in a high stress situation (e.g. fast machining of a hard work piece) the other material would tend to resist deformation and transmit and spread the load to the inner and substrate materials.

Possible failures such as adhesion failure, spalling, cracking, and fatigue failure might occur over a period of time.

However it is not appropriate to consider hardness alone to be the decisive factor. Other properties such as thermal conductivity, thermal expansion, friction coefficient, etc need to be considered to lead to improved performance of coated systems. It may also be necessary to have a compromise between adhesion and the inner layer hardness. More research in this area is required.

Overall, however, despite these uncertainties there is definite evidence that multilayer coating systems offer improved performance over single layer coated tools. Much more systematic research is however clearly needed.

2.2.4 Some other Parameters in Coating Design. Many material properties – chemical, physical and mechanical – as discussed earlier, will be key considerations in the performance of coatings. Trends in physico-chemical and physico-mechanical properties of hard coating materials – borides, carbides and nitrides – have been illustrated by Holleck[28] (Table 7).

It has also been mentioned[7] that in the change from ionic to covalent materials the linear coefficient of thermal expansion decreases. It has been suggested[8] that the thermal expansion coefficient is an important parameter in the performance/failure of a coating. In particular matching thermal expansion coefficients between coating materials (and substrate) would be therefore a major consideration in design. It has been suggested that one way to achieve low coefficient of thermal expansion would be through the use of composite structures.

Poor thermal mismatch could be one of the causes for the occurrence of coating defects such as micro cracking, grain boundary separation, plastic deformation, etc.[50].

Table 6 *Hardness values of selected carbides, nitrides and borides (quoted in(6))*

ELEMENT	BORON	CHROMIUM	HAFNIUM	MOLYBDENUM	NIOBIUM
CARBIDES	3700	1300 - 1600	2270 - 2650	1800	2400 - 2850
NITRIDES	--	2280	1640	--	1396/ 1720
BORIDES	--	1800	2250 - 2900	2350	2100 - 2400

ELEMENT	SILICON	TANTALUM	TITANIUM	TUNGSTEN
CARBIDES	3500	1800 - 2450	2000 - 3200	1450 - 2400
NITRIDES	--	1220	1200 - 2000	--
BORIDES	--	2450 - 2910	2200 - 3500	2400 - 2660

ELEMENT	VANADIUM	ZIRCONIUM
CARBIDES	2460 - 3150	2360 - 2600
NITRIDES	1520 - 1900	1150
BORIDES	2070 - 2800	2250 - 2600

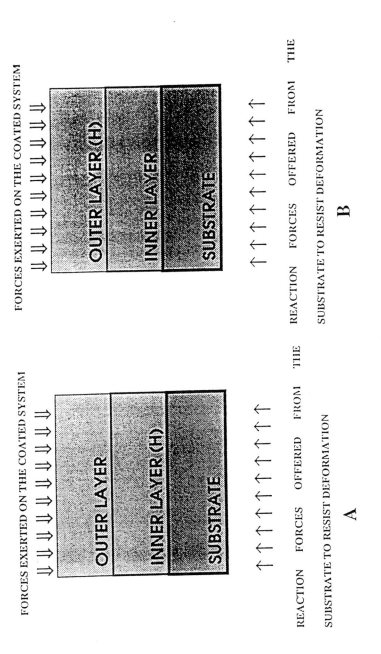

Figure 4 *Hardness sequencing*

Table 7 *Trend in physico-mechanical properties of hard coating materials*[28]

	Trend in chemical bonding ⇑			DECREASING PROPERTY ⇑ Trend in ceramics		
Melting point	M	C	I	CC	BB	NN
Stability	I	M	C	NN	CC	BB
Thermal Expansion Coefficient	I	M	C	NN	CC	BB
Hardness	C	M	I	BB	CC	NN
Toughness	I	C	M	NN	CC	BB
Adherence to metallic substrates	M	I	C	BB	CC	NN
Interaction tendency	M	C	I	BB	CC	NN
Multilayer Compatibility	M	I	C	--	--	--

M = metallic; C = covalent; I = ionic; BB = borides; NN = nitrides

High stresses and strains will tend to develop when large differences exist between the coefficients of thermal expansion existed in coating materials, due to temperatures developed in service and/ or in deposition[44]. The "ideal" coating property requirements would include a combination of high thermal conductivity and low coefficient of thermal expansion. It has been suggested[50] that, in choosing coating materials, it is important to consider the thermal conductivity of the materials, especially the variation of thermal conductivity with temperature. A comparison of various material properties such as thermal expansion coefficient, thermal conductivity, micro hardness and Young's modulus has been made as shown in Figures 5, a–c by Bhushan and Gupta[9]

There is in fact paucity of information in the literature with respect to coated systems' properties and/or the selection criteria required in choosing coating materials with suitable properties such as Young's modulus, coefficient of thermal expansion, thermal conductivity, Poisson's ratio and coefficient of friction. In reviewing some of the bulk properties of coating materials commonly used on cutting tools, Figure 6 describes the combinations of the materials and properties that have been used. In the two layer and multilayer formats a trend is seen in the hardness arrangement which follows a decreasing sequence from inner to outer layer, as discussed earlier. It is evident that hardness **alone** cannot determine the performance of a coated system, as other factors interrelated to adhesion and compatibility should also be met. There is, in general, a lack of systematic information on these perspectives.

3 REPORTED EFFICACY AND PERFORMANCE OF COATED TOOLS

Various coatings in single, duplex, multilayer and/ or in multicomponent formats involving different hard materials have been developed using various process technologies. Different researchers have used a number of techniques in order to evaluate the performance of these coatings/ coating systems[11–35] and understand wear mechanisms[37–40, 51, 52]. Some of the more important observations concerning tribological performance of such entities on various substrates assessed under machining conditions, and the recognised associated wear mechanisms are reviewed.

It is not possible here to give comprehensive detail. Of particular interest are the reported (relative) performances of the different rival surface coating materials (on carbide tools) in single layer format and an appraisal of the generally improved efficacy of these coatings, in different multilayer formats. A fuller review has been provided by Ghopinath[53].

3.1 The Tribological Performance of coated Tools under Machining Conditions

3.1.1 Single Layer Coatings. (a) The performance of TiN coatings and comparisons of TiN coatings with other single layer coating materials.

Machining with single layer TiN coated tools has been reported to yield superior flank wear resistance in comparison with ZrN and HfN coating materials[54]. It has been suggested that the crater wear resistance in ZrN and HfN coated tools was higher than TiN.

In another turning study[22] conducted on coatings of NbN, TiN, (Ti,Al)N and ZrN on HSS inserts it was observed that the ZrN coated inserts performed best followed by the multicomponent (Ti,Al)N coating. Furthermore a comparison of the test results on turning

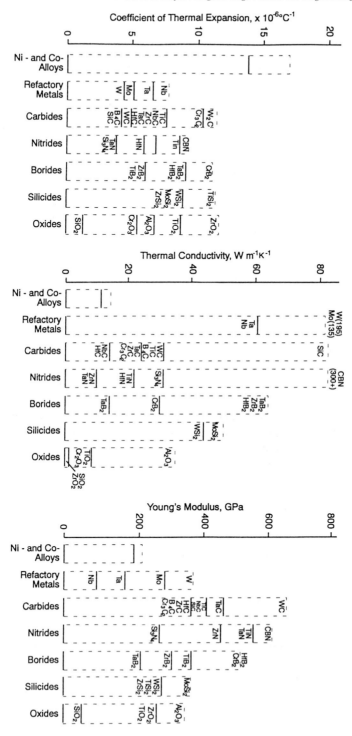

Figure 5 *Properties of various hard coating materials*[9]

with high speed steel inserts reported by Pentinen et al[55] indicated a better wear resistance offered by ZrN and (Ti, Al)N compared with a TiN coating.

It seems that the conflicting evidence concerning the performance of TiN, ZrN, and (Ti,Al)N coatings may be partly due the influence of coating process/ parameters. Thus it has been reported[22] that increasing the current density on the substrate had adversely affected the performance of ZrN, but improved that of (Ti,Al)N and TiN. The reason for the decreased performance in ZrN at higher current density had been attributed to possible high residual stress and hence excessively higher hardness so generated in the coatings. A similar observation has also been cited[56], wherein ceramic coatings of Ti, Zr and Hf produced by the plasma ion nitriding process had indicated the presence of cracks was detected in ZrN coating. It has also been indicated that increasing coating thickness could decrease the residual stresses, presumably within limits.

Several investigations of the performances of cutting tool components coated with TiN (deposited by the Sputter Ion Plating process) where benefits have been reported[57].

Credit here has been given to the high hardness of TiN. Good workpiece surface quality was also produced by the TiN coatings.

Bhat and Woerner, in 1986[58], studied the effect of substrate composition on single layer CVD TiN coated cemented carbides, in comparison with uncoated carbides (VC-5), under machining conditions. The results, indicating the flank wear and deformation resistance of the tools, made clear that VN-8 (CVD TiN coated carbide on a substrate with highest hardness compared with other tools) offered maximum resistance to flank wear and deformation. This observation confirms the importance of the substrate properties as a contribution to the performance of the tool system.

Quinto et al.[59] have described the improved performance of coated tools. In particular a comparison made between TiN (PVD) and TiN (CVD) coated tools with an uncoated standard has been described. It was concluded that PVD TiN was superior to CVD TiN in metal cutting operations. The reason for this has been attributed to the higher abrasive wear resistance of the PVD TiN with high surface fracture strength. On the other hand CVD TiN was reported to increase tool life under turning conditions. It was further suggested that in milling conditions the CVD TiN, by virtue of the presence of tensile residual stress, reduces the transverse rupture strength and the resistance to edge chipping of the carbide.

Byeli et al[60] described the results of turning tests conducted on stainless steel with coated carbide tools. Single layer coatings of TiN at low speeds of 65m/min demonstrated better flank wear resistance than TiC, while the latter has been suggested to be more resistant on the rake face. This observation is contrary however to other results reported in the literature[61,62]. It is believed that the varying cutting conditions used could result in these contradictory observations. These authors[60], also reported that increasing the number of coating layers leads to increased wear resistance. A penta layer coating was reported to have performed the best amongst others with least flank and crater wear. The material properties of TiN and TiC have also been addressed. TiN coatings are known to possess low brittleness but high temperature stability, while TiC has been suggested to support higher diffusion mobility of carbon atoms in the lattice, thus improving adhesion at the interface. The compound also exhibits high hardness (1.5–2 times greater than TiN).

(b) Ti(C,N) coatings. A number of researchers[25, 49, 63, 64] have reported the excellent performance of Ti(C,N) coatings. Observations on the relative performance of single layer coatings of TiC, TiN, Ti(C_x,N_y), HfN and Al_2O_3, under milling conditions, have been made by[64], where the Ti(C,N) coatings performed better than other coatings. It has also been

Two layer coating System

		Hardness	Y's	Coeff. of Thermal expn.	Thermal Conductivity
TiC / TiB$_2$	Inner	Low	High	High	Low
	Outer	High	Low	Low	High
TiN / TiB$_2$	Inner	Low	High	High	Low
	Outer	High	Low	Low	High
TiC / TiN	Inner	High	High	High	High
	Outer	Low	Low	Low	Low
TiC / Al$_2$O$_3$	Inner	High	High	High	Low
	Outer	Low	Low	Low	High
Ti(C,N)/Al$_2$O$_3$		Hardness value of Ti(C,N) not available but it is approximately higher than TiC.			
Ti(C,N)/TiN		N.A			
HfC / HfN	Inner	High	N.A	Low	Low
	Outer	Low	N.A	High	High

Figure 6 *Combinations of materials and properties in two large coating systems*

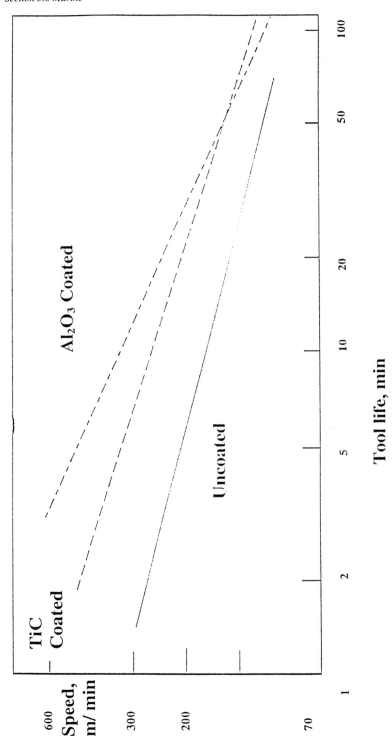

Figure 7 *Tool life comparison between coated carbides and uncoated carbides ($V_b = 0.25mm$, feed $= 0.25mm/rev$)[61]*

mentioned[65] that in machining with drills, inserts and blades, multilayer Ti(C,N) and Ti(B,N) coatings improved productivity. Possible reasons for the improved performance of the Ti(C,N) coatings have been attributed to the high hardness and better adhesion[49].

(c) Single layer TiN, TiC and Al_2O_3 coatings. It has been suggested[61] that some of the factors influencing the performance in machining with coated carbide tools are deposition method, coating materials, coating thickness, and substrate type. Figure 7[61] shows the tests conducted on uncoated, TiC coated and Al_2O_3 coated tools in machining 1045 steel. As can be seen the Al_2O_3 coating performed the best at high cutting speeds but it has been mentioned that, at lower cutting speeds, the TiC coatings, with higher hardness, yielded better performance when compared with Al_2O_3. The main wear mechanism was reported to be abrasion. It was also noted that the thermal wear resistance (ie. wear at elevated temperatures) follows a sequence of (best) Al_2O_3> TiN>TiCN>TiC (worst).

3.1.2 Multilayer Coatings. Comparisons of TiN based multilayer coatings with single layer, multilayer and/ or multicomponent coatings.

There are many references in the literature to the superiority, in general, of multilayer coated tool systems. Tests with two layer, three layer or up to many tens of layers have been reported.

Figures 8 and 9[20] describe the relative performance of PVD and CVD coated carbides involving TiC/TiN, TiN, (Ti,Al)N and (Ti,Zr)N. The superior performance of the (Ti,Zr)N PVD coatings can be seen in Figure 8. Tests conducted on other coatings, such as multicomponent materials with (Ti,Al,V)N and (Ti,Al,Zr)N combinations, when compared with CVD TiN, indicated a relatively better performance of the PVD (Ti,Al,V)N coatings, as shown in Figure 9.

The results of machining tests with multilayer coated tools (coatings shown in Table 8) revealed that flank wear was higher for the "zebra coating" (TiC/TiN/TiC/TiN/TiC/TiN/TiC/TiN) when compared with TiC/TiCN/TiN and TiC/Al_2O_3/TiN at speeds of 190, 260 and 300 m/min[65]. When the speed was increased to 350 m/ min TiC/ Al_2O_3/TiN gave a superior performance. Comparisons made between single layer TiC, multilayered TiC/TiCN/TiN and the "zebra coating" indicated the resistance to flank wear increases in the reverse order and the same trend was also seen in crater wear resistance. It has been suggested that TiC improves adhesive wear, leading to a reduction in flank wear, and that diffusion wear resistance is improved by TiN, thus reducing cratering. Some of the reasons for the superior performance of multilayer coatings have been attributed[65] to the rôle of the multilayer acting as a thermal barrier to the diffusion of the cobalt from the substrate material and also arresting crack propagation. As cutting speed increases the contact temperature increases. It has been stated that contact temperatures of 1200°C are reached at speeds 300 m/min[65].

Goh et al[66] compared the results of machining 1045 steel using cermet tools and tungsten carbide tools coated with TiN, TiC, TiCN and Al_2O_3 in single layer and multilayer formats as shown in Table 9. It was shown that the crater wear resistance in the coated tungsten carbide tools was lower than for the coated cermet tools while on the other hand, the latter had higher flank wear resistance. It has also been mentioned that with increasing speeds (300 m/min at feed rate of 0.25 mm/rev., for 60 min testing time), the cermet tools were found to be superior to the carbide tools. It is worth noting[66] that some of the cermets did not reach the tool life criteria based on ISO 3685 i.e. they exhibited excellent performance, although the cermet tools had higher crater wear.

Coating materials involving a combination of TiC/TiB_2, tested for performance in cutting

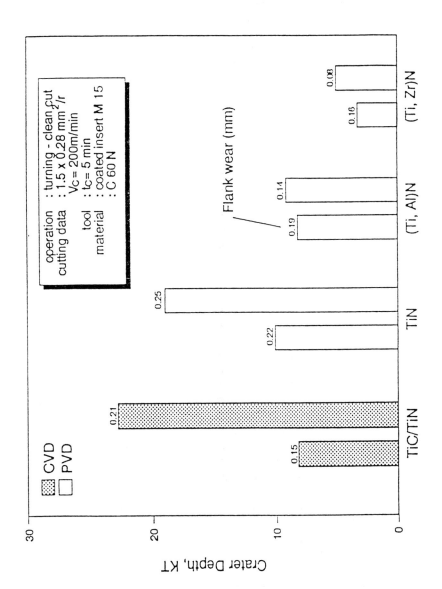

Figure 8 *High flank and crater resistance offered by (Ti Zr)N coatings compared with other coatings*[20]

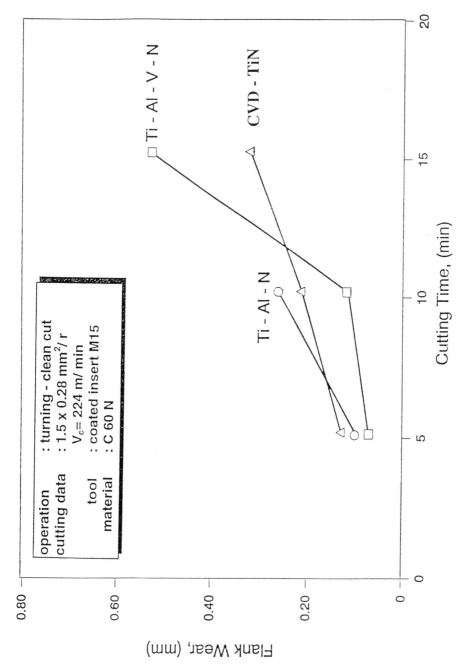

Figure 9 *Comparison between single layer CVD TiN and PVD coated multi-component tools[20]*

conditions, with significant variations in coating layer thickness have been studied[26]. These reports have suggested that such PVD coated inserts with sequential TiC/TiB$_2$ layers involving 100 phase boundaries provided the best wear resistance. It has also been noted[57] that TiC/TiB$_2$ coatings provided **minimum** wear with 500 single layers and also minimum crater wear, as compared with TiC/TiN sequencing.

An improved wear resistance using multilayered coatings with five 1μm layers of TiN/TiC/TiN/TiC/TiN in turning conditions on cemented carbide inserts in machining has also been reported[67]. Here it has also been noted that TiN could be the best choice for the inner and outer layers because of its high temperature stability, low brittleness and also its ability to prevent diffusion of hydrogen during the deposition of carbide into the substrate.

It has been suggested earlier in this review that as a design feature multilayer coating systems should involve materials with some degree of mutual solubility. Thus TiC with materials in combination such as TiN and TiC, Al$_2$O$_3$ and AlN, or like TiC and TiN, and TiC and TiB$_2$, have been used to attempt to obtain good adhesion between layers. It has also been suggested that it is desirable to construct coatings (with hard materials) where the inner layer would be required to provide proper adhesion with the substrate, the intermediate layer(s) provide the overall necessary hardness and strength while the outer layer should provide for less friction and reactivity towards the work piece. It has been quoted[49] that "the hardness of carbonitrides can be correlated with atomic bonding, so that a distinct valance electron concentration of approximately 8.4, maximum hardness occurs". Ceschini et al [49] and Minevich[67] give an indication of the possible use of TiN as a suitable inner layer in the multilayer formats described above. The importance of a low coefficient of friction and less reactivity, most desirable in the outer layer material, has been recognised by[49].

Wick[68] discussed the feasibility, in using various carbide tools (coated and uncoated – on different substrates), of material choice and design for working under specified operating conditions. Thus the suitability of using TiC (inner layer) TiN (outer layer) to improve flank and crater wear resistance at low and moderate cutting speeds has been mentioned. It has been suggested that the cutting performance of a three layered TiC/Al$_2$O$_3$/TiN system is better than a two layered one involving TiC/Al$_2$O$_3$, with the additional bonus of being able to out perform at higher speeds.

The superiority of the multilayered tools is again demonstrated in Figure 10, also from Wick[68]. Coatings with TiC/Al$_2$O$_3$/TiN layers show a distinct improvement under turning conditions of – cutting speed 213 m/min; feed rate 0.28 mm/rev and 1.90 mm depth of cut. (It was also been pointed out incidentally that coating materials such as HfN in a single layered format produced by CVD when tested under turning cast iron showed an improvement of 400% when compared with TiN, while the improved crater wear resistance by TiN is again mentioned[68]).

Despite these positive results obtained by multilayer coatings, some limitations have been suggested. It is noted that in multilayer tools the bonding between layers can cause problems such as adhesive failures at any plane in the many interfaces, in contrast to a single layered coating[69]. Machining of high temperature alloys such as Inconel, Hastelloy and Rene with multilayer coated tools is evidently still not optimised, as it is noted[69] that during the process the "high strength work materials have a tendency to smear rather than to cut".

3.2 Some Aspects of Tool Wear Mechanisms Associated with Uncoated and Coated Tools

Fang[51], has investigated the progressive nature of various types of tool wear of different uncoated tools under cutting conditions. It has been mentioned that the development of tool wear at different regions on a tool is complex. Force variations and surface roughness measurements have been related to tool wear. The different stages in wear at different regions and its complex relationship with surface roughness and forces have been explained and summarised in Table 10. It has been concluded[51] that cutting forces during machining are greatly influenced by the flank and crater wear. Surface roughness in finish turning has been mainly related to the development of minor flank wear, the increase of which has been reported to be consistent with that of associated surface roughness. At high wear stages cutting forces have been reported to change in an erratic fashion. The reason for this is said[70] to be due to the complex influences governing tool wear and forces, and their dependence on work material, cutting parameters, etc.

Some wear mechanisms of various uncoated and coated carbide tools have been assessed by Dearnley and Thompson[37]. The principal wear zones in a cutting tool are shown in Figure 11[37]. It was suggested[37] that coated cemented carbides with coatings of TiC, TiN or Al_2O_3 showed notch wear development. Coated tools, exhibited plastic deformation that caused bulging of the tools.

It has been reported that wear of the coatings in cemented carbides occurs in two ways viz: (a) gradual wear, and (b) catastrophically. The former leads to the formation of a crater, which is suggested[39] to be the region where the rake face temperatures are maximum. It is mentioned that the gradual wear phenomenon on the rake face was a minimum with TiN/TiCN/TiC coated tools, while TiC and Al_2O_3 – TiC coated tools yielded lower resistance. It is also suggested that TiC coatings provided greater crater wear resistance than Al_2O_3 coatings. The appearance of the surfaces at the worn regions has been described[37]. Two distinct features have been observed[37] – (a) smooth, and (b) ridged, or wave-like. The former has been suggested to be caused by a "dissolution/diffusion" mechanism, while the latter has been attributed to discrete plastic deformation. Further support for the discrete plastic deformation and bulk deformation occurring in coatings is due to Kholstedt[71], who has suggested that there is a reduction in hardness and shear strength of TiC, Al_2O_3 and TiN coatings at metal cutting temperatures $> 1000°$ C. It has also been considered that, bearing in mind the solubility limits of TiC, TiN, Al_2O_3 in iron (which fall in TiC (most) > TiN > Al_2O_3 (least), wear occurring by dissolution/diffusion for TiC, TiN and TiCN coatings would also be a possibility.

The performance of coated and uncoated tools in machining has been reported[38] at speeds of 183,244, 305 m/min., feed rate 0.25 mm/rev. and depth of cut 1.27 mm. Under these conditions, the flank and crater wear of the coated tools were estimated to be ~10 and 100 times lower than for the uncoated tools, respectively.

It was noted[38] that the appearance of the worn surface was smooth on the flank face, with no evidence of (workpiece) particle adhesion. The authors have discussed that, as the coatings are harder than the work material, abrasive wear would be unlikely and the most probable wear mechanism has been suggested to be diffusion. Wear mechanisms on the rake face observed in TiC coatings were both diffusion and plastic deformation, while the uncoated and TiN coatings were degraded by diffusion, whereas the Al_2O_3 coatings were worn by plastic deformation. It has been suggested[38] that coated tool wear rate is reduced by low interfacial friction and/or by reduction in temperature at the tool/chip interface. Thus wear resistance by ceramic coatings is associated with good diffusion and oxidation characteristics.

Comparison between PVD and CVD coated cemented carbides tested under turning conditions 250–500 m/min, 0.25 mm/rev feed rate and 2 mm depth of cut with AISI 1046 as work material has been reported[40]. In both coatings progressive wear on the rake face was

Table 8 *Coating materials in multi-layer coated tools tested and the individual layer thicknesses[65]*

(1) **Triple layer coating**, 5.5 μm TiC + 1.5 μm TiCN + 2 μm TiN

(2) **Alumina combined coating**, 4 μm TiC + 6 μm Al_2O_3 + 1 μm TiN

(3) **Zebra coating**, 2 μm TiC + 0.8 μm TiN + 1 μm TiC + 0.8 μm TiN + 1 μm TiC + 0.8 μm TiN + 1 μm TiC + 1.8 μm TiN

Table 9 *Nominal composition and coating thickness of coating materials*[66]

S.No	Insert type	Nominal coating composition	Nominal total coating thickness μm
1.	NX33	Cermet + TiN	3
2.	NX33	Cermet + TiCN	3
3.	T110A	Cermet + TiC	3
4.	T110A	Cermet + TiC + Al_2O_3	3
5.	GC 425	WC + TiC + TiN	8
6.	GC 435	WC + TiC + Al_2O_3 + TiN	8
7.	GC 3015	WC + TiC + Al_2O_3	10

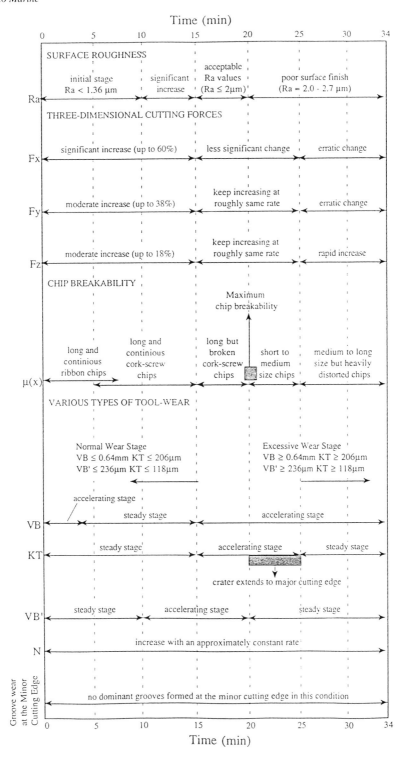

Table 10 *Effects of tool wear on overall machining performance*[51]

Table 11 *Summary of different wear mechanisms occurring in high speed steel and carbide tools*[63]

Wear Location	Mechanisms	Tool Material
Crater	Thermally activated diffusion of elements across chip-crater interface.	HSS
	Chemical dissolution of tungsten carbide grains and diffusion of cobalt binder across chip-crater interface; also mechanical removal of carbide grains.	Carbide
Wearland	At low and normal speeds : abrasive wear of tool matrix by microscopic hard nonmetallic inclusions.	HSS
	At high speeds: thermally activated, but different than crater; possibly fatigue, binder embrittlement, oxidation, etc.	Carbide
Nose	Thermally softened tool matrix plastically deforms and washes away.	HSS, Carbide
Notch or groove	In Ni-based high temperature work materials: plane stress and plane strain plastic flow.	Carbide
	In free - machining steels: grooves on form tools induced by non - uniform BUE and MnS layers.	HSS

observed[37, 40, 72]. It has been concluded that during turning operations the wear mechanism of CVD TiC and PVD TiN were similar and could be chiefly attributed to discrete plastic deformation and dissolution/diffusion[37]. In contrast PVD coatings of TiN were suggested to have sporadic coating failure with likely decohesion at the coating/substrate interface.

It has been concluded[41] that diffusive wear is absent in cutting tools in the coated zones and therefore the wear rate is lower in comparison to an uncoated tool. The reason for the lower wear rate is attributed[41] to lowering of friction and the formation of a barrier to heat dissipation by the coating. It is also noted that coatings in cutting tools cause a decrease in contact length and a fall in compressive stress and the maximum stress and the resultant cutting forces move towards the clearance face and away from the cutting edge on the rake face.

Principal recognised wear mechanisms in high speed steel and carbide tools are summarised in Table 11[63]. However, it should be emphasised out that the wear mechanisms in coated carbides, diamond tools, ceramics and cubic boron nitride tools are still to be completely understood.

4 CONCLUSIONS AND SOME FUTURE RESEARCH DIRECTIONS

The main conclusions to emerge from this summary of a very voluminous and complex literature concerning coatings/coatings systems for cutting tools may be summarised as follows:

1. The advantages in using coatings in various applications, especially in cutting tools, are numerous and clearly evident.
2. The type of coating process, PVD or CVD, has an important effect on the performance of the coated system.
3. The influence of PVD coating process parameters on the performance of the coating is well established in the literature.
4. There is a lack of information in the literature defining the process parameters to be adopted for a particular process to influence coating morphology, composition and microstructure, and coating thickness. There are no standards or literature available at present to systematically evaluate coatings' performance to ensure quality assurance and guarantee consistent production of coatings. More work is required on these perspectives.
5. There is a general lack of information in the literature in the context of carbide tools, regarding details such as chemical composition and microstructure used in their specified grades and their significance, for both uncoated and coated carbide categories. More information in this area is required.
6. The improved performance of coated carbides has been identified in the literature. Although the different behaviour of various coated tools has been reported, there is at present an absence of set standards to evaluate coated tools – especially for coated carbide tools, while at present the ISO standards for testing exist for uncoated carbides.
7. In general, the superior performance of *multilayer coated tools* especially TiC/Al$_2$O$_3$/TiN- and Ti(C,N)-based multilayer coatings relative to *single layer coated tools* is consistently observed.
8. More research is required to identify the wear mechanisms of coated cutting tools – especially of coated carbide cutting tools.
9. With regard to the design of coated tools there is a paucity of information detailing the criteria for determining the optimal coating material properties required to be used in

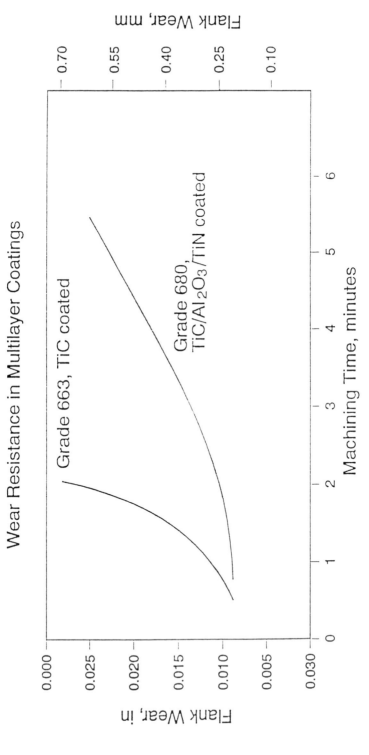

Figure 10 *Wear resistance in multi layer coatings*[68]

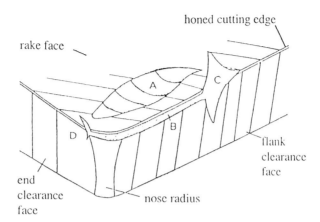

Figure 11 *Principal wear zones on coated carbide tools: A = rake face wear; B = fl. h. face wear; C, D = notch wear*[37]

sequence – multilayer formats, and/or in satisfying compatibility conditions with the substrate. There is a need to establish modelling systems, taking into consideration these properties to improve and develop existing design, concerning which very limited information is available in the literature. This hinders useful comparison, study by study, and the generation of a design rationale.

10. There is a lack of information as to coating morphology, composition and microstructure, coating thickness, etc in most of the tested tools reported by researchers, and also commercially - available coated tools. This makes critical appraisal and comparison of performances difficult, and is an inhibitor to a more systematic approach to coating design.

Acknowledgment

The author would like to thank Dr. C. Subramanian and Mr. M. Ghopinath for useful discussion.

References

1. Sandvik and Coromant, *Metal Working Products – Turning Inserts,* Sweden, 1989.
2. M. Gopinath, *Wear Behaviour of Coated Carbide Tools in Machining,* (Grad. Dip. Dissertation), University of South Australia, 1993.
3. S. Kalpakjian, *Manufacturing Engineering and Technology,* Addison-Wesley, 1989.
4. R. F. Bunshah, *Deposition Technologies for Thin Films and Coatings: Developments and Applications,* Noyes Publication, 1992.
5. Production Engineering Research Association of Great Britain (PERA), *A Survey of*

General Purpose Tools, Report 136, 1965.

6. W. D. Sproul and A. Matthews, *Proc. Conf. Surface and Engineering Technology,* Sydney, 1990.

7. M. G. Hocking, V. Vasantasree and P.S.Sidky, *Metallic and Ceramic Coatings: Production, High Temperature Properties and Applications,* Longman; J. Wiley, Harlow, UK, 1989.

8. D. S. Rickerby and A. Matthews, *Advanced Surface Coatings: A Handbook of Surface Engineering,* Blackie, 1991, New York.

9. B. Bhushan and B.K. Gupta, *Handbook of Tribology: Materials, Coatings and Surface Treatments,* McGraw-Hill, 1991.

10. T. S. Sudarshan, *Surface Modification Technology, An Engineer's Guide,* New York, M. Dekker, 1989.

11. U. Konig, *Surf. Coat. Technol.,* 1987, **33**, 91.

12. H. Tsukada, M. Tobioka, T. Nomura and Y. Saito, *Coated Carbides,* Sumitomo Electric Industries, (1991), Patent - US 5116694.

13. E. O. Ristolainen, J. M. Molarius and S. Vuorinen, *Conf. Electron Microscopy, Vol.11, Kyoto, Japan, TEM and SEM Study of Titanium Nitride/HSS Interface,* Society of Electron Microscopy, 1986, 1365.

14. E. A. Brown, G. L. Sheldon and A. E. Bagoumi, *Wear,* 1990, **138**, 137.

15. H. Holleck, Kuhl and H. Schultz, *J. Vac. Sci. Technol.,* 1985, **A3(b)**, 2345.

16. J. H. Kim and H. A. Jehn, *Journal of the Korean Institute of Metals and Minerals,* 1992, **30**, 1, 75.

17. A. J. Perry, J. P. Schaffer, J. Brunner and W. D. Sproul, *Surf. Coat. Technol.,* 1991, **49**, 188.

18. G. C. Ye and H. J. Shin, *Journal of the Korean Institute of Metals and Minerals,* 1990, **28**, 695.

19. S. L. Rohde and W.D. Munz, in D.S. Rickerby and A. Matthews, *Advanced Surface Coatings: A Handbook of Surface Engineering,* 1991, 92.

20. O. Knotek, M. Bohmer, T. Leylendecker and F. Jungblut, *Mater. Sci. Eng.,* 1988, **A 105/106**, 481.

21. T. Mizuguchi Ko, E. Yoshikawa, T. Tsuji and K. A. Gijutsu, *Journal of the Surface Finishing Society of Japan,* 1990, **41**, 5, 509.

22. J. M. Molarius, *Structure and properties of TiN and ZrN Thin Films Prepared by Reactive Ion Plating,*(Doctral Thesis), 1987.

23. X. Y. Wen and Z. L. Zhang, *Surf. Coat. Technol.,* 1992, **51**, 252.

24. H. S. Legg, K. O. Legg, J. G. Rinker and G. B. Freeman, *J. Vac. Sci. Technol.,* 1986, **A4**, 2844.

25. E. Kubel, *Surf. Coat. Technol.,* 1991, **49**, 268.

26. H. Holleck and H. Schulz, *Surf. Coat. Technol.,* 1988, **36**, 707.

27. R. Fella, H. Holleck and H. Schulz, *Surf. Coat. Technol.,* 1988, **36**, 257.

28. H. Holleck, *J. Vac. Sci. Technol.,* 1986, **A4**, 2661.

29. A. V. Byeli, E. M.Makushok, G. V. Markov, A. A. Minevich and N. N. Popok, *Wear Resistance of Metal Cutting Tools with Multilayer Coatings, in New Cutting Tools of High Performance,* LDNTP, Leningrad, 1989, 61–64 (in Russian).

30. Z. Rengi and L. Ziwei, C. Zhoupping and S. Qi, *Wear,* 1991, **147**, 227.

31. K. H. Habig, *J. Vac. Sci. Technol.,* 1986, **A4**, 2832.

32. Sandvik and Coromant, *Metal Working Products - Turning Inserts,* Sweden, 1993.

33. J. Keem, in T. S. Sudarshan (ed.), *Surface Modification Technologies – An Engineer's Guide*, Marcell Dekker, NY, 1990.

34. A. S. Gates, Jr., *J. Vac. Sci. Technol.*, 1986, **A4**, 2707.

35. P. M. Noaker, *Manuf. Eng.*, Oct., 1991, 63.

36. E. M. Trent, *Metal Cutting 3rd ed.*, Butterworth-Heinemann, 1991.

37. P. A. Dearnley and V. Thompson, *Surf. Eng.*, 1986, **2**, 191.

38. P. A. Dearnley and E. M. Trent, *Met. Technol.*, 1982, **9**, 60.

39. P. A. Dearnley, *Met. Technol.*, 1983, **10**, 205.

40. P. A. Dearnley, R.F. Fowle, N.M. Corbett and D. Doyle, *Surf. Eng.*, 1993, **9**, 312.

41. A. A. Minevich, *Surf. Coat. Technol.*, 1992, **53**, 161.

42. W. Konig, R. Fritsch and D. Kammermeir, *Surf. Coat. Technol.*, 1991, **49**, 316.

43. K.N. Strafford, C. Subramainan and T.P. Wilks, *J. Mater. Process. Technol.*, 1993, **38**, 431.

44. H. W. Grunglig, K. Schneider and L. Singheiser, *Mater. Sci. Eng.*, 1987, **88**, 177.

45. H. Freller and H. Haessler, *Surf. Coat. Technol.*, 1988, **36**, 219.

46. R. D. Arnell, *Surf. Coat. Technol.*, 1990, **43/44**, 674.

47. J. A. Thornton, in R. F. Bunshah (ed.), *Deposition Technologies for Film Coatings – Developments and Application*, Noyes, Publications, 1982.

48. J. M. Blocher, Jr., in (eds.) C.F. Powell, J.H. Oxley and J.M. Blocher, Jr., *Vapor Deposition*, Electrochemical Society, N.Y., Wiley 1966.

49. L. Ceschini, G.L. Garagnani, E. Lanzoni and G. Poli, *Conf. Euromat 91, 1, Advanced Processing*, The Institute of Materials, Cambridge, U.K. 1991.

50. D. S. Rickerby, in D.S. Rickerby and A. Matthews (eds.), *Advanced Surface Coatings*, Blackie, Glasgow, 1991.

51. X. D. Fang, *Wear*, 1994, **173**, 171.

52. V. A. Tipinis, in M.B. Peterson and W. O. Winer (eds.), *Wear Control Handbook*, N. Y. ASME, 1980.

53. M. Gopinath, M. Eng. University of South Australia, 1995.

54. W.D. Sproul, *Surf. Coat. Technol.*, 1987, **33**, 133.

55. I. Penttinen, J.M. Molarius, A.S. Korhonen and R. Lappalainen, *J. Vac. Sci., Technol.*, 1988, **A6**, 2158.

56. Y. Itoh and H. Kashiwaya, *Journal of High Temperature Society of Japan*, 1991, **17**, 317.

57. M. Podob, *Conf. Tool Materials for High Speed Machining*, ASM, Metals Park, USA, Feb. 1987.

58. D. G. Bhat and P. F. Woerner, *J. Metals*, 1986, **32**, 68.

59. D. T. Quinto, A. T. Santhanam and P. C. Jindal, *Mater. Sci. Eng.*, 1988, **A105/106**, 443.

60. A. V. Byeli, E.M. Makushok, G.V. Markov and A.A. Minevich, *Int. Tribology Conf.* Brisbane, 1990, 54.

61. S. Takatsu, *High Temp. Materials and Processes*, 1990, **9**, (2–4), 175.

62. K. Dreyer and J. Kolaska, *Proceedings of the International Conference – Towards Improved Performance of Tool Materials*, The National Physical Laboratory and the Metals Society, London, April 1981.

63. K. Keller, *Industrie-Anzieger*, 1990, **112**, 30.

64. P. Hedenqvist, M. Olsson and G. Hogmark, *Surf. Eng.*, 1992, **8**, 30.

65. S. Kalpakjian, *Manufacturing Processes for Engineering Materials*, Addison –

Wesley, 1984.

66. T. N. Goh, M. Rahman, K. H. W. Seah and C. H. Lee, *J. Mater. Process. Technol.*, 1993, **37**, 655.
67. A.A. Minevich, *Surf. Coat. Technol.*, 1992, **53**, 161.
68. C. Wick, *Manuf. Eng.*, 1987, **98**, 45.
69. J. R. Coleman, *Manuf. Eng.*, 1990, **104**, 38.
70. L. Dan and J. Matthew, *Int. J. Mach. Tools Manufact.*, 1990, **30**, 579.
71. D. L. Kohlstedt, *J. Mater. Sci.*, 1973, **8**, 777.
72. P. A. Dearnley, *Surf. Eng.*, 1985, **1**, 43.

Contributor Index

This is a combined index for all three volumes. The volume number is given in roman numerals, followed by the page number.

A

Ager F. J. I 19
Ahmed R. I 229
Arnell R. D. I 119, II 19
Audy J. III 167

B

Baldwin K. R. I 119
Bao Y. II 132, II 222
Barker D. II 296
Baron R. P. I 158
Bates J. R. I 307
Bates R. I. I 119
Bedingfield P. B. I 238
Bell T. I 260, II 111
Benham R. I 168
Bienk E. J. III 218
Binfield M. L. I 203
Blunt L. III 167
Botella J. I 19
Brooks J. S. II 31
Bryar J. C. I 53
Bull S. J. I 274, II 48, III 209
Burnell-Gray J. S. I 1, III 70, III 78
Bushby A. J. I 250

C

Calzavarini R. III 197
Carré A. I 292
Carroll M. W. III 115
Cawley J. II 31, III 99
Chan W. Y. III 78
Chen F. II 111
Chisholm C. U. I 95, I 111

D

Daadbin A. III 70

Damborenea J. J. II 211
Datta P. K. I 1, I 53, I 217, I 238,
 III 70, III 78
Dayal R. K. III 38
D'Errico G. E. III 179, III 197
DeSilva A. M. I 111
Di-Agieva J. D. II 290
Ding G. II 179
Dodds B. E. II 280
Dong H. II 111
Dou Z. II 132
Dowson A. L. I 238
Du H. L. I 53
DuPont J. N. II 156

E

El-Gammal M. M. III 315
El-Sharif M. R. I 95, I 111
Erickson L. C. I 195
Escourru L. I 180
Eyre T. S. I 203

F

Fouad M. I 286
Feng L. I 95
Flamant G. II 59
Flis J. I 132, I 260
Flitris J. II 72
Fujikawa H. I 41

G

Gabe D. R. II 265
Gauthier D. II 59
Gawne D. T. II 132, II 222, II 245
Green J. III 99
Griffin D. III 70
Guglielmi E. III 179

H

Hadfield M. I 229
Hammersley J. G. II 199
Hao He-Sheng II 179
Harrison K. III 131
Hawthorne H. M. I 195
Hocking M. G. II 168
Hodgkiess T. II 88, III 277
Hull J. B. II 233

I

Ibrahim S. K. II 245
Ives M. II 31

J

Jacobs M. H. I 53
Jian L. I 76
Jiang J. I 143
Jiangping T. I 76
Jokinen T. II 186
Jones A. M. II 48
Jones A. R. II 233
Jones R. E. III 99

K

Kalantary M. R. II 265
Kariofillis G. II 72
Kathirgamanathan P. I 307
Kelly P. J. II 19
Kempster A. I 168
Kerr C. II 296
Khalil S. I 286
Kimura Y. II 39
Kiourtsidis G. I 180
Kipkemoi J. I 32
Knotek O. III 147
Krajewski A. III 19
Krishnan K. III 38

L

Lanzoni E. III 187
Latha G. III 247, III 303
Lawson K. J. I 83
Leggett J. I 53
Lehmberg C. E. II 274
Lewis D. B. II 31, II 274, II 280
Löffler F. III 147, III 224

Lugscheider E. II 146, III 147
Luna C. I 19

M

Ma Xiong-dong II 179
Mankowski J. I 132, I 260
Maoshen L. I 76
Marder A. R. I 158, II 156, III 51
Marshall G. W. II 274, II 280
Martini C. III 187
Mazza G. II 59
McCabe A. R. II 48, III 209
McMillan W. III 325
Meuronen I. II 186
Mikkelsen N. J. III 218
Miles R. W. I 307
Mo Z. II 111
Möller M. III 147
Mudali U. K. III 26, III 38
Münz W-D. II 31

N

Nair K. G. M. III 26
Narita T. I 67
Neville A. II 88, III 277
Ni Z-Y. II 81
Nicholls J. R. I 53, I 83
Nicoll A. R. III 131
Nitta S. II 39

O

Odriozola J. A. I 19

P

Palombarini G. III 187
Paúl A. I 19
Pavlikianova A. I. II 290
Pen Q. II 111
Pena D. III 260

Q

Qu J-X. II 179

R

Rajendran N. III 247, III 303
Rajeswari S. III 26, III 247, III 303
Ravagliok A. III 19
Remer P. II 146

Respaldiza M. A. I 19
Reymann H. II 146
Rice-Evans P. II 48
Rickerby D. S. I 83, III 99
Rodríguez G. P. II 211

S

Saleh A. II 48
Salpistis C. II 72
Sanchez I. II 59
Sarwar M. III 159
Shan L-Y. II 81
Shanahan M. E. R. I 292
Shandong J. P. II 111
Shida Y. I 41
Silva M. F. I 19
Skolianos S. M. I 180
Smith A. B. I 168
Smith C. J. E. I 119
Smith J. I 168
Sneddon D. I 111
Soares J. C. I 19
Stack M. M. I 143, III 260
Stapountzis H. II 72
Stott F. H. I 143
Stoyanov V. A. II 290
Strafford K. N. I 189, III 167, III 325, III 364
Subbaiyan M. III 26, III 38
Subramanian C. I 189, III 325
Sundararajian T. III 26
Sut Y. J. I 95
Swain M. V. I 250

T

Tabaghina T. M. I 286
Thampi N. S. III 26, III 38
Troczynski T. I 195
Tsipas D. I 32, II 72

V

Vázquez A. J. II 211
Veerabadran K. M. III 38
Vicenzi B. III 197
Vincenzini P. III 19

W

Walsh F. II 296

Wang C. II 111
Wang P-Z. II 81, II 179
Ward L. P. I 217
Watson A. II 245
Wellman R. G. III 294
Wilcox G. D. II 265
Wilks T. P. III 325
Wood III 115
Wright N. P. II 121

Y

Yang S. II 31
Yang Y-S. II 179
Ye P. II 81
Yoshioka T. I 67

Z

Zakroczymski T. I 260

Subject Index

This is a combined index for all three volumes. The volume number is given in roman numerals, followed by the page number.

Symbols

(Ti,Al)N III 187
(Ti,Nb)N III 187
WC-Co coatings I 229
2D weave structure III 212
316L stainless steel I 223, III 38, III 303
3D weave III 212

A

ABAQUS III 71
abrasion II 81, III 240, III 384
abrasion of MMCs III 260
abrasion wear I 180, I 205, II 147, III 325
abrasive flow machining II233
abrasive particles III 294
abrasive-rolling II 147
ABS III 221
absorption effects III 282
acetone II 253
acidification of the solution III 251
acoustic emission I 190
activating properties II 121
activator I 34
active site III 251
actuators II 28
adhesion I 77, I 191, I 194, I 218, II 81, II 247, III 237, III 240
adhesion failures III 373
adhesive III 325
adhesive failure I 225, I 276
adhesive force III 231
adhesive ridge I 300
adhesive strength III 226, III 238
aero gas turbine engine III 99
aerodynamic flow III 294
aerospace I 32, II 210, III 159
Ag and Cu wires II 174

aged II 128
ageing I 103
aggregation II 256
air plasma spraying III 99
AISI 410 stainless steel I 206
AISI 4150 I 189
AISI M35 tool steel III 188
AISI-304 stainless steel I 19
Al extrusion II 72
Al-2024 composites I 180
Al_2O_3/Al metal matrix composites III 70
alloy coatings I 238
alloy steels III 179
alloys and refractories III 209
alumina coatings III 216
alumina crucibles III 213
alumina filler II 231
aluminium I 250
aluminium coatings I 119
aluminium nitride III 160
aluminium/magnesium alloys II 28
anions II 290
anisotropic behaviour I 229
annealing II 53
anodic currents I 135
anodic cyclic polarization III 304
anodic dissolution III 306
anodic polarisation I 135, I 268, II 272, II 298, III 283
anodic vacuum arc III 227
anodising II 292
anomalous co-deposition mechanism II 265
anti-oxidation I 67
apex angle I 232
application of inhibitors III 247
arc II 156

arc current II 156
arc discharges II 23
arc evaporation (AE) III 187
arc power II 166
arc spray III 296
arc-height of curvature II 200
argon shrouded plasma III 100
argon-treated duplex stainless steel III 289
Armco iron I 133
aspect ratio I 244
ASTM B117 I 126
ASTM G5-82 II 133
ASTM G65 I 205
atmospheric pressure non-equilibrium
    plasma II 121
atomic absorption spectroscopy III 41,
    III 231
audit I 8
Auger electron spectroscopy I 95, I 260
austenitic Cr18Ni9Ti steel I 260
austenitic lattice I 265
austenitic stainless steel III 38, III 277,
    III 278, III 279, III 286
austenitic steels III 179
auto-catalytic process III 306
automobile I 111
automotive II 210, III 159
automotive industry III 325
axial ball bearing III 151
axial magnetic field II172

B

back-scattered electron image I 234
bacterial corrosion III 315
ball bearing test III 243
ball bearings III 147
ball-cratering I 170
ball-on-disc I191
barrier I124
barrier to oxygen III 211
basic sputtering process II 19
be heat-treated I 119
bearing rings III 154
bench-marking I 8
bending stresses I 276
bending test I 112
beta-ray backscatter method III 231,
    III 232

bias-voltage III 148
binding energy III 249
biocompatibility I 217, III 38
biomaterials I 217
biomedical III 159
biomedical applications III 38
blow moulding dies II 233
boiler III 296
boiler tubes I 34, III 294
bore III 144
boric oxide glass III 210
borides III 211
boriding II 82
boron carbide II 238
boron-containing compounds II 72
break-away I 162
breakdown potential III 248
breaking-in I 176
brittle fracture I 210, III 271
BS 5466 I 112
Bückle's rule III 236
buckling failure I274
built-up-edge III 183
bulk materials and coatings II55

C

C/N ratio I 265
C/SiC ratio III 211
cadmium replacements II 28
calibration III 136
capacitance of storage capacitor II 169
carbide tool tips III 163
carbides III 211
carbon steels III 179
carburizing/boriding I I81
cast tungsten carbide II 179
cathode current efficiencies II 281
cathode diffusion layer II 251
cathode potential II 290
cathodic arc PVD process III 180
cathodic reaction rate II 272
cationic ion-exchange II 248
cations II 290
cavitation III 240
cavitation measurement III 241
cell potential II 247
cemented carbide tips III 163
cemented carbides III 364

cementite I 266
centrifugal spray deposition I 55
ceramic matrix composites III 209
ceramic-base coatings II 88
ceramics III 277
cermet II 88
cermet cutting tools III 179
cermet insert III 183
characterization of as-deposited films
    III 189
chemical and mineral processing II 88
chemical degreasing II 292
chemical reactivity III 252
chemical vapour deposition II 48, II 88,
    III 364
chipless machining II 234
chloride electrolytes II 248, II 295
chloride ions III 303
chlorination I 76
chromate conversion I 95, I 119
chrometan/malate/sulphate-based electrolyte
    II 250
chromic III 280
chromising I 158, I 168
chromium (VI) baths I I245
chromium and nickel ions III 38
chromium plating II 290
chromium (III) II 262
chromium (III) ligand exchange II 251
chronoamperometry II 133
circular saw III 164
cladding II 156, II 172
clean remelting I 58
clearance control I 229
closed-reactor chemical vapour deposition
    I 168
Co-Cu, FeNi-Cu/Cu II 265
co-deposition II 28, I 32
coefficiency of friction III 164
CO_2 laser III 278
coal III 294
coal utilisation I 32
coal-fired power III 247
coated ball bearings III 147
coating adhesion II 226
coating deposition III 213
coating detachment I 274

coating manufacturing II 180
coating microstructure II48, III 216
coating system III 364
coating thermal conductivity I 85
coating thickness I 83, II 168
coating/substrate I 218, I 274
coating/substrate adhesion I 39
coatings III 133
coatings of ceramic III 210
cobalt-chromium alloys III 38
cohesive III 238
cohesive energy density II 257
cohesive failure I 225
cold forging III 160
cold working II 199
collaborative development I 10
composite I297
composite coatings I 111, II 179, II 280
compound zone I 132
compound/diffusion zone I 265
compressive strength of coating I 230
compressive stresses I 275, II 199, II 296
computer integrated manufacturing III 146
conductive heat flux II 229
conductivity II 253
constant amplitude loading I 239
consumable electrode processes II 156
contact angle I 292
contact fatigue II 290
contact pressure I 253
conversion coating I 95, I 103
cooling systems III 303
corona discharges II 123
corrosion II 265, II 274,
    II 280, III 239, III 299
corrosion and wear resistance II 280
corrosion current I 112
corrosion initiation III 289
corrosion potential I112, III 248
corrosion propagation III 291
corrosion protection I95
corrosion resistance I 217, I 260, II 222,
    III 26, III 132, III 277, III 290
corrosion resistant coatings II 19
corrosion response I 247
corrosion tests III 278
corrosive environments III 209

cost I 33, III 131
cost effective III 299
coulometry III 231
counter electrode III 248
coverage II 201
crack propagation I 233, III 75
cracking and detachment III 211
crater wear III 183, III 381
crater wear rate I 174
craters I 263
crevice corrosion II 97, III 248, III 305
critical crevice potential III 251, III 305
critical load I 194, I 222, I 274, III 238
critical nodal stress III 75
critical pitting potential III 251
critical process conditions III 290
CrN III 42
CrN, NO_3^- and NH_4^+ III 311
crystal lattice II 293
crystallisation I 239
crystallite size I 241
crystallographic orientation I 224
current II 266
current density I 112, I 158, II 42, II 275
    II 290
current efficiency II 245
cutting edge geometry III 162
cutting edges III 161
cutting energy parameter III 165
cutting tools II 122, III 364
CVD I 83, I 189, II 59, III 159
CVD diamond I 86
CVD SiC III 212
cyclic oxidation I 34

D

D/max-RB type X-ray, diffractometer
    II 180
DC anodic polarisation potentiodynamic
    II 90
DC-anodic polarisation III 280
dc-sputtering III 226
decoration III 226
decorative I 119, III 159
deep groove rolling element I 231
defects III 38
defocusing laser III 278

deformable coatings III 210
degree of overlap III 281, III 290
delamination I 230
dendrites II 174
density measurement I 77
deposit morphology II 270
deposition kinetics I 239
deposition parameters I 247
depth of craters III 183
depth of failure I 236
depth of transition regions I 264
depth sensing I 250
design audit III 362
design of coated tools III 365
detonation gun I 229, II 89
dewetting I 295
diamond jet II 146
diamond-on-plate I 206
dielectric II 255
dies II 72, II2 33, III 325, III 364
diffusion coatings I 32
diffusion layer II 251
diffusion zone I 132
diffusion zones of nitrided chromium steels
    I 260
dimensional changes II 222
dimethylamineborane (DMAB) I 238
dimethylformamide II 248
dimples II 201
DIN 4768 III 234
DIN 4772 III 234
DIN 50 103 part 1 III 238
DIN V ENV 1071 III 232, III 238
dislocations II 52, II 118
dispersion hardening II 287
donor power of solvent II 257
Doppler-broadening S-parameter II 50
dose rate III 41
downtime III 294
drill life III 165
droplet velocity II 168
dry powder II 222
dry sliding wear II 84, II 183, II 185
duplex III 277
duplex coating systems I194

E

economics III 138
economiser III 296
E$_{cor}$ II 298
eddy current I 120, III 231
edge cracks I 234
edge sharpness III 165
EDTA I 77
EDXS I 111
effects of oxidation III 275
elastic limit II 202
elastic/plastic behaviour II 201
elasto-hydrodynamic lubrication I 232
elastomers I 295
electrolyte temperature II 290
electric arc spraying II 225
electrical insulation III 132
electro-discharge II 233
electrochemical cell assembly III 248
electrochemical measurements II 42
electrochemistry II 245
electrode agitation/rotation II266
electrode extension II157
electrode preparation III 247
electrodeposition I 111, I 158, II 265
electrodeposition apparatus II 246
electrodeposition of chromium (III) II 245
electrodischarge machining II 234
electroforming II 290
electrogaphical indentation III 235
electroless deposition I 238
electroless nickel I 189
electroless nickel bath composition I76
electroless nickel coatings III 326
electroless nickel-phosphorus III 350
electrolyte chemistry II 245
electron beam III 227
electron beam evaporation I 253
electron beam hollow cathode discharge (EB-HCD) III 187
electron beam physical vapour deposition III 99
electron collision frequency II 125
electron probe microanalysis I 234, III 188, III 280
electronics III 159
electronics industry I 83

electrostatic spraying II 223
elongation at fracture II 202
energy dispersive X-ray spectroscopy I 111
engine block III 144
ENV draft 38 III 233
environment III 145
environmental disadvantages II 146
Environmental Protection Act II 128
epitaxial layers II 290
epoxy II 127
equilibrium vapour pressures of iron chlorides I 81
erosion I 196, I 201, I 203, II 265
erosion mechanism III 262, III 271, III 275
erosion mechanism map III 275
erosion-corrosion apparatus III 262
erosion-corrosion resistance III 271
erosive abrasion II 180
erosive wear of metal matrix composite III 260
exhaust gas emission III 144
exhaust gas sensor III 138
explosion shock wave II 171
explosion spraying II 168.
explosive ablation II 169
extruded metal and oxide III 271

F

failure criterion III 75
failure modes I 229, I 274, III 165
Faraday's laws II 281
fast explosion II 169
fasteners I 119
fatigue life I 230, I 242
fatigue response I 243
fatigue strength II 296
fatigue wear II 83
FCC I 78
Fe-based alloys III 271
Fe- and Ni- base Alloys I 32
FeCl$_2$ I 78
FeCl$_3$ I 78
feed rate II 156
fibre bridging III 71
fibre/matrix interface III 75
fibre/matrix interface bond strength III 70
filler metal II 156

filler wire II 164
film deposition rate II 21
film thickness standards III 229
finishing teeth III 164
finite element analysis (FEA) I 274, III 71
finite element modelling I 278, III 70
finite element numerical techniques I 251
fireside corrosion III 294
fixation III 38
flame hardenable cast steel III 326
flame spraying II 168, III 262
flank wear III 164, III 179, III 181
flares II 28
flock spraying II 224
flue gas desulphurization III 247
flue gas III 298
fluidized bed dipping II 223
fluidized bed of erodent particles III 262
fluorescence III 282
fly ash erosion III 294
focused collision sequences II 55
formation of chlorides I 76
formation of concentration cell III 309
fracture analysis III 71
fracture energy I 279
fractures III 38
free corrosion potentials II 101
Frenkel pairs II 118
friction III 221, III 325
frictional force I 190
furnace fused coatings II 88
fusion II 156

G

γ'-phase (Fe$_4$N) I132
γ-TiAl I 58
galling II 81
galvanic corrosion III 315
galvanic hard chromium platings II 146
gas carburising II 88
gas combustion III 209
gas industry II 88
gas nitriding II 72
gas permeability of coating III 215
gas scattering II 21
gas turbines I 32, II 265
gels I 304

general corrosion III 315
geometrical accuracy III 165
glass-polyamide composite coating II 231
glow II 19
glow discharge optical emission spectroscopy
    I 133, I 260, II 275, II 281
glow discharges II 123
gradation of hard reinforcement particulates
    III 260
graded coatings III 260
graded material III 274
grain boundary precipitates III 283
grey cast iron III 179
grit blasting III 145, III 213
growth stress relief I 39
gullet geometry III 162

H

H13 tool steel II 72
hafnia III 211
halogen corrosion I 76
hangers III 296
hard chromium plating III 326, III 350
hardness I 191, I 253, I 260, II 290
hardness measurement III 236
heat conduction resistance II 229
heat input III 290
heat transfer II 124, II 230
heat treatment I 117, II 75, III 145
heat wave microscopy III 235
heliostats II 214
Hertz contact stress I 233
Hertzian I 252
Hertzian cracks I 220, I 274
Hertzian pressure III 151
hexavalent chromate I 128
hexavalent chromium electrolytes II 245
Hf-modified coatings I38
Hf-plasma etching III 148
high frequency II 209
high speed steel tips III 163
high temperature composites III 216
high temperature X-ray diffraction II 48,
    II 51
high velocity oxy-fuel I 229, II 88, II 146,
    III 297
high velocity plasma spraying I 229

high voltage corona field II 224
higher wear, corrosion, and heat resistance
    II 179
hole-drilling techniques II 201
hot corrosion I 32, I 35
hot topping I 54
hot working tool steels II 72
HSS Drills III 165
Hull cell II 275
HVOF coated I 229
hydrated ferric oxide II 142
hydrogen evolution II 259
hydrophobicity II 122

I

IEC test specification 587 II 128
igniters II 28
impact angle II 75
impact test III 241
impedance analysis I 95
implanted argon II 54
in-flight polymer particles II 226
INC718 I 34
inclined plane tracking test II 128
including fasteners II 28
inclusions III 42
increased ionization II 21
induction heating II 169
industries I 32
implantable ceramics technology III 19
injection mould inserts III 220
inlet nozzles III 294
inter-splat I 195
interface III 161
interfacial crack I 275
interfacial defects I 275
interfacial failure I 276
interfacial fracture energies I 283
interfacial tension I 292
intergranular corrosion II 102, II 199
intermetallic I 53
intermetallic phase formation III 278
intermittent thermal shock III 183
Internal coating II 168
internal combustion engines II 168
internal oxidation I 37, I 164
internal stresses II 290

internal surface spraying III 144
interrupted cut III 242
intersplat bonding II144
intersplat slippage III 210
interstitial atom II 53
inward diffusion of chlorine I 76
ion bombardment II 23, II 53, II 118
ion density II 126
ion implantation I 19, III 26, III 159,
    III 218
ion nitriding III 278
ion-plating I 217
ionization efficiency II 21
iron III 43
iron chlorides I 76
ISO 4287 III 234
ISO 4288 III 234
ISO 9000 III 224
isothermal conditions III 210
isothermal exposure III 213

J

joints and nuts III 309
just-in-time III 325

K

kinetics processes I 78
Knoop III 237
knowledge-based systems I 11
KONTRON II 149
Kushner's stressometer II 297

L

lamella structure I 229
laminar boundary layer I 80
lanthanum implanted I 19
laser II 179
laser-assisted deposition III 159
laser cladding II 180, III 277
laser coating II 179
laser cutting III 277
laser-induced chemical/physical vapour
    deposition II 179
laser ionisation III 280
laser irradiation III 290
laser PVD III 228
laser run III 282
laser surface melting III 278

laser surface modification III 277
laser transformation hardening II179
laser treated layer III 277
laser treated layer microhardness III 290
laser treatment I 195
laser/electron beam III 159
lattice parameter II 48
lattice parameter relaxation II53
leaching III 41
lean manufacturing III 325
ligand exchange reactions II 247
line broadening I 240
line-of-sight factor I 33
liquid droplet erosion III 294, III 299
load-bearing ability II 222
localised attack III 38
localised corrosion initiation III 289
low pressure centrifugal spray deposition
  I 56
low pressure plasma spraying III 99,
  III 100
LP turbines III 299
lubrication I 304, II 222
Luggin capillary II 298

M

M-Cr-Al-Y I 32
m-hydroxo bridged II 261
machine tooling III 144
magnesia III 211
magnetoinduction III 231
magnetron sputter ion plating III 147,
  III 226
man-made SO_2 emission III 247
management issues I 7
management of change I 8
manipulator III 141
manufacturing II 72
manufacturing industry III 159
manufacturing standards III 165
Marble's reagent I 260
marine environment I 111
martensitic Cr17Ni2 steel I 261
masking III 145
mass analyser III 280
master/slave surface concept III 71
masteralloy powder I 34
maximum shear stress I 257

measuring techniques III 228
mechanical cleaning II 292
mechanical damage I 82
mechanical interlock I 232
melting efficiency II 160
melting points I 81
melting pool II 183
metal cutting III 160
metal finishing industry I 111
metal ions II 290
metal matrix composites I 180, II 179,
  II 184, III 277
metal oxide debries III 38
metal screens III 296
metallic passivity III 252
metallurgical bond II 108
metallurgical transformation III 290
metastable phases II 290
microstructure of the coating I 238, II 183
micro-abrasion test I 197, III 210
microcracks I 236, II 247
microcrystalline I 240
microhardness I 229, I 241, II 75,
  II 281 III 188, III 229
microhardness profile III 279
microhenrys II 169
microphase II 293
microscopy of eroded surfaces III 271
microstrain broadening II 53
microstructure III 277
microstructure and microhardness III 284
microwave plasma III 228
mild abrasion I 199
mild steel I 189
mild steel cathodes I 95
mining II 88
mirrored configurations II 23
MMCs I 111
mode I cracking III 71
model test methods III 239
modified laser treatment III 284
molten iron III 209
molybdenum carbides III 180
molybdenum salt film III 251
molybdenum spectra III 255
moulds II 233
mullite and zircon coatings III 213

mullite and zirconium silicate coatings
III 211
multilayer pyrotechnic coatings II 28
multilayer coating system III 376
multilayer Zn-Ni coatings II 265
multiple fatigue cracks I 247

N

nano-indenter III 228, III 237
Nb I 217
NBA II 180
near-surface region I 271
neutral salt fog tests I 124
new drill design III 165
Ni I 78
Ni-P and Ni-B electroless coatings I 238
$Ni_{12}P_5$ II 286
Ni_5P_2 II 286
nickel-based, self fusing alloys II 179
nickel ions III 43
$NiCl_2$ I 78, I 79
Niklad 740 I 239
NiO and $Ni(OH)_2$ III 255
nitrided II 290
nitrided layer I 265
nitrides III 211
nitriding I 132, I 189, II 88, III 26
nitrogen II 290
nitrogen alloying I 260
nitrogen atmosphere III 289
nitrogen-containing steel III 289
nitrogen implanted III 38
nitrogen interstitials II 54
nitrogen supersaturated solid solution
I 260
nodal failure stresses III 75
nodular defect I 244
non-conducting lacquer II 246
non-consumable electrode processes II 156
non-equilibrium structures II 290
non-linear ordinary differential equations
II 64
"non-thermal" (non-equilibrium) plasma
II 122
non-traditional machining II 234
nuclear power III 303
nuclear reaction analysis (NRA) I 24

O

OCP-time measurement III 40
octahedral nitrogen II 54
octahedral titanium sites II 54
oil and gas industry II 88
oil film II2 90
oilfield II 88
open reactor CVD I 168
open-circuit potential I 123, III 33, III 41
operator III 133
optical III 159, III 226
optical mirrors I 250
optical roughness measuring methods
III 234
optimum dose III 40
osprey processing I 55
out-of-pack cementation I 168
outward vapour transport I 76
oxidation I 32, I 53, III 294
oxidation testing III 213
oxide spinels III 100
oxides III 211
oxy-acetylene thermal spray II 104
oxygen concentration I 271

P

P – asymptote hardening modulus II 207
pack aluminising I 32
pack cementation I 168, II 75
pack chromising I 32
pack chromising-aluminising I 32
pack surface treatment I 158
packaging I 250
paint I 304
parabolic oxidation I 19
parallel gate valves I 203
partially yttria stabilised zirconia III 99
particle angularity III 260
particle flux III 274
particle size I 77
particulate deposition II 224
passive film III 41, III 303
passive region I 269
peening hammer II 199
PEENSTRESS software II 202
peg formation III 100
periodically varied current II 265

phosphorus II 297
physical vapour deposition
    II 48, II 88, II 265, III 147
physical vapour nitridation III 26
physiological saline solution III 40
pile-up I 256
pin-on-disc geometry I 205
pin-on-disk II 147, III 240
pinhole-ratio II 42
pit cracks II 247
pit formation III 306
pit growth III 305
pit morphology III 306
pit protection III 248
pit protection potential E_p III 40, III 248
pits II 132
pitting attack II 107
pitting corrosion III 38, III 248, III 304,
    III 315
pitting resistance III 251
plant networks I 9
plasma arc cold-hearth melting I 54
plasma confinement II 21
plasma electron frequency II 125
plasma model II 126
plasma nitriding I 261, II 111 III 278,
    III 364
plasma physics II 129
plasma polymerisation II 122
plasma spraying I 195, III 133, III 213
plasma-assisted chemical vapour deposition
    II 89
plastic deformation I 210, I 236
plastic flow I 235
plastic pile-up I 275
plastic strains I 286
plastics moulding industry III 218
plating bath I 158, II 275
plating of nickel phosphorous II 291
P_o – initial hardening modulus II 207
pointed indenters I 251
Poisson contraction III 72
polar solvent II 257
polarisation curves I 262
polarization II 133
polarization overpotential I 124
polyamide particle II 229

polymeric adhesive I 292
polymethyl methacrylate II 127
polytetrafluoroethylene III 352
porosity I 236, II 89, II 108, II 148,
    II 247
positive ions II 20
positron annihilation spectroscopy II 55
post-coating finishing I 168
post-coating surface processing III 145
post-deposition heat treatment I 239
post-plating processes I 119
potentiodynamic polarisation I 95, I 124,
    II 133, III 304
potentiotransient studies III 27
powder coating II 222
power industry III 294
power input rate II 183
pre-hardening II 209
preoxidized I 19
press forming tools III 325
pretreatment II 296
process parameters II 48
processing route I 53
properties of thin films II 48
protection potential E_p III 41
protective silica scales III 213
pseudo-cyclic II 201
PTFE III 40, III 326
pulp and paper manufacture II 88
pulse/periodic current modulation II 266
PVC III 221
PVD I83, I 119, I 189, I 223,
    III 159, III 219

Q

quadruply bonded dimeric chromium(II)
    II 251
quality III 131, III 364
quality and rate of deposition II 247
quality assurance III 238
quality control III 132, III 299
quality management I 8
quality system III 224

R

radiographic fine structure analysis III 235
rake face III 162
random arc III 227

ranking order of erosion resistance III 270
rare earth elements II 111
rare earth oxide deposition I 19
rare-earth modified Cr/Al coatings I 32
rates of chlorination I 76
RE plasma nitriding II 112
reactive element effect I 33
red rust I 112, I 124, II 271
reducing agent II 293
refractory metals and ceramics II 226
Rene 80 I 34
repassivation II 132
repassivation potential III 248, III 308
residual or internal stresses II 296
residual stresses I 238, II 53, II 144
residual surface stresses II234
resistance to passivity III 283
resistance to sliding wear I229
resonant frequency II 169
reversed transferred arc III 132
rf-arc III 227
rf-sputtering III 226
Ringer's solution III 27, III 40
road conditions I 111
robot III 141
Rockwell indentation test III 238
roller design I 230
rolling contact fatigue I 229
rolling element bearings I 229
root bending fatigue of a gear II 208
rotodynamic equipment II 88
roughened surface III 290
rubber wheel abrasion I 205
rust II 142
Rutherford backscattering spectrometry
    I 20

S

σ phase I 260
S-parameter II 52
sacrificial
sacrificial coating I 95, I 111
sacrificial protection I 123
safety III 145
salt spray I 112
saturated calomel electrode I 262,
    III 40, III 248, III 280, III 304

saturation point II 200
scale plasticity I 39
scoring III 325
Scotch test III 240
scratch adhesion testing I 274
scratch testing I 191, I 218, I 274, III 238
scratch width I 277
scuffing II 81, III 354
sealing II 89
sealing coat II 144
sealing surfaces I 203
seawater II 89
Seebeck arc welding calorimeter II 159
self-aligned bearings III 262
semi-open-reactor CVD process I 168
semiconductors I 250
Senderoff's spiral contractometer II 297
shock wave energy II 170
short-circuit diffusion I 80, II 119
shot velocity II 202
shot-peening II 88
shrouding gas III 278
SiC particles III 260
SiC/SiC CMCs III 212
silica III 216
silicates III 211
silicon carbide III 148, III 210
silicon carbide matrix composites
    III 209, III 210
silicon nitride III 147
silicon phase transformation I 256
silver paste III 40
simulated combustion III 213
simulated SO_2 Scrubber Environment
    III 247
simulation of failure mechanisms III 70
single pass welds II 158
sintered ceramics III 209
skin effect II169
skin thickness effect II 170
slapper plate II 171
slip properties III 221
slurry abrasion tests II 185
slurry conditions III 260
smoothing cubic spline funcrion I 288
S_o – initial micro-plasticity II 207
SO_2 I 58

soft metallic coating III 210
sol-gel infiltration I 195
solar Energy II 211
solar energy materials and solar cells
     II 221
solid lubricants III 371
solid particle erosion III 260, III 299
solution concentration II 266
solution hardening II 182
solvation II 256
solvent-based paints II 222
sonotrode III 241
sorbitic Cr3Mo steel I 133, I 261
spallation I 189, III 100
special welding procedures III 38
specific cutting energy parameter III 165
specific wear rate I 206
spectrographic analysis I 77
spherical cup grinding III 231
spherical indenters I 252
spherical shots II 199
spin casting I 292
splats II 224
sprayability III 107
spreading I 292
sputter ion plating I 277
sputter plasma III 148
sputtering I 119, I 221, II 118,
     III 219, III 225
squeeze cast SiC particulates reinforced
     Al-2024 C I 180
stainless steel (41C grade) II 132
stainless steel-PA II 231
stainless steels I 19, II 50
standard calomel electrode (SCE) II 246
standardized hardness test blocks III 229
steady state abrasive wear rate I 173
steel substrates I 119
steered arc III 227
stellite powders III 145
stoichiometric Al_2O_3 II 26
stored electrical energy II 170
strain energy I 223
strain rate sensitivity I286
strain tolerance III 215
strain-to-failure I 209
strategic alliances I 9

strategic decision-making I 10
strategic management I 8
strategy I 9
strength I 260
strengthening by fibre reinforcement III 71
stress corrosion cracking II 199, II 209
stress measurements I 229
stresses perpendicular to interface
     I 282
strip deflection residual stress I 239
strip turning test III 242
structural laser layer uniformity III 290
structure analyses and hardness test
     II 180
subsea pipelines I 203
substrate I 86, III 161
substrate bond strength I 204
substrate condition II 296
substrate damage II 123
substrate-to-target separation II 21
sulphamate bath II 280
sulphate electrolytes II 250
sulphato bridged oligomers II 250
sulphidation I 53
sulphur penetration I 37
Super D-Gun™ I 212
super austenitic stainless steels III 303
superduplex stainless steel III 283
superheated plasma II 171
superimposed pulsed current II 288
superplastic forming I 286
superplastic materials I 286
supersaturated solid solution I 241, I 265
surface activation III 132
surface engineering I 1, II 19, II 72, II 88
     II 111, II 122, II 289, III 159,
     III 260, III 277
Surface Engineering Research Group I 12
surface fatigue cracking III 240
surface films III 247
surface finish II 225, II 233, III 165,
     III 183
surface irradiation III 278
surface microhardness I 246
surface modification I 32, II 122
surface morphology II 266
surface roughness II 296

surface tension I 292
surface treatments I 4
surfacing II 156
synergistic influence III 306
synthetic diamond I 83
system approach III 159

T

t~butanol II 253
Ta I 217
Taber abraser III 240, III 241
Tafel plots I 124, II 133, II 298, II 303
tantalum carbide III 180
target II 19
target voltage II 50
Teflon III 352
tensile cracking I 274
tetrahedral sites II 54
texture I 84
the interface I 247
thermal barrier I 229, III 209
thermal conductivity I 85
thermal efficiency II 158
thermal expansion coefficient II 222
thermal expansion mismatch III 210
thermal insulation III 132
thermal plasmas II 122
thermal power III 247
thermal spray II 88, III 294
thermal spray coatings II 88
thermal spraying I 4, II 231, III 100, III 159
thermal spraying of polymer coatings II 226
thermal strains III 144
thermal treatment II 290
thermally grown oxide III 99
thermally sprayed coatings I 204
thermionic arc evaporation III 227
thermochemical treatments II 111
thermogravimetric experiments I 77, I 78
thin coatings I 274
thin strip deflection I 239
Thornton's model I 83
threaded parts I 119
three-body wear I 235
three-body abrasion I 203, III 331

three-point bend loading I 239
threshold level of applied stress II 209
through thickness cracking I 274
Ti(B,N) III 384
Ti(C,N) III 187
Ti-48Al-2Nb-2Mn I 53
Ti_2N coating III 181
Ti_2N monolayers III 179
Ti_3Al I 67
TiAl I 67
$TiAl_3$ I 67
TiAlN III 218
TiB_2 coating III 210
tightly knitted splats II 231
TiN III 278, III 367
titanium II 54, III 26, III 179, III 278
titanium and its alloys III 38
titanium carbonitride III 179
titanium dioxide II 127
titanium nitride coating I 189, II 50, III 160
tools II 72
tooth height III 165
total systems approach III 166
toxicity II 245
Toyota diffusion process III 326, III 327
transition region I 263
transpassive region I 137
tribiological behaviour III 26
tribo system III 239
tribological behaviour I 189, III 190
tricone bits II 85
TRIM calculations II 54
triple line I 292
tube leaks III 294
tube wastage III 297
tubular spindle III 262
tungsten and alumina particles II 229
tungsten carbide LW45 I 203
tungsten carbide-cobalt I 229
turbine inlet nozzles III 299
turbines III 299

U

ultimate tensile stress II 202
ultra micro-indentation I 250
ultrasonic machining II 234

unbalanced magnetron sputtering I 119, II 20
unbalanced sputtering III 226
undercoat II 144
unduloids II 169, II 174
urea II 248

V

vacancies II 55
vacancies and interstitials II 50
vacuum annealing II 89
vacuum chamber II 19
vacuum furnace fused coatings II 88
vapour phase deposition III 224, III 364
vapour pressure of nickel chloride I 76
various species (bands) II 248
VDI strip turning test III 242
vertical CNC machine III 185
Vickers indenter I 252, III 237
voids I 77, I 80
volatile organic compounds II 126, II 128
volatilization of ferric chloride I 81
volume fraction III 271

W

washing II 292
water desalination II 211
water droplet contact angle II 129
water vapour attack III 211
water-cooled target II 20
waterwall III 297
wavelength dispersive spectroscopy III 101
WC volume fraction III 262
wear I 119, I 180, I 194, I 206, II 75, II 152, II 265, II 290, III 38, III 165, III 325, III 364, III 365
wear and corrosion resistance II 245
wear debris III 144
wear environments III 262
wear mechanisms III 179
wear protectiveness II 290
wear resistance I 238, I 260, II 122, II 280, III 132
wear scar I 171, I 206
wear tests II 180
wear track I 234
wear volume I 206

wearing in III 354
weathering II 128
wedge cracks I 278
Weibull statistics I 278
weight of deposit II 247
weld surfacing III 159
wet lime/limestone scrubbing III 247
wetting III 209
wetting ridge I 293
wheel-on-wheel configuration II 113
white layer III 218
white rust I 95
Winger effect II 118
wire explosion II 168
Wohlfahrt's Hertzian pressure II 201
work cell III 141
working electrode III 248, III 304
world-class competitive advantage I 7

X

X-ray diffraction I 77, I 78, I 239, I 260, II 201, III 188, III 280
X-ray fluorescence spectrometer I 96, I 112, III 231, III 232
X-ray photoelectron spectroscopy I 95, III 188, III 249, III 252, III 305
XRD, neutron diffraction II 201

Y

Y-modified coatings I 39
Young's equation I 292
Young's modulus of elasticity II 202
yttrium I 34

Z

zinc I 111
zinc-nickel electrodeposits II 274
zirconia III 211
Zn-Ni II 265
zone-1 structure I 223
zone-2 structure I 223
zone-3 structure I 221
zone-T region in Messier model III 189

T
1 Month